FIREFIGHTER

Robert Andriuolo
Former Deputy Chief, N.Y.C. Fire Department

Macmillan • USA

Eleventh Edition

Macmillan General Reference
A Simon & Schuster Macmillan Company
1633 Broadway
New York, NY 10019-6785

An Arco Book

MACMILLAN is a registered trademark of Macmillan, Inc.
ARCO is a registered trademark of Prentice-Hall, Inc.

Library of Congress Cataloging-in-Publication Data

Andriuolo, Robert.
 Firefighter / by Robert Andriuolo. — 11th ed.
  p. cm.
 "An Arco book"—T.p. verso
 ISBN 0-02-860316-8
 1. Fire extinction—Examinations, questions, etc. 2. Civil service—
United States—Examinations. I. Title.
 TH9157.A37 1995   95-30118
 628.9'25'076—dc20   CIP

Manufactured in the United States of America

10 9 8 7 6 5 4 3 2 1

CONTENTS

WHAT THIS BOOK WILL DO FOR YOU

ARCO has followed testing trends and methods ever since the firm was founded in 1937. We specialize in books that prepare people for tests. Based on this experience, we have prepared the best possible book to help you score high.

To write this book we carefully analyzed every detail surrounding firefighter examinations:

- the job itself

- official and unofficial announcements concerning firefighter examinations

- previous examinations, many not available to the public

- related examinations

- technical literature

Can You Prepare Yourself for Your Test?

You want to pass this test. That's why you bought this book. Used correctly, your "self-tutor" will show you what to expect and will give you a speedy brush-up on the subjects tested in your exam. Some of these are subjects not taught in schools at all. Even if your study time is very limited, you should:

- become familiar with the type of examination you will have

- improve your general examination-taking skill

- improve your skill in analyzing and answering questions involving reasoning, judgment, comparison, and evaluation

- improve your speed and skill in reading and understanding what you read—an important part of your ability to learn and an important part of most tests

This book will help you to prepare by presenting many different types of questions that have appeared on actual firefighter exams.

This book will help you find your weaknesses. Once you know what subjects you're weak in, you can get right to work and concentrate on those areas. This kind of selective study yields maximum test results.

This book will give you the *feel* of the exam. All of the questions in the practice exams and in the chapters giving instruction in answering reading-based questions, in answering questions of reasoning and judgment, and in dealing with spatial orientation questions were taken from actual previous firefighter exams given at various times in various places. This book will give you experience in answering real questions.

This book will give you confidence now, while you are preparing for the test. It will build your self-confidence as you proceed and will prevent the kind of test anxiety that causes low test scores.

Every city creates its own firefighter exam. They are not all alike. Number of questions varies; timing varies; actual content of the exam varies. However, nearly all firefighter exams have in common the multiple-choice question format and a few basic subject areas like reading-based questions, reasoning and judgment, use of maps and diagrams, and observation and memory. This book will give you lots of practice with questions in all of these areas.

A special feature of ARCO books is an explanation for every answer. You can learn immediately why a correct answer is correct, and why the wrong answers are wrong. There is a great deal of information in the answer explanations. Study of answer explanations for all questions, even those you got right, will serve as a course in itself.

THE KIND OF WORK YOU WILL BE DOING

Nature of the Work

Every year, fires destroy thousands of lives and property worth millions of dollars. Firefighters help protect the public against this danger. This book is concerned with paid career firefighters; it does not cover the many thousands of volunteer firefighters in communities across the country.

During duty hours, firefighters must be prepared to respond to a fire and handle any emergency that arises. Because firefighting is dangerous and complicated, it requires organization and teamwork. At every fire, firefighters perform specific duties assigned by a company officer such as a lieutenant, captain, or other department officer. They may connect hoselines to hydrants, operate a pump, or position ladders. Because their duties may change several times while the company is in action, they must be skilled in many different firefighting activities, such as rescue, ventilation, and salvage. Some firefighters operate fire apparatus, emergency rescue vehicles, and fire-boats. In addition, they help people to safety and administer first aid.

Most fire departments also are responsible for fire-prevention activities. They provide specially trained personnel to inspect public buildings for conditions that might cause a fire. They may check building plans, the number and working condition of fire escapes and fire doors, the storage of flammable materials, and other potential hazards. In addition, firefighters educate the public about fire prevention and safety measures. They frequently speak on this subject before school assemblies and civic groups, and, in some communities, they inspect private homes for fire hazards.

Between alarms, they have practice drills and classroom training, and are responsible for cleaning and maintaining the equipment they use.

Working Conditions

Firefighters spend much of their time at fire stations, which usually have facilities for dining and sleeping. When an alarm comes in, firefighters must respond rapidly, regardless of the weather or hour. They may spend long periods outdoors fighting fires in adverse weather.

Firefighting is among the most hazardous occupations. The job of a firefighter involves risk of death or injury from sudden cave-ins of floors or toppling walls and danger from exposure to flames and smoke. Firefighters also may come in contact with poisonous, flammable, and explosive gases and chemicals.

In some cities, firefighters are on duty for 24 hours, then off for 48 hours, and receive an extra day off at intervals. In other cities, they work a day shift of 10 hours for three or four days, a night shift of 14 hours for three or four nights, have three or four days off, and then repeat the cycle. Although in many large cities, particularly in the East, firefighters work a standard 40-hour week, many firefighters average as many as 56 hours per week. In addi-

1

tion to scheduled hours, firefighters often must work extra hours when they are bringing a fire under control. Fire lieutenants and fire captains work the same hours as the firefighters they supervise. Duty hours may include some time when firefighters are free to read, study, or pursue other personal interests.

Places of Employment

More than 280,000 persons work as firefighters. Nine out of ten work in municipal fire departments. Some very large cities have several thousand firefighters on the payroll while many small towns have fewer than 25. Some firefighters work in fire departments on federal installations; others work at airports and in large manufacturing plants.

Training, Other Qualifications, and Advancement

Applicants for municipal firefighting jobs must pass a written test, a medical examination, and tests of strength, physical stamina, and agility, as specified by local regulations. These examinations are open to persons who are at least 18 years of age and have a high school education or the equivalent. Those who receive the highest scores on the examinations have the best chances for appointment. Extra credit usually is given for military service. Experience gained as a volunteer firefighter or through training in the Armed Forces also may improve an applicant's chances for appointment.

As a rule, beginners in large fire departments are trained for several weeks at the department's training center. Through classroom instruction and practical training, the recruits study firefighting techniques, fire prevention, hazardous materials, local building codes, and emergency medical procedures; also, they learn how to use axes, saws, chemical extinguishers, ladders, and other firefighting and rescue equipment. After completing this training, they are assigned to a fire company, where they are evaluated during a probationary period.

A small but growing number of fire departments have accredited apprenticeship programs lasting three to four years. These programs combine formal, technical instruction with on-the-job training under the supervision of experienced firefighters. Technical instruction covers subjects such as firefighting techniques and equipment, chemical hazards associated with various combustible building materials, emergency medical procedures, and fire prevention and safety.

Many fire departments offer continuing in-service training to members of the regular force. This training may be offered during regular on-duty hours at an easily accessible site that is connected to the station house alarm system. Such courses help firefighters maintain certain seldom-used skills and permit thorough training on any new equipment that the department may acquire.

Most experienced firefighters continue to study to improve their job performance and prepare for promotion examinations. To progress to higher level positions, firefighters must acquire expertise in the most advanced firefighting equipment and techniques and in building construction, emergency medical procedures, writing, public speaking, management and budgeting procedures, and labor relations. Fire departments frequently conduct training programs, and some firefighters attend training sessions sponsored by the National Fire Academy on a variety of topics such as executive development, anti-arson techniques, and public fire safety and education. Most states also have extensive firefighter training programs.

Many colleges and universities offer courses such as fire engineering and fire science that are helpful to firefighters, and fire departments often offer firefighters incentives such as tuition reimbursement or higher pay to pursue advanced training. Many fire captains and other supervisory personnel have college training.

Among the personal qualities firefighters need are mental alertness, courage, mechanical aptitude, endurance, and a sense of public service. Initiative and good judgment are extremely important because firefighters often must make quick decisions in emergencies. Because members of a crew eat, sleep, and work closely together under conditions of stress and danger, they should be dependable and able to get along well with others in a group. Leadership qualities are assets for officers, who must establish and maintain discipline and efficiency as well as direct the activities of firefighters in their companies.

Opportunities for promotion are good in most fire departments. As firefighters gain experience, they may advance to a higher rank. After three to five years of service, they may become eligible for promotion to the grade of lieutenant. The line of further promotion usually is to captain, then battalion chief, assistant chief, deputy chief, and finally to chief. Advancement generally depends upon scores on a written examination, performance on the job, and seniority. Increasingly, fire departments are using assessment centers—which simulate a variety of actual job performances tasks—to screen for the best candidates for promotion. However, many fire departments require a master's degree—preferably in public administration or a related field—for promotion to positions higher than battalion chief.

Employment Outlook

Employment of firefighters is expected to increase about as fast as the average for all occupations through the year 2000 due to the increase in the nation's population and fire protection needs. Employment should rise as new fire departments are formed and as others enlarge their fire prevention sections. However, little employment growth is expected in large, urban fire departments. Much of the expected increase will occur in smaller communities with expanding populations that augment volunteers with career firefighters to better meet growing, increasingly complex fire protection needs. A small but increasing number of local governments are expected to contract for firefighting services with private companies.

Turnover of firefighter jobs is unusually low, particularly for an occupation that requires a relatively limited investment in formal education. Nevertheless, most job openings are expected to result from the need to replace those who retire or stop working for other reasons, or who transfer to other occupations.

Firefighting attracts many people because a high school education is sufficient, earnings are relatively high, a pension is guaranteed upon retirement, and promotion is possible to progressively more responsible positions on the basis of merit. In addition, the work is frequently exciting and challenging and affords an opportunity to perform a valuable public service. Consequently, the number of qualified applicants in most areas generally exceeds the number of job openings, even though the written examination and physical requirements eliminate many applicants. This situation is expected to persist through the year 2000. Opportunities should be best in smaller communities.

Layoffs of firefighters are not common. Fire protection is an essential service, and citizens are likely to exert considerable pressure on city officials to expand or at least preserve the level of fire-protection coverage. Even when budget cuts do occur, local fire departments usually cut expenses by postponing equipment purchases or the hiring of new firefighters, rather than by laying off staff.

Earnings

The majority of career firefighters are members of the International Association of Fire Fighters (AFL-CIO). This union is a highly effective bargaining unit that is able to gain for its members wages and other benefits commensurate with the efforts they make, the responsibility they shoulder, and the hazards they face. In addition, the union is often able to negotiate pay parity with police officers, who are also effectively represented.

According to a 1985 survey by the International Personnel Management Association, nonsupervisory firefighters had a median salary of about $20,900 a year, but their earnings varied considerably depending on city size and region of the country. Earnings for firefighters are lowest in the South and highest in the West, and generally are higher in large cities than in small ones. Entrance salaries for beginning full-time firefighters averaged about $18,600 a year, while maximum salaries averaged $24,100 a year.

Fire lieutenants had a median annual salary of about $25,800 in 1985. Their starting annual salaries averaged $23,900, and maximum salaries about $28,400.

Fire captains had a median salary of about $28,500 a year in 1985. They started at an average annual salary of $26,700, and could advance to an average maximum salary of $32,000. Some fire captains had salaries in excess of $57,000 a year.

Salaries for firefighters have kept pace with rises in the cost of living. By 1993, New York City firefighters were earning a starting salary of $26,000 with a maximum of $39,000 to be reached after five years on the job. Salaries were proportionately higher at the promotional levels.

The law requires that overtime be paid to those firefighters who average 53 hours or more a week during their work period—which ranges from 7 to 28 days.

Practically all fire departments provide protective clothing (helmets, boots, and coats) and many also provide dress uniforms.

Firefighters generally are covered by liberal pension plans that often provide retirement at half pay at age 50 after 25 years of service or at any age if disabled in the line of duty.

Related Occupations

Firefighters work to prevent fires and to save lives and property when fires and other emergencies such as explosions and chemical spills occur. Related fire-protection occupations include wild-land firefighters and fire-protection engineers, who identify fire hazards in homes and workplaces and design prevention programs and automatic fire detection and extinguishing systems. Other occupations in which workers respond to emergencies include police officers and emergency medical technicians.

Sources of Additional Information

Information on obtaining a job as a firefighter is available from local civil service offices or fire departments.

Information about a career as a firefighter may be obtained from:

International Association of Fire Chiefs, 1329 18th St. NW, Washington, DC 20036.

International Association of Fire Fighters, 1750 New York Ave. NW, Washington, DC 20006.

Information about firefighter professional qualifications may be obtained from:

National Fire Protection Association, Batterymarch Park, Quincy, MA 02269.

Additional information on the salaries and hours of work of firefighters in various cities is published annually by the International City Management Association in its *Municipal Yearbook*, which is available in many libraries.

TECHNIQUES OF STUDY AND TEST-TAKING

You want to become a firefighter. That's great. Every city, every town, every village needs qualified and enthusiastic new recruits to maintain staffing at a full level. You have made a good start toward joining the firefighting force by buying this book.

This book has been carefully researched and has been written to help you through the qualifying process. The information in this book will prepare you for the written exam and will give you valuable tips towards preparing yourself for the physical performance exam.

It is important that you allow the book to help you. Read it carefully. Do not skip over any information. You must know what to expect and then prepare yourself for it. If you are prepared, you can feel self-confident. If you feel confident, you can answer questions quickly and decisively, finish the exam, and earn a high score. If you feel confident, you can enter the physical performance course without hesitation and can prove that you are fit for the job.

Preparing for the Exam

1. **Make a study schedule.** Assign yourself a period of time each day to devote to preparation for your firefighter exam. A regular time is best, but the important thing is daily study.

2. **Study alone.** You will concentrate better when you work by yourself. Keep a list of questions you find puzzling and points you are unsure of to talk over with a friend who is preparing for the same exam. Plan to exchange ideas at a joint review session just before the test.

3. **Eliminate distractions.** Choose a quiet, well-lighted spot as far as possible from telephone, television, and family activities. Try to arrange not to be interrupted.

4. **Begin at the beginning.** Read. Underline points that you consider significant. Make marginal notes. Flag the pages that you think are especially important with little **Post-it**® Notes. We have highlighted some pages with colored paper. Read these pages more than once.

5. **Concentrate on the information and instruction chapters.** Read the Glossary of Firefighting Terminology, page 13. Learn the vocabulary of the job. Get yourself psyched for the whole world of firefighting. Learn how to handle reading-based questions. Focus on the approach to elimination of wrong answers. This information is important to answering all multiple-choice questions, but is especially vital to questions of reasoning and judgement.

6. **Answer the practice questions in each chapter.** Then study the answer explanations. You can learn a great deal from answer explanations. Even when you have answered correctly, the explanation may bring out points that had not occurred to you. This same suggestion—read *all* the explanations—applies throughout this book, to the exams as well as to the instructional chapters.

7. **Try the practice exams.** When you believe that you are well prepared, move on to the exams. If possible, answer an entire exam in one sitting. If you must divide your time, divide it into no more than two sessions per exam.

When you do take the practice exams, treat them with respect. Consider each as a dress rehearsal for the real thing. Time yourself accurately and do not peek at the correct answers. Remember, you are taking these for practice; they will not be scored; they do not count. So learn from them. Learn to think; learn to reason like an effective firefighter; learn to pace yourself so that you can answer all the questions. Then learn from the explanations.

IMPORTANT: Do not memorize questions and answers. Any question that has been released will not be used again. You may run into questions that are very similar, but you will not be tested with any of these exact questions. These questions will give you good practice, but they will not give you the answers to any of the questions on your exam.

How to Take an Exam

1. **Get to the examination room about ten minutes ahead of time.** You'll get a better start when you are accustomed to the room. If the room is too cold, too warm, or not well ventilated, call these conditions to the attention of the person in charge.

2. **Make sure that you read the instructions carefully.** In many cases, test-takers lose points because they misread some important part of the directions. (An example would be reading the *incorrect* choice instead of the *correct* choice.)

3. **Don't be afraid to guess.** The best policy is, of course, to pace yourself so that you can read and consider each question. Sometimes this does not work. Most civil service exam scores are based only on the number of questions answered correctly. This means that a wild guess is better than a blank space. There is no penalty for a wrong answer, and you just might guess right. If you see that time is about to run out, mark all the remaining spaces with the same answer. According to the law of averages, some will be right.

   However, you have bought this book with entire exams for practice answering firefighter questions. Part of your preparation is learning to pace yourself so that you need not answer randomly at the end. Far better than a wild guess is an educated guess. You make this kind of guess not when you are pressed for time but when you are not sure of the correct answer. Usually, one or two of the choices is obviously wrong. Eliminate the obviously wrong answers and try to reason among those remaining. Then, if necessary, guess from the smaller field. The odds of choosing a right answer increase if you guess from a field of two instead of from a field of four. When you make an educated guess or a wild guess in the course of the exam, you might want to make a note next to the question number in the test booklet. Then, if there is time, you can go back for a second look.

4. **Reason your way through multiple-choice questions very carefully and methodically.** Here are a few samples that we can "walk through" together:

   I. On the job, your supervisor gives you a hurried set of directions. As you start your assigned task, you realize you are not quite clear on the directions given to you. The best action to take would be to

   (A) continue with your work, hoping to remember the directions and do the best you can

(B) ask a co-worker in a similar position what he or she would do

(C) ask your supervisor to repeat or clarify certain directions

(D) go on to another assignment

In this question you are given four possible answers to the problem described. Though the four choices are all *possible* actions, it is up to you to choose the *best* course of action in this particular situation.

Choice (A) will likely lead to a poor result; given that you do not recall or understand the directions, you would not be able to perform the assigned task properly. Keep choice (A) in abeyance until you have examined the other alternatives. It could be the best of the four choices given.

Choice (B) is also a possible course of action, but is it the best? Consider that the co-worker you consult has not heard the directions. How could he or she know? Perhaps his or her degree of incompetence is greater than yours in this area. Of choice (A) and choice (B), the better of the two is still choice (A).

Choice (C) is an acceptable course of action. Your supervisor will welcome your questions and will not lose respect for you. At this point, you should hold (C) as the best answer and eliminate (A).

The course of action in (D) is decidedly incorrect because the job at hand would not be completed. Going on to something else does not clear up the problem; it simply postpones your having to make a necessary decision.

After careful consideration of all choices given, choice (C) stands out as the best possible choice of action. You should select (C) as your answer.

Every question is written about a fact or an accepted concept. The question above indicates the concept that, in general, most supervisory personnel appreciate subordinates questioning directions which may not have been fully understood. This type of clarification precludes subsequent errors on the part of the subordinates. On the other hand, many subordinates are reluctant to ask questions for fear that their lack of understanding will detract from their supervisor's evaluation of their abilities.

The supervisor, therefore, has the responsibility of issuing orders and directions in such a way that subordinates will not be discouraged from asking questions. This is the concept on which the sample question was based.

Of course, if you were familiar with this concept, you would have no trouble answering the question. However, if you were not familiar with it, the method outlined here of eliminating incorrect choices and selecting the correct one should prove successful for you.

We have now seen how important it is to identify the concept and the key phrase of the question. Equally, or perhaps even more important, is identifying and analyzing the key word—the qualifying word—in a question. This word is usually an adjective or adverb. Some of the most common key words are *most, least, best, highest, lowest, always, never, sometimes, most likely, greatest, smallest, tallest, average, easiest, most nearly, maximum, minimum, only, chiefly, mainly, but,* and *or.* Identifying these key words is usually half the battle in understanding and, consequently, answering all types of exam questions.

Now we will use the elimination method on some additional questions:

II. On the first day you report for work after being appointed as a firefighter, you are assigned to routine duties which seem to you to be very petty in scope. You should

(A) perform your assignment perfunctorily while conserving your energies for more important work in the future

(B) explain to your superior that you are capable of greater responsibility

(C) consider these duties an opportunity to become thoroughly familar with the firehouse

(D) try to get someone to take care of your assignment until you have become thoroughly acquainted with your new associates

Once again we are confronted with four possible answers from which we are to select the *best* one.

Choice (A) will not lead to getting your assigned work done in the best possible manner in the shortest possible time. This would be your responsibility as a newly appointed firefighter, and the likelihood of getting to do more important work in the future following the approach stated in this choice is remote. However, since this is an (A) choice, we must hold it aside because it may turn out to be the best of the four choices given.

Choice (B) is better than choice (A) because your superior may not be familiar with your capabilities at this point. We therefore should drop choice (A) and retain choice (B) because, once again, it may be the best of the four choices.

The question clearly states that you are newly appointed. Therefore, would it not be wise to perform whatever duties you are assigned in the best possible manner? In this way, you would not only use the opportunity to become acquainted with firehouse procedures, but also to demonstrate your abilities. Choice (C) contains a course of action which will benefit you and the firehouse in which you are working because it will get needed work done. At this point, we drop choice (B) and retain choice (C) because it is by far the better of the two.

The course of action in choice (D) is not likely to get the assignment completed and it will not enhance your image to your fellow firefighters. Choice (C), when compared to choice (D), is far better and therefore should be selected as the *best* choice.

Now let us take a question that appeared on a police officer examination:

III. An off-duty police officer in civilian clothes riding in the rear of a bus notices two teenage boys tampering with the rear emergency door. The most appropriate action for the officer to take is to

(A) tell the boys to discontinue their tampering, pointing out the dangers to life that their actions may create

(B) report the boys' actions to the bus operator and let the bus operator take whatever action is deemed best

(C) signal the bus operator to stop, show the boys the officer's badge, and then order them off the bus

(D) show the boys the officer's badge, order them to stop their actions, and take down their names and addresses

Before considering the answers to this question, we must accept that it is a well known fact that a police officer is always on duty to uphold the law even though he or she may be technically off duty.

In choice (A), the course of action taken by the police officer will probably serve to educate the boys and get them to stop their unlawful activity. Since this is only the first choice, we will hold it aside.

In choice (B), we must realize that the authority of the bus operator in this instance is limited. He can ask the boys to stop tampering with the door, but that is all. The police officer can go beyond that point. Therefore, we drop choice (B) and continue to hold choice (A).

Choice (C) as a course of action will not have a lasting effect. What is to stop the boys from boarding the next bus and continuing their unlawful action? We therefore drop choice (C) and continue to hold choice (A).

Choice (D) may have some beneficial effect, but it would not deter the boys from continuing their actions in the future. When we compare choice (A) with choice (D), we find that (A) is the better one overall, and therefore it should be your selection.

The next question illustrates a type of question which has gained popularity in recent examinations and which requires a two-step evaluation. First, the reader must evaluate the condition in the question as being "desirable" or "undesirable." Once the determination has been made, we are then left with making a selection from two choices instead of the usual four.

IV. A visitor to an office in a city agency tells one of the office aides that he has an appointment with the supervisor of the office who is expected shortly. The visitor asks for permission to wait in the supervisor's private office, which is unoccupied at the moment. For the office aide to allow the vistor to do so would be

(A) desirable; the vistor would be less likely to disturb the other employees or to be disturbed by them
(B) undesirable; it is not courteous to permit a visitor to be left alone in an office
(C) desirable; the supervisor may wish to speak to the visitor in private
(D) undesirable; the supervisor may have left confidential papers on the desk

First of all, we must evaluate the course of action on the part of the office aide of permitting the visitor to wait in the supervisor's office as being *very* undesirable. There is nothing said of the nature of the visit; it may be for a purpose that is not friendly or congenial. There may be papers on the supervisor's desk that he or she does not want the visitor to see or to have knowledge of. Therefore, at this point, we have to make a choice between (B) and (D).

This is definitely not a question of courtesy. Although all visitors should be treated with courtesy, permitting the visitor to wait in the supervisor's office in itself is not the only act of courtesy. Another comfortable place could be found for the visitor to wait.

Choice (D) contains the exact reason for evaluating this course of action as being undesirable, and when we compare it with choice (B), choice (D) is far better.

One word of caution: Once you have made a determination that a course of action is desirable or undesirable, confine your thoughts to the situations illustrating your choice. Do not revert to again considering the other extreme or you will flounder and surely not select the correct answer.

5. **Mark the answer sheet clearly**. Unless you have applied to a fire force in a very small locality, your exam will be machine scored. You will mark your answers on a separate answer sheet. Each exam in this book has its own answer sheet so that you can practice marking all your answers in the right way. Tear out the answer sheet before you begin each exam. Do not try to flip pages back and forth. Since your answer sheet will be machine scored, you must fill it out clearly and correctly. You cannot give any explanations to the machine. This means:

• Blacken your answer space firmly and completely. ● is the only correct way to mark the answer sheet. ◑, ⊠, ⊘, and ∅ are all unacceptable. The machine might not read them at all.

• Mark only one answer for each question. If you mark more than one answer you will be considered wrong even if one of the answers is correct. (See note, page 12.)

• If you change your mind, you must erase your mark. Attempting to cross out an incorrect answer like this ✖ will not work. You must erase any incorrect answer completely. An incomplete erasure might be read as a second answer.

A Typical
Notice of
Examination

The general provisions of the notice of examination and the general examination regulations of the New York City Department of Personnel are part of this notice of examination. They are posted, and copies are available, at the Application Section of the Department of Personnel at 18 Washington Street, New York, NY 10004.

This position is open to both men and women.

Candidates appointed to this position may be required to work rotating tours or shifts including Saturdays, Sundays, and holidays.

Promotion Opportunities: Employees in the title of Firefighter are accorded promotion opportunities, when eligible, to the title of Lieutenant, Fire Department.

Requirements

MINIMUM REQUIREMENTS: By the date of appointment, you must be a United States citizen and possess a 4-year High School Diploma or its equivalent.

LICENSE REQUIREMENT: On the date of appointment as a firefighter, possession of a valid New York State Driver License is required.

AGE REQUIREMENT: Pursuant to Section 15-103 of the Administrative Code, only persons who have passed their eighteenth birthday but not their twenty-ninth birthday on the date of filing qualify for this examination. However, only persons who have reached their twenty-first birthday shall be appointed.

In accordance with Section 54 of the New York Civil Service Law, no applicant who is within six months of the minimum age for filing can be prohibited from taking the written test. In summary, the age requirement means that you must not have reached your twenty-ninth birthday by September 2, but you must be at least 17 1/2 years of age by September 22, in order to take the December 12 exam.

EXCEPTION TO THE AGE REQUIREMENT: All persons who were engaged in military duty, as defined in Section 243 of the Military Law, may deduct from their actual age the length of time spent in such military duty, provided the total deduction for military duty does not exceed six years, of which not more than four years may be for voluntary service after May 7, 1975.

INVESTIGATION: At the time of appointment and at the time of investigation, candidates must present all the official documents and proof required to qualify. At the time of investigation, candidates will be required to present proof of birth by transcript of records from the Bureau of Vital Statistics, or other satisfactory evidence. Proof of military service and similar documents may be required. Failure to present required documents, including proof of educational or license requirements, will result in disqualification for appointment. Any willful misstatement or omission will be cause for disqualification.

CHARACTER AND BACKGROUND: Proof of good character and satisfactory background will be an absolute prerequisite to appointment. In accordance with the provisions of law, persons convicted of a felony are not eligible for positions in the uniformed force of the Fire Department. In addition, the rules of the Department of Personnel provide that no person who has been dishonorably discharged from the Armed Forces shall be certified for appointment as a firefighter.

MEDICAL TEST: Candidates will be required to pass a medical test prior to appointment.

RESIDENCY REQUIREMENT: The New York Public Officers Law requires that any person employed as a firefighter in the New York City Fire Department be a resident of the City of New York, or of Nassau, Westchester, Suffolk, Orange, Rockland or Putnam counties.

PROOF OF IDENTITY: Under the Immigration Reform and Control Act of 1986, you must be able to prove your identity and your right to obtain employment in the United States prior to employment with the City of New York.

Job Description

Duties and Responsibilities: Under supervision, assists in the control and extinguishment of fires and in the enforcement of laws, ordinances, rules, and regulations regarding the prevention, control, and extinguishment of fires; performs inspectional, investigation, and regulative duties connected with fire prevention, control, and extinguishment of fires; performs all additional functions for the rank prescribed by orders, directives, or regulations of the Fire Department and performs special duties or assignments as directed by the Fire Commissioner.

Test Information

Tests: There will be a written test of the multiple-choice type and a physical test. Only candidates who achieve the passing mark of 70% on the written test will be summoned for the physical test. The weight of the written test will be announced by the date of the written test.

The written test will be of the multiple-choice type and will be designed to test the candidates' ability to read and understand English and capacity to learn and perform the work of a firefighter. It may include questions on understanding job information; applying laws, rules, and regulations to job situations; recognizing appropriate behavior in job situations; understanding mechanical devices; and remembering details of a floor layout.

The physical test will be designed to test the candidates' capacity to perform the physical aspects of the work of a firefighter.

Medical and Physical Standards: Eligibles must meet the medical and physical standards to be established and posted on the Department of Personnel Bulletin Board. Candidates must pass the physical test and a qualifying medical test. Eligibles will be rejected for any physical, mental, or medical condition which will prevent the eligible from performing in a reasonable manner the duties of a firefighter.

The Application

The application form which you must file may be as simple as the New York City application form reproduced on the next page or may be considerably more complex and more detailed than the Experience Paper which follows it.

What is important is that you fill out your application form neatly, accurately, completely and truthfully.

Obviously, neatness counts. If your application is misread it is likely to be misfiled or misinterpreted to your detriment. If your application is extremely sloppy, it brands you as a slob. No one wants to hire a slob.

Accuracy is important. A single wrong digit in your social security number or exam number can misroute your application forever. If your inaccuracy is discovered and corrected, you nevertheless carry with you the label of "careless."

Completeness is a requirement. An incomplete application gets tossed into the wastebasket. No one will take the time to chase you down to fill in the blanks.

The need to tell the truth should be self-evident. Certainly you should describe your duties and responsibilities in the best possible light, but do not exaggerate. Do not list a degree you did not receive. Do not give a graduation date if you did not graduate. If your application form asks for reasons for leaving previous employment give a very brief, but truthful reason. Any statement you make on an application form is subject to verification. There are staff members at the Civil Service office whose job it is to check with schools and with former employers. These investigators do ask questions about your job performance, attendance record, duties and responsibilities.

Any misstatements which are not picked up by background investigators are likely to be unmasked at an interview. The last step of the screening process is likely to be a personal interview before one person or before a panel of interviewers. The interviewer(s) will sit with a copy of your application form and will ask you some questions based upon it. This will be your opportunity to explain unfavorable items such as dismissal from a job or frequency of job changes. If you have made false claims or have stretched the truth too far, this is the point at which you are likely to trip yourself up. It is difficult to maintain a lie under the pressure of pointed questioning. The interviewers are prepared and are at ease. You are in an unfamiliar situation and are nervous about qualifying for a job. Do not make it harder for yourself. Be open and truthful on the application; defend yourself at the interview.

APPLICATION FOR OPEN COMPETITIVE EXAMINATION
(See Instructions on Reverse Side)

CITY OF NEW YORK
DEPARTMENT OF PERSONNEL
APPLICATION SECTION

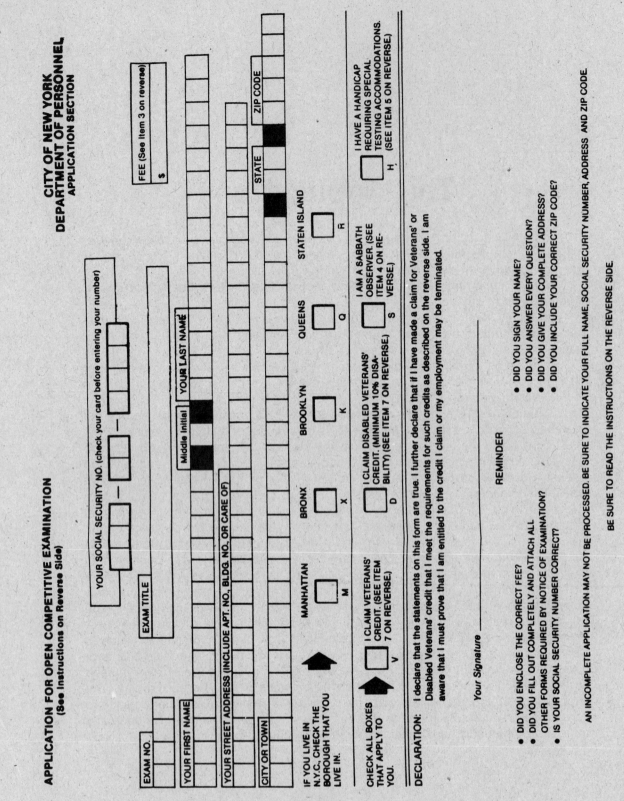

YOUR SOCIAL SECURITY NO. (check your card before entering your number)

FEE (See Item 3 on reverse)

$

EXAM NO.

EXAM TITLE

YOUR FIRST NAME

Middle Initial   YOUR LAST NAME

YOUR STREET ADDRESS (INCLUDE APT. NO., BLDG. NO., OR CARE OF)

CITY OR TOWN

STATE   ZIP CODE

IF YOU LIVE IN N.Y.C., CHECK THE BOROUGH THAT YOU LIVE IN.

MANHATTAN M   BRONX X   BROOKLYN K   QUEENS Q   STATEN ISLAND R

CHECK ALL BOXES THAT APPLY TO YOU.

I CLAIM VETERANS' CREDIT. (SEE ITEM 7 ON REVERSE.) V

I CLAIM DISABLED VETERANS' CREDIT. (MINIMUM 10% DISABILITY) (SEE ITEM 7 ON REVERSE.) D

I AM A SABBATH OBSERVER. (SEE ITEM 4 ON REVERSE.) S

I HAVE A HANDICAP REQUIRING SPECIAL TESTING ACCOMMODATIONS. (SEE ITEM 5 ON REVERSE.) H

DECLARATION: I declare that the statements on this form are true. I further declare that if I have made a claim for Veterans' or Disabled Veterans' credit that I meet the requirements for such credits as described on the reverse side. I am aware that I must prove that I am entitled to the credit I claim or my employment may be terminated.

Your Signature _____

REMINDER

- DID YOU ENCLOSE THE CORRECT FEE?
- DID YOU FILL OUT COMPLETELY AND ATTACH ALL OTHER FORMS REQUIRED BY NOTICE OF EXAMINATION?
- IS YOUR SOCIAL SECURITY NUMBER CORRECT?

- DID YOU SIGN YOUR NAME?
- DID YOU ANSWER EVERY QUESTION?
- DID YOU GIVE YOUR COMPLETE ADDRESS?
- DID YOU INCLUDE YOUR CORRECT ZIP CODE?

AN INCOMPLETE APPLICATION MAY NOT BE PROCESSED. BE SURE TO INDICATE YOUR FULL NAME, SOCIAL SECURITY NUMBER, ADDRESS AND ZIP CODE.

BE SURE TO READ THE INSTRUCTIONS ON THE REVERSE SIDE.

DP-848C (R. 1/86)

4A

INSTRUCTIONS FOR FILLING OUT THIS APPLICATION FORM

Be sure to read the Notice of Examination carefully before completing this form.

1. **FORMS**
If the Notice of Examination calls for Experience Form A and/or other forms, these forms must be included with your application, or your application cannot be accepted.

2. **ADDRESS**
Give your full mailing address, including building, apartment number, or "in care of" information where needed. (If you change your address after applying, write to the Examining Service Division of the Department of Personnel, Room 216, 220 Church Street, New York, N.Y. 10013, with your new address, plus your Social Security Number and the number and title of the examination for which you applied.)

3. **FEE**
Enclose the correct fee made out to the N.Y.C. Department of Personnel. The amount of the fee is stated in the Notice of Examination for this examination. DO NOT MAIL CASH. If paying by check: write your name, social security number, home address, and the examination number on the front of the check. A fee is not charged if you are a N.Y.C. resident receiving public assistance from the N.Y.C. Department of Social Services. To have the fee waived you must present or enclose a clear photocopy of your current Medicaid Card. The photocopy must accompany the application, even if it is filed in person.

4. **SABBATH OBSERVER**
If because of religious belief you cannot take a test on the scheduled date, you must come or write to: Examining Service Division (Room 216); N.Y.C. Department of Personnel, 220 Church Street, New York, N.Y. 10013 no later than five work days prior to the test date to request an alternate date. All requests must be accompanied by a signed statement on letterhead from your religious leader certifying to your religious observance.

5. **HANDICAP**
If you have a handicap which will interfere with your ability to take this test without special accommodations, amanuensis, or other assistance, you must submit a written request for specific special accommodations to the Examining Service Division (Room 216); N.Y.C. Department of Personnel, 220 Church Street, New York, N.Y. 10013. The request must be received no later than 15 work days prior to the test date. A physician or agency authorized for this purpose must corroborate the specific nature of your handicap and must justify the need for the special accommodations you request. For more details, consult Regulation E.8 of the General Examination Regulations, which is available from the Application Section of the Department of Personnel.

6. **MAIL**
Check the Notice of Examination to see if applications are accepted by mail. If you mail your application, use a 4¼" x 9½" (legal size) envelope, and address to:

    N.Y.C. Department of Personnel
    Application Section

7. **VETERANS' CREDIT**
Your application **MUST BE RECEIVED IN THIS DEPARTMENT BY THE LAST DAY OF FILING.**

FOR VETERANS' OR DISABLED VETERANS' CREDIT YOU MUST:

a. Have served on active duty, other than for training purpose, in the Armed Forces of the United States during:
   - December 7, 1941 - September 2, 1945; or
   - June 26, 1950 - January 31, 1955; or
   - January 1, 1963 - May 7, 1975; and

b. Be a resident of New York State at the time of list establishment; and

c. Have an honorable discharge or have been released under honorable conditions.

d. For Disabled Veterans' Credit, in addition to a, b, and c above, at the time the examination list is established you must receive or be entitled to receive at least 10% compensation from the V.A. for a disability incurred in time of war. The V.A. must also certify that the disability is permanent. If it is not permanent, you must have been examined by the V.A. within one year of the establishment of the examination list.

NOTES

1. You may use Veterans' or Disabled Veterans' Credit only once after January 1, 1951 for appointment or promotion from a City, State or County civil service list.

2. The above is only a summary of necessary conditions. The complete provisions are contained in statutory and/or decisional law.

SOCIAL SECURITY NO.	EXPERIENCE PAPER	FORM A
___/___/___	Exact Title of Examination	EXAMINATION NO.

DO NOT WRITE IN THIS SPACE

Not qualified under terms of advertisement	QUALIFIED: Admitted for further examination pending final determination of qualifications.	RATING

To Investigator: Verify items checked and also data mentioned here ....
....

☐ If this box is checked, submit a report with full details to ....
....

FOLLOW THESE INSTRUCTIONS CAREFULLY:

a. Type, print or write (in blue or black ink) the information requested.

b. Be COMPLETE, CONCISE AND ACCURATE. The information you enter below will be used to determine your rating on training and experience. If statements of material facts are found to be false, exaggerated or misleading, you may be disqualified.

c. DO NOT REVEAL YOUR IDENTITY in any way on this experience paper. If you are or have been in business for yourself or have been employed by a relative of the same name, write "self" or the relationship of the relative in the space headed "Name of Employer".

1. LICENSE OR REGISTRATION: If a license or registration is required, answer the following.

a. Title of lic. or reg. you possess which is valid in this State ....   License or Registration No. ....

b. Name of issuing agency...

c. Date of original issue..................   Date last renewed or registered...........................   Renewal number.............................   Date of expiration ....

2. EDUCATION: **3. DATE OF BIRTH*** ....

Give Name and Locations of Schools Below	Day or Night	From Mo.	Yr.	To Mo.	Yr.	Were you Graduated (Yes or No)	Degree Received	Total Credits Completed	Major Subject	No. of Credits in Major
High School or Trade School										
College or other School										

4. List below any courses which you have passed which: (a) are necessary to meet the minimum requirements for the position; (b) provide training appropriate to the position; (c) are in related fields.

NAME OF COLLEGE OR INSTITUTION	Course Number	EXACT TITLE OF COURSE (Place "G" after graduate courses)	No. of Credits	Date Completed

*The Human Rights Law prohibits discrimination on the basis of age, creed, color, national origin, sex, disability or marital status of any individual. The law allows certain age or sex specifications if based upon a bona fide occupational qualification or statutory authorization.

QUALIFYING EMPLOYMENT. List in chronological order those positions you have held which tend to qualify you for the position sought. Begin with the present or most recent position. Use a separate block for each position. List as a separate employment every material change of duties although with the same employer. Include pertinent experience or training in the armed forces.

1

Dates of Employment (give month and year)	Length of Employment	Exact Title of Your Position	Starting Salary _____ per
From To	Yrs. Mos.		Last Salary _____ per

Name and Address of Employer	Nature of Employer's Business
	Title (not name) of your Immediate Superior
Number and Titles of Employees You Supervise	If with the City or State was this a provisional appointment? Yes ☐ No ☐ / No. of Hours Worked per Week

Describe Your Duties

2

Dates of Employment (give month and year)	Length of Employment	Exact Title of Your Position	Starting Salary _____ per
From To	Yrs. Mos.		Last Salary _____ per

Name and Address of Employer	Nature of Employer's Business
	Title (not name) of your Immediate Superior
Number and Titles of Employees You Supervise	If with the City or State was this a provisional appointment? Yes ☐ No ☐ / No. of Hours Worked per Week

Describe Your Duties

3

Dates of Employment (give month and year)	Length of Employment	Exact Title of Your Position	Starting Salary _____ per
From To	Yrs. Mos.		Last Salary _____ per

Name and Address of Employer	Nature of Employer's Business
	Title (not name) of your Immediate Superior
Number and Titles of Employees You Supervise	If with the City or State was this a provisional appointment? Yes ☐ No ☐ / No. of Hours Worked per Week

Describe Your Duties

4

Dates of Employment (give month and year)	Length of Employment	Exact Title of Your Position	Starting Salary _____ per
From To	Yrs. Mos.		Last Salary _____ per

Name and Address of Employer	Nature of Employer's Business
	Title (not name) of your Immediate Superior
Number and Titles of Employees You Supervise	If with the City or State was this a provisional appointment? Yes ☐ No ☐ / No. of Hours Worked per Week

Describe Your Duties

OTHER EXPERIENCE: List here only the other employments you have had which were not included before.

Dates of Employment		Name and Address of Employer	Business of Employer	Title of Your Position and Highest Salary	Brief Description of Duties You Performed
From	To				

Any additional qualifications you have for the position may be listed here; for example: field work in connection with training, authorship or co-authorship of books or of articles in recognized journals, special certification from professional boards, other professional attainments, etc. (Caution: Do not mention your name.) Additional courses or experience may also be listed here.

IF YOU HAVE ADDITIONAL QUALIFYING EXPERIENCE, ATTACH SEPARATE PAPER USING THE SAME FORMAT

Medical Standards

Firefighting is a physically stressful occupation. Firefighters who spend much of their time doing relatively undemanding physical tasks in the firehouse and who spend a good deal of the time in sedentary activity like card playing and watching television are suddenly called to put forth major physical exertion. The rapid and extreme shifts in demands on the body of the firefighter result in real punishment. This is why the selection process places such emphasis on medical condition and upon physical fitness and capacity.

The actual firefighting activity involves heavy lifting and carrying while encumbered by heavy protective gear under conditions of extreme temperature, water exposure, and poor air quality. This work must by conducted at great speeds and with constant awareness of personal danger. No wonder there is such concern for hiring potential firefighters who are physically and emotionally suited for the job.

Federal law prohibits discrimination against job applicants on the basis of age. However, where youth is a bona fide prerequisite for effective performance of the duties of the job, employers are permitted to prescribe age limits. There is no question that the age limitation on firefighter applicants is justified. The work requires young, agile bodies. There is a big investment of time, effort, and money in the training of firefighters. The goal is to hire and train healthy, capable young men and women who will serve many productive years as effective firefighters.

In fact, medical standards are a matter of official policy:

This position is physically demanding and affects public health and safety. Therefore, the Department requires candidates to meet appropriately high standards of physical and mental fitness. The object of the preemployment examination is to obtain personnel fit to perform in a reasonable manner all the activities of the position. Appointees may be subject to reexamination at any time during their probationary periods.

A candidate will be medically disqualified upon a finding of physical or mental disability which renders the candidate unfit to perform in a reasonable manner the duties of this position, or may reasonably be expected to render the candidate unfit to continue to perform in a reasonable manner the duties of this position. (For example, a latent impairment or a progressively debilitating impairment may, in the judgment of the designated medical officer, reasonably be expected to render the candidate unfit to continue to perform the duties of the position.)

The fitness of each candidate is determined on an individual basis, in relation to this position. The designated medical officer may utilize diagnostic procedures, including the use of scientific instruments or other laboratory methods which, in his or her discretion, would determine the true condition of the candidate before he or she is accepted. The judgment of the designated medical officer, based on his or her knowledge of the activities involved in the duties of this position and the candidate's condition, is determinative.

The following definitions apply:

- Physical or Mental Disability: A physical, mental, or medical impairment resulting from an anatomical, physiological, or neurological condition which prevents the exercise of a normal bodily function or is demonstrable by medically accepted clinical or laboratory diagnostic techniques. Such impairment may be latent or manifest.

- Accepted Medical Principles: Fundamental deduction consistent with medical facts and based upon the observation of a large number of cases. To constitute an accepted medical principle, the deduction must be based upon the observation of a large number of cases over a significant period of time and be so reasonable and logical as to create a moral certainty that they are correct.

- Impairment of Function: Any anatomic or functional loss, lessening, or weakening of the capacity of the body, or any of its parts to perform that which is considered by accepted medical principles to be the normal.

- Latent Impairment: Impairment of function which is not accompanied by signs and/or symptoms but which is of such a nature that there is reasonable and moral certainty, according to accepted medical principles, that signs and/or symptoms will appear within a reasonable period of time or upon change of environment.

- Manifest Impairment: Impairment of function which is accompanied by signs and/or symptoms.

- Medical Capability: General ability, fitness, or efficiency (to perform duty) based on accepted medical principles.

As you can see, the medical standards for prospective firefighters are extremely strict and rigid. Conditions which would not deter a person's performance in a less strenuous occupation are absolutely disqualifying for firefighters. Some of these conditions are temporary, like pregnancy, and others, like the use of drugs or obesity (overweight) may be correctable. Temporary or correctable conditions may cause the candidate to be barred from the physical performance test at a given time and so may effectively keep the person from competition and employment. However, once the temporary or correctable conditions are overcome, the candidate may, if still within the age limit, compete at a later date.

A number of medical conditions, depending upon their severity, may bar a candidate from employment as a firefighter even if the candidate is able to score high on the physical performance test and is otherwise highly qualified. In fact, a person with any of these conditions may cause him or herself injury or permanent complications of the condition as a result of undergoing strenuous physical exercise, or participating in conditioning programs, or in the process of competing in the physical performance test. Some of these conditions are:

- Alcoholism
- Anemia
- Asthma
- Chronic Gastro-Intestinal Disorders
- Diabetes
- Drug Dependence
- Epilepsy or other Seizure Disorder
- Heart Disease
- High Blood Pressure
- Joint Disease
- Kidney Disease
- Liver Disease

- Lung Disease
- Muscular Disorders
- Obesity (greater than simple overweight)
- Proneness to Heat Illness such as Heat Stroke or Heat Exhaustion
- Sickle Cell Disease
- Ulcers

The above list represents conditions that are likely to be permanently disqualifying. Other conditions may militate against taking part in physical training or taking the physical performance test until they are removed by passage of time or by active curative measures. Examples of such conditions include:

- Acute Gastro-Intestinal Disorder
- Dehydration
- Drug Use
- Hernia
- Infections
- Overweight
- Pregnancy
- Severe Underweight

The cooperation and assistance of a physician can help a candidate to attain the health and medical status to permit participation in fitness programs and in the test itself. A person who has suffered from any one of these conditions or similar should remain under the supervision of a doctor while undergoing vigorous physical training and should not attempt to participate in a physical performance test until given approval by the doctor.

Still other physical/medical conditions are disqualifying to the firefighter candidate simply by virtue of their presence and without regard to physical fitness. Considering the requirements of firefighting, it should be obvious that persons with visual impairments (nearsightedness, color blindness, poor depth perception, etc.) cannot function as firefighters. Severe hearing deficit also is disqualifying. Firefighters must not be dependent on glasses or contact lenses or on hearing aids which may be lost, broken, or rendered dysfunctional under heat stress. Persons with these conditions may be in excellent health and physically strong and able, but they cannot serve as firefighters. Other medical conditions such as severe allergies, cancer, or presence of the HIV virus do not augur well for long-term effectiveness as a firefighter and so are automatically disqualifying to the candidate.

You are probably very much aware of any disabilities or major problems that would keep you from serving as a firefighter. If you are considering yourself as a firefighting candidate, you evidently think of yourself as a healthy, physically fit person. Even so, it would be wise to consult with your own doctor before proceeding to invest time and money in the application process and in preparing for the written exam. It is also wise to get the approval of your doctor before you embark on a vigorous physical conditioning program. Tell your doctor about the type of work you have in mind, describe the physical demands, and ask for an assessment of your potential to withstand these rigors. If your doctor foresees any potential problems, either in passing the exams or in facing the demands of the job, discuss corrective measures and remedial programs right now. Follow the medical advice you receive. While speaking with your doctor, you might describe the physical performance test as well. You may be able to pick up special tips to prepare yourself to earn the top score on the performance test itself. Your doctor may even have a physical conditioning program to recommend. Since so much of the hiring decision is based upon the physical status of the applicant, it makes sense to devote at least as much attention to preparing your body for the physical test as to preparing your mind for the written exam.

The exam for which you are now preparing yourself is the first step toward your career in firefighting. If all goes well, you are planning to spend the bulk of your working life in this career. You cannot expect to prepare in a few weeks for a career which will last a lifetime. Plan ahead. Start now.

The book that you have before you is a good starting place for preparation for the written test. Now consider preparing for the physical test at the same time.

Step one: Go to your doctor for a full medical workup. If you get the go-ahead from your doctor, you can begin an aggressive physical fitness training program. If your doctor has any reservations, take care of your health problems. Follow your doctor's advice for improving your general health, for curing any specific problems, and for building strength and stamina at a rate consistent with your overall condition.

Step two: Stop polluting your body. This means: If you smoke, stop smoking now. Medical authorities do not agree on all the details, but they all acknowledge that some harm is caused by tobacco smoke. Smoking has been linked to lung cancer, breathing disorders, heart disease, and circulatory diseases. Smoking begins to affect the body long before actual diseases develop. Smokers generally do not have the stamina of non-smokers. Smoking reduces the power of the lungs to distribute oxygen. With less oxygen to the muscles of the body, smokers cannot run or climb as long, cannot carry as much, cannot sustain effort to the degree that non-smokers can. For a top score on the physical performance test and for a long career in firefighting, stop smoking now.

"Stop polluting your body" also means: If you use drugs—inhaled, injected, ingested, or smoked—stop now. Different drugs affect the body in different ways, but all alter functioning in some way. Regardless whether the drug you use affects your vision, your hearing, your emotions, or your thinking, it creates an unnatural condition which does not enhance your chances of top scores on examinations nor your chances of being selected for a firefighter position. In fact, drug testing is part of the screening process for firefighters. Persons who test positive for drug use can not be hired. And medical authorities warn that the intense activity of the physical performance test can cause permanent damage to the body of a drug user.

Excess alcohol and excessive caffeine are also body pollutants. Moderate amounts of these stimulants is probably not harmful, though experts do disagree as to the definition of "moderate." Let common sense be your guide. Do not overdo.

Step three: Control your weight. Excess weight requires muscles, heart, and lungs to work harder. The overweight person tends to have less speed and less stamina than the well-proportioned person and tends to have a shorter productive life span as well. But, rapid weight reduction constitutes a shock to the system. Quick weight loss is unhealthy. Weight must be lost gradually through a combination of calorie cutback and sensible exercise.

Very low body weight can also be a problem for the prospective firefighter. The person who is severely underweight is unlikely to be able to carry heavy loads for any length of time and may tire easily. If underweight does not stem from disease, the underweight person should be able to gain slowly by eating more food and by adding muscle weight through exercise and weight lifting.

Each department sets its own standards, but the tables below can serve as a general guide to weight requirements.

ACCEPTABLE WEIGHT IN POUNDS ACCORDING TO FRAME
MALE

A Height (in bare feet)		B Small Frame	C Medium Frame	D Large Frame
Feet	Inches			
5	2	128–134	131–141	138–150
5	3	130–136	133–143	140–153
5	4	132–148	135–145	142–153
5	5	134–140	137–148	144–160
5	6	136–142	139–151	146–164
5	7	138–145	142–154	149–168
5	8	140–148	145–157	152–172
5	9	142–151	148–160	155–176
5	10	144–154	151–163	158–180
5	11	146–157	154–166	161–184
6	0	149–160	157–170	164–188
6	1	152–164	160–174	168–192
6	2	155–168	164–178	172–197
6	3	158–172	167–182	176–202
6	4	162–176	171–187	181–207

ACCEPTABLE WEIGHT IN POUNDS ACCORDING TO FRAME
FEMALE

A Height (in bare feet)		B Small Frame	C Medium Frame	D Large Frame
Feet	Inches			
4	10	102–111	109–121	118–131
4	11	103–113	111–123	120–134
5	0	104–115	113–126	122–137
5	1	106–118	115–129	125–140
5	2	108–121	116–132	128–143
5	3	111–124	121–135	131–147
5	4	114–127	124–138	134–151
5	5	117–130	127–141	137–155
5	6	120–133	130–144	140–149
5	7	123–136	133–147	143–163
5	8	126–139	136–150	146–167
5	9	129–142	139–153	149–170
5	10	132–145	142–156	152–173
5	11	135–148	145–159	155–176
6	0	138–151	148–162	158–179

NOTE: Although the above tables commence at specified heights, no minimum height requirement has been prescribed. The Civil Service examining physician may determine that weight in excess of that shown in the tables (up to a maximum of twenty pounds) is lean body mass and not fat. Decision as to frame size of a candidate shall be made by the examining physician.

Step four: Now embark on a positive program to build up your strength, agility, speed, and stamina. The chapter that follows will give you some suggestions.

Preparing for the Physical Test

Defining Fitness

Physical fitness is to the human body what fine tuning is to an engine. It enables us to perform up to our potential. Fitness can be described as a condition that helps us look, feel, and do our best. More specifically, it is:

"The ability to perform daily tasks vigorously and alertly, with energy left over for enjoying leisure-time activities and meeting emergency demands. It is the ability to endure, to bear up, to withstand stress, to carry on in circumstances where an unfit person could not continue and is a major basis for good health and well-being."

Physical fitness involves the performance of the heart and lungs and the muscles of the body. And, since what we do with our bodies also affects what we can do with our minds, fitness influences to some degree qualities such as mental alertness and emotional stability.

As you undertake your fitness program, it's important to remember that fitness is an individual quality that varies from person to person. It is influenced by age, sex, heredity, personal habits, exercise, and eating practices. You can't do anything about the first three factors. However, it is within your power to change and improve the others where needed.

Knowing the Basics

Physical fitness is most easily understood by examining its components, or "parts." There is widespread agreement that these four components are basic:

CARDIORESPIRATORY ENDURANCE–the ability to deliver oxygen and nutrients to tissues and to remove wastes over sustained periods of time. Long runs and swims are among the methods employed in measuring this component.

MUSCULAR STRENGTH–the ability of a muscle to exert force for a brief period of time. Upper-body strength, for example, can be measured by various weight-lifting exercises.

MUSCULAR ENDURANCE–the ability of a muscle, or a group of muscles, to sustain repeated contractions or to continue applying force against a fixed object. Pushups are often used to test endurance of arm and shoulder muscles.

FLEXIBILITY–the ability to move joints and use muscles through their full range of motion. The sit-and-reach test is a good measure of flexibility of the lower back and backs of the upper legs.

BODY COMPOSITION is often considered a component of fitness. It refers to the makeup of the body in terms of lean mass (muscle, bone, vital tissue, and organs) and fat mass. An optimal ratio of fat to lean mass is an indication of fitness, and the right type of exercises will help you decrease body fat and increase or maintain muscle mass.

A Workout Schedule

How often, how long, and how hard you exercise, and what kinds of exercises you do should be determined by what you are trying to accomplish. Right now your goal is to prepare your body so that it can withstand the rigors of the physical performance test and make a top showing. Your goal should also be solid preparation for the physical demands of firefighting.

Your exercise program should include something from each of the four basic fitness components described previously. Each workout should begin with a warm up and end with a cool down. As a general rule, space your workouts throughout the week and avoid consecutive days of hard exercise.

Here are the amounts of activity necessary for the average, healthy person to maintain a minimum level of overall fitness. Included are some of the popular exercises for each category.

WARM UP–5–10 minutes of exercises such as walking, slow jogging, knee lifts, arm circles or trunk rotations. Low intensity movements that simulate movements to be used in the activity can also be included in the warm up.

MUSCULAR STRENGTH–a minimum of two 20-minute sessions per week that include exercises for all the major muscle groups. Lifting weights is the most effective way to increase strength.

MUSCULAR ENDURANCE–at least three 30-minute sessions each week that include exercises such as calisthenics, pushups, situps, pullups, and weight training for all the major muscle groups.

CARDIORESPIRATORY ENDURANCE–at least three 20-minute bouts of continuous aerobic (activity requiring oxygen) rhythmic exercise each week. Popular aerobic conditioning activities include brisk walking, jogging, swimming, cycling, rope-jumping, rowing, cross-country skiing, and some continuous action games like racquetball and handball.

FLEXIBILITY–10–12 minutes of daily stretching exercises performed slowly, without a bouncing motion. This can be included after a warm up or during a cool down.

COOL DOWN–a minimum of 5–10 minutes of slow walking, low-level exercise, combined with stretching.

A Matter of Principle

The keys to selecting the right kinds of exercises for developing and maintaining each of the basic components of fitness are found in these principles:

SPECIFICITY–pick the right kind of activities to affect each component. Strength training results in specific strength changes. Also, train for the specific activity you're interested in. For example, optimal swimming performance is best achieved when the muscles involved in swimming are trained for the movements required. It does not follow that a good runner is a good swimmer.

OVERLOAD–work hard enough, at levels that are vigorous and long enough to overload your body above its resting level, to bring about improvement.

REGULARITY–you can't hoard physical fitness. At least three balanced workouts a week are necessary to maintain a desirable level of fitness.

PROGRESSION–increase the intensity, frequency and/or duration of activity over periods of time in order to improve.

Some activities can be used to fulfill more than one of your basic exercise requirements. For example, in addition to increasing cardiorespiratory endurance, running builds muscular endurance in the legs, and swimming develops the arm, shoulder and chest muscles. If you select the proper activities, it is possible to fit parts of your muscular endurance workout into your cardiorespiratory workout and save time.

Measuring Your Heart Rate

Heart rate is widely accepted as a good method for measuring intensity during running, swimming, cycling, and other aerobic activities. Exercise that doesn't raise your heart rate to a certain level and keep it there for 20 minutes won't contribute significantly to cardiovascular fitness.

The heart rate you should maintain is called your **target heart rate**. There are several ways of arriving at this figure. One of the simplest is: **maximum heart rate** (220 – age) x 70%. Thus, the target heart rate for a 40-year-old would be 126.

Some methods for figuring the target rate take individual differences into consideration. Here is one of them:

1. Subtract age from 220 to find **maximum heart rate**.
2. Subtract resting heart rate (see below) from maximum heart rate to determine **heart rate reserve**.
3. Take 70% of heart rate reserve to determine **heart rate raise**.
4. Add heart rate raise to resting heart rate to find **target rate**.

Resting heart rate should be determined by taking your pulse after sitting quietly for five minutes. When checking heart rate during a workout, take your pulse within five seconds after interrupting exercise because it starts to go down once you stop moving. Count pulse for 10 seconds and multiply by six to get the per-minute rate.

A Measure of Your Progress

You will, of course, be able to observe the increase in your strength and stamina from week to week in many ways.

In addition, there is a 2-minute step test you can use to measure and keep a running record of the improvement in your circulatory efficiency, one of the most important of all aspects of fitness.

The immediate response of the cardiovascular system to exercise differs markedly between well-conditioned individuals and others. The test measures the response in terms of pulse rate taken shortly after a series of steps up and down onto a bench or chair.

Although it does not take long, it is necessarily vigorous. Stop if you become overly fatigued while taking it. You should not try it until you have completed a number of weeks of conditioning exercises.

The Step Test

Use any sturdy bench or chair 15–17 inches in height.

Count 1—Place right foot on bench.
Count 2—Bring left foot alongside of right and stand erect.
Count 3—Lower right foot to floor.
Count 4—Lower left foot to floor.

REPEAT the 4-count movement 30 times a minute for two minutes.

THEN sit down on bench or chair for two minutes.

FOLLOWING the 2-minute rest, take your pulse for 30 seconds.

Double the count to get the per-minute rate. (You can find the pulse by applying middle and index finger of one hand firmly to the inside of the wrist of the other hand, on the thumb side.)

Record your score for future comparisons. In succeeding tests—about once every two weeks—you probably will find your pulse rate becoming lower as your physical condition improves.

Three important points:

1. For best results, do not engage in physical activity for at least 10 minutes before taking the test. Take it about the same time of day and always use the same bench or chair.

2. Remember that pulse rates vary among individuals. This is an individual test. What is important is not a comparison of your pulse rate with that of anybody else—but rather a record of how your own rate is reduced as your fitness increases.

3. As you progress, the rate at which your pulse is lowered should gradually level off. This is an indication that you are approaching peak fitness.

Staying Fit

Once you have reached the level of conditioning you have chosen for yourself, you will wish to maintain your fitness.

While it has been found possible to maintain fitness with three workouts a week, ideally, exercise should be a daily habit. If you can, by all means continue your workouts on a five-times-a-week basis.

Broadening Your Program

There are many other activities and forms of exercise which, if you wish, you may use to supplement a basic program.

They include a variety of sports; water exercises you can use if you have access to a pool; and isometrics—sometimes called exercises without movement—which take little time (6–8 seconds each).

Isometrics

Isometric contraction exercises take very little time, and require no special equipment. They're excellent muscle strengtheners and, as such, are valuable supplements.

The idea of isometrics is to work out a muscle by pushing or pulling against an immovable object such as a wall...or by pitting it against the opposition of another muscle.

The basis is the "overload" principle of exercise physiology—which holds that a muscle required to perform work beyond the usual intensity will grow in strength. And research has indicated that one hard, 6- to 8-second isometric contraction per workout can, over a period of six months, produce a significant strength increase in a muscle.

The exercises illustrated and described in the following pages cover major large muscle groups of the body.

They can be performed almost anywhere and at almost any time.

There is no set order for doing them—nor do all have to be completed at one time. You can, if you like, do one or two in the morning, others at various times during the day whenever you have half a minute or even less to spare.

For each contraction, maintain tension *no more than eight seconds.* Do little breathing during a contraction; breathe deeply between contractions.

And start easily. Do *not* apply maximum effort in the beginning.

For the first three or four weeks, you should exert only about one-half what you think is your maximum force.

Use the first three or four seconds to build up to this degree of force—and the remaining four or five seconds to hold it.

For the next two weeks, gradually increase force to more nearly approach maximum. After about six weeks, it will be safe to exert maximum effort.

Pain indicates you're applying too much force; reduce the amount immediately. If pain continues to accompany any exercise, discontinue using that exercise for a week or two. Then try it again with about 50 percent of maximum effort and, if no pain occurs, you can go on to gradually build up toward maximum.

Neck

Starting position: Sit or stand, with interlaced fingers of hands on forehead.
Action: Forcibly exert a forward push of head while resisting equally hard with hands.
Starting position: Sit or stand, with interlaced fingers of hands behind head.
Action: Push head backward while exerting a forward pull with hands.
Starting position: Sit or stand, with palm of left hand on left side of head.
Action: Push with left hand while resisting with head and neck. Reverse using right hand on right side of head.

Upper Body

Starting position: Stand, back to wall, hands at sides, palms toward wall.
Action: Press hands backward against wall, keeping arms straight.
Starting position: Stand, facing wall, hands at sides, palms toward wall.
Action: Press hands forward against wall, keeping arms straight.
Starting position: Stand in doorway or with side against wall, arms at sides, palms toward legs.
Action: Press hand(s) outward against wall or doorframe, keeping arms straight.

Arms

Starting position: Stand with feet slightly apart. Flex right elbow, close to body, palm up. Place left hand over right.
Action: Forcibly attempt to curl right arm upward, while giving equally strong resistance with the left hand. Repeat with left arm.

Arms and Chest

Starting position: Stand with feet comfortably spaced, knees slightly bent. Clasp hands, palms together, close to chest.
Action: Press hands together and hold.
Starting position: Stand with feet slightly apart, knees slightly bent. Grip fingers, arms close to chest.
Action: Pull hard and hold.

Abdominal

Starting position: Stand, knees slightly flexed, hands resting on knees.
Action: Contract abdominal muscles.

Lower Back, Buttocks and Back of Thighs

Starting position: Lie face down, arms at sides, palms up, legs placed under bed or other heavy object.
Action: With both hips flat on floor, raise one leg, keeping knee straight so that heel pushes hard against the resistance above. Repeat with opposite leg.

Legs

Starting position: Sit in chair with left ankle crossed over right, feet resting on floor, legs bent at 90 degree angle.
Action: Forcibly attempt to straighten right leg while resisting with the left. Repeat with opposite leg.

Inner and Outer Thighs

Starting position: Sit, legs extended with each ankle pressed against the outside of sturdy chair legs.

Action: Keep legs straight and pull toward one another firmly. For outer thigh muscles, place ankles inside chair legs and exert pressure outward.

Water Activities

Swimming is one of the best physical activities for people of all ages—and for many of the handicapped.

With the body submerged in water, blood circulation automatically increases to some extent; pressure of water on the body also helps promote deeper ventilation of the lungs; and with well-planned activity, both circulation and ventilation increase still more.

Weight Training

Weight training also is an excellent method of developing muscular strength—and muscular endurance. Both barbells and weighted dumbbells—complete with instructions—are available at most sporting goods stores. A good rule to follow in deciding the maximum weight you should lift is to select a weight you can lift six times without strain. If you have access to a gym with sophisticated equipment, take advantage of the advice of professional trainers in establishing weight training programs and goals.

Sports

Soccer, basketball, handball, squash, ice hockey, and other sports that require sustained effort can be valuable aids to building circulatory endurance.

But if you have been sedentary, it's important to pace yourself carefully in such sports, and it may even be advisable to avoid them until you are well along in your physical conditioning program. That doesn't mean you should avoid all sports.

There are many excellent conditioning and circulatory activities in which the amount of exertion is easily controlled and in which you can progress at your own rate. Bicycling is one example. Others include hiking, skating, tennis, running, cross-country skiing, rowing, canoeing, water skiing, and skindiving.

Games should be played with full speed and vigor only when your conditioning permits doing so without undue fatigue.

On days when you get a good workout in sports you can skip part or all of your exercise program. Use your own judgment.

If you have engaged in a sport which exercises the legs and stimulates the heart and lungs—such as skating—you could skip the circulatory activity for that day, but you still should do some of the conditioning and stretching exercises for the upper body. On the other hand, weight-lifting is an excellent conditioning activity, but it should be supplemented with running or one of the other circulatory exercises.

Whatever your favorite sport, you will find your enjoyment enhanced by improved fitness. Every weekend athlete should invest in frequent workouts.

New York City
Physical Test for Firefighters

The physical demands upon the bodies of firefighters are far greater than demands made upon the bodies of those in most other occupations. Firefighters alternate long periods of relative inactivity with periods of intensive strenuous effort. In their active periods, firefighters are burdened with heavy protective equipment under conditions of oppressive heat and smoke, time pressure, and an emotional component as well, often in cramped, confining spaces for punishingly long stretches. To perform effectively under these conditions, a firefighter must be healthy and strong and must be in excellent condition. Because top physical condition and stamina are so important, the New York City physical performance test carries as much weight as the written exam in qualifying the firefighter applicant. The physical performance test is scored, and the applicant must score high to remain in competition for hiring. In most civil service testing, the written exam is the sole component of the competitive score; the physical performance test, if any, is only a qualifying test. However, for New York City firefighting candidates, half of the competitive score is based on the score on the physical performance test. Firefighting candidates in other locations may also find themselves facing physical performance tests that enter into calculation of the competitive score or they may take pass/fail style physical performance exams. Either way, passing the physical performance segment is a vital step along the way to earning a place on a firefighting force.

The physical performance test most recently administered by the City of New York challenged candidates to perform a series of activities which are often performed in the course of firefighting, encumbered by some of the clothing that firefighters must wear and carrying some equipment that firefighters must carry. Even the time pressure of the firefighter physical performance test is similar to the time pressure created by the requirements of an actual fire. The only difference is that the test is taken under controlled temperature and air-quality conditions rather than under the high temperature, heavy smoke, and low oxygen conditions of actual firefighting.

Each fire department prepares its own exam and each includes some features that are not found on the exams given by other departments. However, the similarities far exceed the differences. All seek to measure strength, agility, and stamina. Your own exam will not be identical to the New York City exam described here (indeed the next New York City exam will probably not be identical to the last one), but if you are in good shape and are prepared for this exam, you will be prepared for whatever physical performance exam you must take.

You should report to the physical performance test site on the date and at the time assigned to you wearing long pants, comfortable clothing, and sneakers. Kneepads are not permitted; leave them home even if you have them. At the test site you will be issued a weighted vest, a firefighter's air tank, and gloves. You will be required to wear these for all events except for the ceiling hook event in which you must pull down a ceiling. You may remove the air tank for the ceiling hook event. The firefighter's gear weighs about forty pounds.

The eight events of the physical performance test follow one right after the other, with rest periods only as noted, between events. You may drink water during the rest periods. The following are descriptions of the events in the order in which they are administered.

The Test

1. Stepmill:

Fully clad in firefighter's gear including weighted vest, airpack, and gloves, the candidate must walk on a stepmill at the rate of 59 steps per minute for five minutes and twelve seconds. The stepmill is not a conventional stairmaster, but rather resembles a short, 3-4 step escalator.

Five-minute rest period.

2. Ladder event:

This event is in two parts. Both involve activities typically engaged in with respect to ladders. In the first part, the candidate, fully laden with vest, gloves, and airpack, must lift a 20-foot straight aluminum ladder that is lying flat on the ground and that is hinged to the floor at the butt end. The candidate lifts the ladder by starting at the top end and "walking" the ladder up until it is positioned against a wall. In the second part of the event, the candidate must raise the extension on a 24-foot aluminum extension ladder that is mounted to a wall. The entire ladder event must be completed in 15.81 seconds.

Twenty seconds to get to next event.

3. Hose feed:

In this event the candidate must feed a total of 50 feet of a 150-foot, 1¾ inch diameter hose line (weighted as if charged) across a circular roller bed within 18.19 seconds by using a hand over hand motion. The initial effort to lift and start the hose on its way is very strenuous. Once the hose begins moving across the roller bed, the movement of the rollers makes the task easier and the candidate can concentrate on speed.

Twenty seconds to get to next event.

4. Hose advance:

In this event the candidate must perform the actions necessary to pull a hose to a fire. As this event begins, ten feet of a 150-foot fully charged (water filled) hose is resting on a roller bed. The remainder of the hose is coiled behind the roller bed. The candidate must take the end of the hose from the roller bed and pull it through a 50-foot course with two turns. The course through which the hose must be pulled is only four feet high, so candidates cannot stand upright while pulling the hose. The turns can present a problem; the hose can get caught on the corners. The candidate should take the turns very wide so as to avoid getting caught in the bends. This event must be completed within 46.33 seconds.

Twenty seconds to get to next event.

5. Forcible entry:

In the forcible entry exercise, the candidate must use a 12-pound sledgehammer and muscle power in striking a rubber pad mounted on a device which measures force. This event simulates the activity used by firefighters to force open a locked door. The candidate must generate enough force within 10.95 seconds to light a bulb connected to the device on which the rubber pad is mounted.

Twenty seconds to get to next event.

6. Search:

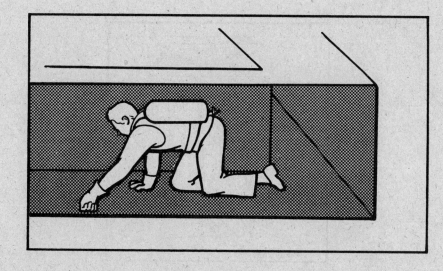

In this event the candidate must perform the actions necessary to crawl through a smoke-filled structure, but without the smoke. Wearing weighted vest, gloves, and air pack, the candidate must crawl through an 80-foot, totally dark, enclosed tunnel which includes a number of obstructions and turns. At its roomiest the tunnel is three feet high and five feet wide. In two separate locations, the width of the tunnel is reduced to three feet. It is easy to lose your way or your sense of direction in the dark tunnel. A wise maneuver is to feel for the walls making turns so as not to turn full circle. The search event must be completed within 95.66 seconds.

Twenty seconds to get to next event.

7. Rescue:

In this event, the candidate must demonstrate the strength and agility to drag a victim to safety. The candidate must drag a 130-pound dummy 30 feet along a U-shaped course to a finish line. The course through which the dummy must be pulled is only four feet high, so the candidate must perform the task in a semi-crouch position. Speed is vital in firefighting and rescue, so speed is an important factor throughout the physical performance test. The rescue event must be completed within 23.80 seconds.

Five-minute rest period.

8. Ceiling pull:

For this final event, which simulates firefighter activities after a fire, candidates may remove the air pack. This event requires four one-minute intervals of pushing and pulling. The candidate begins by entering the test area and removing a hooked pole from a canister. The candidate must push up a 60-pound hinged lid, resembling a manhole cover, about one foot until a light goes on. The candidate then runs to another hinged ceiling grate, this one about 80 pounds, from which a pole is hanging down. The candidate pulls the grate from the ceiling and allows it to pop back up five times. Since the grate springs back, the candidate will find the task much easier if he or she hangs on tightly to the pole and allows body weight to assist in the next pull-down in a bouncing type of action. After five pull-downs, the candidate returns to the push-up to begin another set. Each set consists of one push-up and five pull-downs. A candidate is permitted to do up to seven sets in one minute. After seven sets or one minute, the candidate must rest for 30 seconds. Candidates are required to perform 20 such sets within four one-minute periods in order to pass this event.

24A

A Word of Caution

Without a doubt, this is a physically demanding test. However, a consulting expert in sports medicine and exercise physiology has concluded that the test course does not appear to cause greater average changes in the body than would be expected from other forms of similar exercise. It poses no health hazard to physically fit candidates. It is a physically exerting test as it reflects the strenuous job firefighters are required to do, and only those who are physically fit should attempt to take it.

Development of this test included many trials by current firefighters. These firefighters agreed that the exam reasonably approximates the physical demands upon firefighters. The timing requirements of the individual events are based upon the performance of the active firefighters who participated in the development of the test and the test site.

It is suggested that candidates avoid eating or drinking for several hours before taking the test. Further, candidates are informed that those who have had a cold, flu-like illness, nausea or vomiting two weeks prior to the test should reschedule the test; strenuous exercise should be avoided 48 hours before and 24 hours after the test; alcohol should be completely avoided for 24 hours before and after the test; the use of any over-the-counter medication such as aspirin or aspirin-containing products as well as non-steroidal, anti-flammatory agents such as Advil, Nuprin, Medipren, Haltran, Midol and Trendar should be avoided for 24 hours before and after the test; food consumption from any source for which the sanitary nature of its preparation cannot be verified should be avoided. Individuals who are users of drugs such as cocaine, crack, heroin and marijuana are at an increased risk of developing medical complications and should not take the test.

Candidates are required to bring a doctor's note and will be informed of medical disorders that may place them at increased risk of serious medical complications. Electrolyte solutions and other liquids are available at the test site. Emergency Medical Technicians, paramedics and an ambulance are stationed at the test site to care for those candidates who may have disregarded these precautions.

Scoring

Each of the eight events is scored separately on a pass/fail basis. There is no numerical score for any single event, nor is there partial credit. The candidate who does not pass at least six of the eight events fails the entire examination. Since the score on the physical performance test constitutes 50% of the firefighter candidate's total examination score, the candidate who fails the physical performance exam fails the firefighter exam. Candidates passing six of the eight events receive a score of 75% on the physical test; those passing seven of the eight events receive 87.5%; and those passing all eight events score 100% on the physical half of the firefighter exam. Written score and physical performance score are converted to "standard scores" according to Personnel Department formula. The two scores are then combined to yield a total score on the New York City firefighter exam. Passing scores are reported in a range from 70% to 100%. Where applicable, veterans credits and credit for New York City residency may be added to passing scores so that some candidates may achieve final reported scores higher than 100%.

Another Physical Fitness Test

Testing methods vary. Your written test may be quite different from any of the written tests described in this book, though all will be intended to assess your capacity to learn the job of the firefighter and to do it well. Likewise, while all firefighter physical fitness tests are designed to measure physical strength and stamina, the procedures used may be very different from one fire department to the next. The following physical fitness test is very different from the New York City exam just described, but you must be equally fit to pass it.

The candidates who qualify on the medical examination will be required to pass the qualifying physical fitness test.

A total score of 20 is required for passing this test; the scores attained on the five individual tests are added together to obtain your final score.

Requirements for the Qualifying Physical Fitness Test

Test 1—Trunk Flexion Test—3 Chances

Candidates will assume a sitting position on the floor with the legs extended at right angles to a line drawn on the floor. The heels should touch the near edge of the lines and be 5 inches apart. The candidate should slowly reach with both hands as far forward as possible on a yardstick that is placed between the legs with the 15-inch mark resting on the near edge of the heel line.

The score is the most distant point (in inches) reached on the yardstick with fingertips.

Rating	Trunk Flexion (inches)	Points
Excellent	22 and over	6
Good	20–21	5
Average	14–19	4
Fair	12–13	3
Poor	10–11	2
Very poor	9 and under	1

Test II—Hand Grip Strength Test—3 Chances

The candidate places the dynamometer (hand grip tester) at the side and without touching the body with any part of the arm, hand or the dynamometer, should grip the dynamometer as hard as possible in one quick movement. The best of the 3 tries will be recorded.

Rating	Hand Grip in Kg.	Points
Excellent	65 and above	6
Good	57–64	5
Average	44–56	4
Fair	37–43	3
Poor	30–36	2
Very poor	29 and under	1

Test III—Standing Broad Jump—3 Chances

Candidates will be permitted 3 chances in consecutive order, and the longest distance will be credited.

Candidates will be required to jump from a standing position, both feet together. Distance of jump will be recorded from starting point to back of heels. It is each candidate's responsibility to have a non-skid surface on the soles of his or her sneakers.

Rating	Distance	Points
Excellent	7'10" or better	6
Good	7' to 7'9"	5
Average	6'1" to 6'11"	4
Fair	5'6" to 6'0"	3
Poor	5'0" to 5'5"	2
Very poor	Less than 5'	1

Test IV—One-Minute Situp Test

The candidate will start by lying on the back with the knees bent so that the heels are about 18 inches away from the buttocks. An examiner will hold the ankles to give support. The candidate will then perform as many correct situps (elbows alternately touching the opposite knee) as possible within a one minute period. The candidate should return to the starting position (back to floor) between situps.

Rating	Situps in 1 minute	Points
Excellent	35	6
Good	30–34	5
Average	20–29	4
Fair	15–19	3
Poor	10–14	2
Very poor	9 and under	1

Test V—Three-Minute Step Test

The candidate will step for 3 minutes on a 12-inch bench at a rate of 24 steps per minute. The time will be maintained by a metronome. Immediately after the 3 minutes of stepping, the subject will sit down and relax without talking. A 60-second heart rate count is taken starting 5 seconds after the completion of stepping.

Rating	Pulse	Points
Excellent	75–84	6
Good	85–94	5
Average	95–119	4
Fair	120–129	3
Poor	130–139	2
Very poor	140 and over	1

The Interview

An interview can take place at any time during the screening process, but where there are many applicants, the employment interview tends to be the last step. An interview takes a great deal of the interviewer's time so is offered only to candidates who apear fully qualified on the basis of all other measures—written test, background investigation, medical examination, and physical fitness performance test.

If you have passed all the cuts up to the interview, you are very close, but you do not yet have a place on the fire force. The interview is a very important step in the firefighter screening process.

The purpose of the interview is to gather information along a number of dimensions and to fill you out as a whole person. Some of the aims of the interview are to:

1. **Supplement the application form**. The interviewer will ask about childhood, education and prior employment. He, she, or they will give you an opportunity to explain employment gaps and abrupt terminations. If your record is anything less than perfect, this is your chance to indicate that you have learned from your prior experience. Now is the time to let the interviewer know how you have matured. Do not blame others for your mistakes. Take responsibility for impetuous behavior, personality clashes, brushes with the law; then make clear that you have developed the self-control to keep from such misbehaviors in the future.

   You may also have an opportunity to expand upon the kinds of work you did at prior jobs. Where the application form gave you limited space in which to describe duties and responsibilities, you can now fill in details. You can convey enthusiasm for those tasks which you especially liked or at which you were notably skilled.

2. **Learn of your motivation**. Firefighting is not easy work. The interviewer wants to know why you want to be a firefighter. Is your motivation strictly financial or do you have an unhappy home life that you want to escape for long stretches? Do you care about people or are you just seeking excitement? How do your interests coincide with those of current firefighters? Do your patterns of interest and motivation match those of successful, satisfied firefighters?

3. **Assess your stability and personality**. This is a tall order, and your interview may or may not do it successfully. The firefighter lives and works under stress. In the firehouse the firefighter lives in close quarters with other men and women, sharing household and maintenance chores, sacrificing privacy. The firefighter must be able to cooperate with all members of the shift without showing undue irritation at the person who may prove annoying. Likewise, the firefighter must not display habits or mannerisms which might prove irritating to others and create dis-

cord in the firehouse. In short, the firefighter must "get along" and must "fit in." The stress which firefighters face is even greater when they fight fires. Interview questions will try to gauge how well you can follow orders and how you will react under conditions of real physical pressure. The interviewer will look for signs as to how you can juggle order-taking with initiative. Some of the questions may be hard to answer. The interview is itself a stressful situation. Consider alternatives and give a decisive answer. A firefighter must think before acting, but must not spend too much time thinking about next steps. Try to convey this same balance between deliberation and quick thinking as you answer the tough questions. Give the impression that you have the self-confidence both to sustain difficult activity under pressure and to size up situations and deviate from prescribed routine when warranted by extreme emergency. Some fire departments supplement the interview with psychological tests to further screen out candidates who give signs of a tendency to "crack" when in real danger or when faced with frustrating decisions.

Since this interview is the final step, the moment of decision as to whether or not you will be accepted on the force, you will want to come across at your very best. There are bound to be some surprises, but to a large extent you can prepare yourself for an interview.

First of all, write down the date, time, and place on a wall calendar at home and in your pocket calendar. Fold the notice of interview into your wallet so that you cannot make a mistake. If you are not sure of how to get to the interview site, look into public transportation or automobile routes ahead of time. Do a dry run if necessary. On interview day, leave enough time for the unexpected. A three-hour pile-up on the highway will be reported in the newspaper and is an acceptable reason for missing an appointment, but a twenty-minute bus breakdown is no excuse. If you arrive much too early, go for a cup of coffee. Err on the side of time to spare. Actually, you will want to go to the floor of your interview about ten minutes ahead of time so as to be able to go to the restroom to wash your hands and comb your hair.

Next, dress for the occasion. You are not being interviewed for an executive position. You need not wear a three-piece suit. However, your clothing should be businesslike, clean and neat. You should be well groomed. You want to impress the interviewer as someone he or she would find pleasant to spend a lot of time with. You will add to this impression with a firm handshake, an easy smile, and frequent eye contact with the interviewer or interviewers. Do not chew gum and do not smoke even if you are offered a cigarette.

Prepare yourself with answers to the questions. You may wonder how you can prepare the answers to questions that you have not seen. Actually, this is easy. The questions are quite predictable. Begin your preparation by looking over the application forms that you filled out and any other papers that you were required to file. You should be able to pick out the points that an interviewer will want you to amplify or to explain.

Questions arising from the information you have already given might be:

- Why did you choose your area of concentration in school?
- What particularly interests you about _____ (particular subject)?
- Why did you transfer from x school to y school?
- How did you get the job with _____?
- Which of the duties descibed in your second job did you like best? Which least? Why?
- What did you do during the nine months between your second and third jobs?
- Please explain your attendance pattern at your first job.

- Explain the circumstances of your leaving a particular job.
- Please clarify: armed forces service, arrest record, hospitalization record, etc., as applicable.

Other questions are almost routine to all interviews. They can be anticipated and prepared for as well.

- Why do you want to leave the kind of work you are doing now?
- Why do you want to be a firefighter?
- How does your family feel about your becoming a firefighter?
- What do you do in your leisure time?
- Do you have hobbies? What are they? What do you particularly like about _____?
- What is your favorite sport? Would you rather play or watch?
- How do you react to criticism? If you think the criticism is reasonable? If you consider the criticism unwarranted?
- What is your pet peeve?
- Name your greatest strengths. Weaknesses.
- What could make you lose your temper?
- Of what accomplishment in your life are you most proud?
- What act do you most regret?
- If you could start over, what would you do differently?
- What traits do you value most in a co-worker? In a friend?
- What makes you think you would make a good firefighter?

Still other questions may be more specific to a firefighter interview. You should have prepared answers to:

- How much sleep do you need?
- Are you afraid of heights?
- What is your attitude toward irregular hours?
- Do you prefer working alone or on a team?
- Are you afraid of dying?
- What would you do with the rest of your life if your legs were crippled in an accident?
- How do you deal with panic? Your own? That of others?
- What is your attitude toward smoking? Drinking? Drugs? Playboy magazine? Gambling?
- What is your favorite TV program? How do you feel about watching news? Sports? Classical drama? Rock music? Opera? Game shows? Etc.

Now make a list of your own. The variety of interview questions is endless, but most can be answered with ease. Preparation makes the whole process much more pleasant and less frightening.

There is one question that strikes terror into the heart of nearly every job candidate. This question is likely to be the first and, unless you are prepared for it, may well throw you off your guard. The question is, "Tell me about yourself." For this question you should have a prepared script (in your head, not in your hand). Think well ahead of time about what you want to tell. What could the interviewer be interested in? This question is not seeking information about your birthweight or your food preferences. The interviewer wants you to tell about yourself with relation to your interest in and qualifications for firefighting. Think of how to describe yourself with this goal in mind. What information puts you into a good light with reference to the work for which you are applying? Organize your presentation. Then analyze what you plan to say. What is an interviewer likely to pick up on? To what questions will your little speech lead? You must prepare to answer these questions to which you have opened yourself.

The temptation to talk too much is greatest with this open-ended question, but it exists with many questions. Resist the urge to ramble on and on. Give full, complete answers, but stop when you have answered the questions. Do not tell the interviewer more than he or she needs to know. Do not volunteer extra, superfluous information. Do not get anecdotal, chatty or familiar. Remember that the interview is a business situation, not a social one.

Towards the end of the interview, the interviewer will most likely ask if you have any questions. You undoubtedly will have had some before the interview and should have come prepared to ask. If all of your questions have been answered in the course of the interview, you may tell this to the interviewer. If not, or if the interview has raised new questions in your mind, by all means ask them. The interview should serve for your benefit; it is not just to serve the purposes of the Personnel Division or the Fire Department.

The invitation of your questions tends to be the signal that the interview is nearly over. The interviewer is satisfied that he or she has gathered enough information. The time allotted to you is up. Be alert to the cues. Do not spoil the good impression you have made by trying to prolong the interview.

At the end of the interview, smile, shake hands, thank for the opportunity to have the interview, and leave. A brief thank-you note to the interviewer or to the chairperson of the interviewing panel would not be out of order. Such a note brands you as a courteous, thoughtful person and indicates your continued interest in appointment.

- All of your answers should be in the form of blackened spaces. The machine cannot read English. Do not write any notes in the margins. If you have done any figuring in the margins of the test booklet itself, be sure to mark the letter of your answer on the answer sheet. Correct answers in the test booklet do not count. Only the answer sheet is scored.

- MOST IMPORTANT: Answer each question in the right place. Question 1 must be answered in space 1; question 52 in space 52. If you should skip an answer space and mark a series of answers in the wrong places, you must erase all those answers and do the questions over, marking your answers in the proper places. You cannot afford to use the limited time in this way. Therefore, as you answer *each* question, look at its number and check that you are marking your answer in the space with the same number. The risk involved in slipping out of line on the answer sheet is the reason we recommend that you answer every question in order, even if you have to guess. Make it an educated guess if you can. If not, make a wild guess. Just mark the question number in the test booklet so that you can try again if there is time.

TEST DAY STRATEGY

On the examination day assigned to you, allow the test itself to be the main attraction of the day. Do not squeeze it in between other activities. Arrive rested, relaxed, and on time. In fact, plan to arrive a little bit early. Leave plenty of time for traffic tie-ups or other complications which might upset you and interfere with your test performance.

In the test room the examiner will hand out forms for you to fill out. He or she will give you the instructions that you must follow in taking the examination. The examiner will tell you how to fill in the grids on the forms. Time limits and timing signals will be explained. If you do not understand any of the examiner's instructions, ASK QUESTIONS. Make sure that you know exactly what to do.

At the examination, you must follow instructions exactly. Fill in the grids on the forms carefully and accurately. Filling in the wrong grid may lead to loss of veterans' credits to which you may be entitled or to an incorrect address for your test results. Do not begin until you are told to begin. Stop as soon as the examiner tells you to stop. Do not turn pages until you are told to. Do not go back to parts you have already completed. Any infraction of the rules is considered cheating. If you cheat, your test paper will not be scored, and you will not be eligible for appointment.

Once the signal has been given and you begin the exam, READ every word of every question. Be alert for exclusionary words that might affect your answer—words like "not," "most," "least," "all," "every," and "except."

READ all the choices before you mark your answer. It is statistically true that most errors are made when the last choice is the correct answer. Too many people mark the first answer that seems correct without reading through all the choices to find out which answer is *best*.

The following list summarizes the suggestions we have already given for taking this exam. Read the suggestions now and before you attempt the exams in this book. Review them before you take the actual exam. You will find them all useful.

1. Mark your answers by completely blackening the answer space of your choice.

2. Mark only ONE answer for each question, even if you think that more than one answer is correct. You must choose only one. If you mark more than one answer, the scoring machine will consider you wrong even if one of your answers is correct.*

3. If you change your mind, erase completely. Leave no doubt as to which answer you have chosen.

4. If you do any figuring on the test booklet or on scratch paper, be sure to mark your answer on the answer sheet.

5. Check often to be sure that the question number matches the answer space number and that you have not skipped a space by mistake. If you do skip a space, you must erase all the answers after the skip and answer all the questions again in the right places.

6. Answer every question in order, but do not spend too much time on any one question. If a question seems to be "impossible," do not take it as a personal challenge. Guess and move on. Remember that your task is to answer correctly as many questions as possible. You must apportion your time so as to give yourself a fair chance to read and answer all the questions. If you guess at an answer, mark the question in the test booklet so that you can find it easily if time allows.

7. Guess intelligently if you can. If you do not know the answer to a question, eliminate the answers that you know are wrong and guess from among the remaining choices. If you have no idea whatsoever of the answer to a question, guess anyway. Choose an answer other than the first. The first choice is generally the correct answer less often than the other choices. If your answer is a guess, either an educated guess or a wild one, mark the question in the question booklet so that you can give it a second try if time permits. If time is about to run out and you have not had a chance to answer all the questions, mark all the remaining questions with the same answer. According to the law of averages, you should get 25 percent of them right, and you can't lose.

8. If you happen to finish before time is up, check to be sure that each question is answered in the right space and that there is only one answer for each question. Return to the difficult questions that you marked in the booklet and try them again. There is no bonus for finishing early so use all your time to perfect your exam paper.

* Note: A recent administration of the Firefighter Exam in New York City experimented with a new method of marking answers. Instead of marking only one answer for each question as is done in conventional multiple-choice exams, test-takers were told to mark first choice, second choice, and third choice for each question. Partial credit was then awarded to questions for which a test taker marked the correct answer as a second or third choice. This change in the marking system was reported in the newspapers only days before the exam and caught everyone by surprise. Detailed instructions for marking the answer sheet accompanied the exam and answer sheet. Even so, many applicants had difficulty with the complicated system. Test makers do experiment. The lesson to you is: READ CAREFULLY. Do not assume that all rules and procedures will remain as they have always been. Read all instructions. Ask questions before the exam begins. Be certain that you understand exactly what you are to do. Then, follow through as directed.

GLOSSARY OF FIREFIGHTING TERMINOLOGY

access stairs—In a high-rise building, a stairway, usually open, serving a number of floors of a common tenant. Also known as convenience stairs.

aerial ladder—A large metal extension ladder mounted on a truck, having a bed ladder with three sliding sections. The overall length is generally 100 feet.

apparatus—The term given to the rolling stock of a fire department, for example, ladder truck, pumpers, rescue vans.

attack stairway—A fire stair used by the fire department to gain access to the fire area, where the door between the stairway and the fire area is being maintained in an open position.

attic—The space between the ceiling framing of the top story and the underside of the roof.

backdraft—A smoke explosion. For example, if a fire starts in a sealed room, the fire will smolder for lack of oxygen. Heat and fuel are available, but oxygen is lacking. When a door or window is opened, allowing air to rush in, the fire will accelerate with explosive rapidity.

basement—A story partly underground with less than one-half its height below the curb line.

blind shaft elevators—Elevators serving the upper areas of a high-rise building in a shaft that is not equipped with hoistway doors on the lower floors.

bulkhead—An enclosed structure on or above the roof of a building not designed or used for human occupancy. For example, a stair bulkhead or an elevator bulkhead.

cellar—A story partly underground with more than one-half its height below the curb line.

charged—The term used to indicate that the hoseline has water under pressure. This term is also used to indicate that an area or building is filled with smoke.

chocks—Wooden wedges carried by firefighters to keep doors open when stretching hoselines.

cockloft—The area in a building between the ceiling of the top floor and the roof boards.

compartmentation—The subdividing of floor areas by fire-resistive separations into smaller spaces or compartments.

core type building—A building in which the elevators, stairways, and building support systems are grouped together in one area of the building. This area could be in the center of the building as in a center core building or on one of the sides of the building as in a side core building.

damper—A device to seal off or to control air flow in a heating, ventilation, air conditioning system.

distributor—A revolving nozzle that is connected to a hose and used for underpier or cellar fires.

drafting—Taking water from a river, pond, lake, etc., by suction through a hose.

dropladder—A vertical iron ladder that is generally found on the first balcony of a fire escape. It is lowered to the ground before using.

elevating platform, tower ladder, or **cherry picker**—A fire department apparatus used in rescue and firefighting situations. Generally 75 to 100 feet is the full extension of this equipment.

engine company—That branch of the fire service that is responsible for the stretching of hoselines and extinguishment of the fire.

evacuation stairway—A fire tower or a fire stairs that is remote from the fire area and is used for the evacuation of the building occupants. A fire tower is the preferred evacuation stairway.

exposures—The structures or areas that border the four sides of the fire.

extinguisher or **"can"**—The "can" is the pressurized water exinguisher used to extinguish fire in its early stages. Generally carried to the fire by the ladder company first to arrive.

fire damper—A damper used to restrict the passage of heat.

fire safety director—In a high-rise building, a designated employee holding a certificate of fitness from the fire department qualifying him or her to perform fire safety duties as required.

fire tower—An enclosed stairway connected at each story by an outside balcony or fire-proof vestibule vented to the outside.

forcible entry—Gaining entry to a locked building, store, or apartment using special "forcible entry" tools.

gooseneck—The vertical iron ladder connecting the uppermost fire escape with the roof.

halligan tool—A steel bar approximately 1¼ inches thick and 30 inches long with a slightly bent fork at one end and an adz (chisel) and point at the other. It is generally used in forcible entry but has many uses.

hook or **pike pole**—A wooden or fiberglass pole with a steel point and hook that is used to pull ceilings, strip plaster from walls, and many other uses. The poles have lengths of 6, 10, 15, and 20 feet.

hose strap—A short length of strong rope approximately 3 feet long with an iron hook at one end and a loop at the other. It is used to secure a hoseline to a ladder or fire excape but has many other uses.

ladder company or **truck company**—That branch of the fire service that is responsible for the raising of ladders, rescue, forcible entry, and ventilation of the fire building.

ladderpipe—A large caliber nozzle with sufficient hose that is mounted on the upper rungs of an aerial ladder in order to direct a stream of water into the upper floors of a building.

leeward—That side of a fire that is away from the wind. The smoky side of a fire is generally the leeward side.

locked door fail safe system—A system in which the lock mechanism is controlled electrically from a remote location.

outriggers or **tormentors**—Steel jacks that are used to stablize aerial ladders and tower ladders when they are in the elevated position.

overhauling—After a fire has been extinguished, firefighters will generally open walls, floors, and ceilings in the immediate vicinity of the main body of fire to locate any hidden fire.

parapet—The vertical extension of a wall above the roof of a building.

plenum—An air compartment or chamber to which one or more ducts are connected and which forms a part of an air distribution system. In a high-rise building, the space between the suspended ceiling and the underside of the floor above is used as a plenum for the collection of the return air.

"Q" decking—A type of composite floor construction in which corrugated steel is used to support the concrete floor.

reducer—A fitting, generally made of brass or aluminum, that is used to couple hose of two different sizes. A $2^1/_2 \times 1^1/_2$-inch reducer is used to attach a $1^1/_2$-inch line to a $2^1/_2$-inch line.

rolled-up or **donut**—A length of hose that is rolled and tied. Rolled-ups are carried into buildings for firefighters on the upper floors of high buildings.

scissor stairs—Two stairways constructed side by side in the core of a building with doors alternating the point of exit to opposite sides of the core.

scuttle—An opening or hatch in the roof generally covered by a wooden hatch cover.

scuttle ladder—A vertical iron ladder providing access to the roof from the top floor of the building.

search—The action taken by firefighters to find people that may be victims of fire.

sky lobby—An elevator terminal point on an upper floor of a high-rise building where passengers can change from one bank of elevators to another.

skylight—Glass in a metal frame on the roof of a building to allow daylight to enter the room or area. The skylight has louvers that permit ventilation of the room.

stretch—The term used for stretching hoselines at a fire. Example: The lieutenant ordered the firefighters to *stretch* into the fire building.

ventilation—As used in the fire service, ventilation means the opening of the roof, doors, windows, and any other opening in a structure to allow the escape of toxic gases and smoke from a fire.

water hammer—The sudden pressure rise that may accompany a sudden change in the velocity of water flowing in a line. When a nozzle discharging water under pressure is shut down quickly, a water hammer develops that can cause the hose to burst.

windward—That side of the fire that is into the wind. If you are facing the fire with the wind at your back, you are on the windward side.

READING-BASED QUESTIONS

A recent survey of firefighter examinations given nationwide indicates that there is a wide variation in the subject matter of these exams. The single topic that is common to all exams is reading. Some exams include classic reading comprehension questions that present a passage and then ask questions on the details of the passage and, perhaps, on its meaning. Other exams require candidates to indicate proper behavior based on their reading of printed procedures and regulations. Still another type of reading-based question requires candidates to reason and predict next steps on the basis of information presented in a reading passage. Of course, questions of judgment in emergency and non-emergency situations rely heavily on reading as well. Actually, there are nearly as many variations of the reading-based question as there are test-makers.

Before you begin to devote attention to strategies for dealing with reading-based questions, give some thought to your reading habits and skills. Of course, you already know how to read. But how well do you read? Do you concentrate? Do you get the point on your first reading? Do you notice details?

Between now and the test day, you must work to improve your reading concentration and comprehension. Your daily newspaper provides excellent material to improve your reading. Make a point of reading all the way through any article that you begin. Do not be satisfied with the first paragraph or two. Read with a pencil in hand. Underscore details and ideas that seem to be crucial to the meaning of the article. Notice points of view, arguments, and supporting information. When you have finished the article, summarize it for yourself. Do you know the purpose of the article? The main idea presented? The attitude of the writer? The points over which there is controversy? Did you find certain information lacking? As you answer these questions, skim back over your underlinings. Did you focus on important words and ideas? Did you read with comprehension?

As you repeat this process day after day, you will find that your reading will become more efficient. You will read with greater understanding, and will "get more" from your newspaper.

You can't sit down the night before a test involving reading comprehension and cram for it. The only way you can build up on your reading skill is to practice systematically. The gains you make will show up not only in an increased score on the test, but also in your reading for study and pleasure.

Trying to change reading habits that you have had for a long time can be difficult and discouraging. Do not attempt to apply *all* of the suggestions we have given to *all* your reading *all* at once. Try to follow a program like the one below.

1. Set aside 15 minutes a day to practice new reading techniques.

2. Start off with a short, easy-to-read article from a magazine or newspaper. Time yourself. At the end of the your practice session, time yourself on another short article, and keep a record of both times.

3. Select a news story. Read it first, and then practice an eye-scan exercise. Work towards reducing your eye fixations to no more than two per line the width of a newspaper column.

4. Read an editorial, book review, or movie or drama review in a literary magazine or newspaper. This type of article always expresses the author's (or the paper's) point

of view and is therefore good practice for searching out the main idea. After you read, see whether you can write a good title for the article and jot down in one sentence the author's main idea. Also, you can try making up a question based on the article with four choices. This is excellent practice for determining main ideas, and you can use the questions to test your friends.

5. Find one new word and write the sentence in which it appears. Guess at its meaning from the context. Then look up the definition in a dictionary. Try to make up a sentence of your own, using the word. Then try to use the word in your conversation at least twice the following day.

If you follow this program daily, you will find that your test score will show the great gains you have made in reading comprehension.

A major aspect of your daily reading that deserves special attention is vocabulary building. The most effective reader has a rich, extensive vocabulary. As you read, make a list of unfamiliar words. Include in your list words that you understand within the context of the article, but that you cannot really define. In addition, mark words that you do not understand at all. When you put aside your newspaper, go to the dictionary and look up *every* new and unfamiliar word. Write the word and its definition in a special notebook. Writing the words and their definitions helps seal them in your memory far better than just reading them, and the notebook serves as a handy reference for your own use. A sensitivity to the meaning of words and an understanding of more words will make reading easier and more enjoyable even if none of the words you learn in this way crops up on your exam. In fact, the habit of vocabulary building is a good lifetime habit to develop.

Success with reading-based questions depends on more than reading comprehension. You must also know how to draw the answers from the reading selection, and be able to distinguish the best answer from a number of answers that all seem to be good ones, or from a number of answers that all seem to be wrong.

Strange as it may seem, it's a good idea to approach reading comprehension questions by reading the questions—not the answer choices, just the questions themselves—before you read the selection. The questions will alert you to look for certain details, ideas, and points of view. Use your pencil. Underscore key words in the questions. These will help you direct your attention as you read.

Next skim the selection very rapidly to get an idea of its subject matter and its organization. If key words or ideas pop out at you, underline them, but do not consciously search out details in the preliminary skimming.

Now read the selection carefully with comprehension as your main goal. Underscore the important words as you have been doing in your newspaper reading.

Finally, return to the questions. Read each question carefully. Be sure you know what it asks. Misreading of questions is a major cause of error on reading comprehenson tests. Read *all* the answer choices. Eliminate the obviously incorrect answers. You may be left with only one possible answer. If you find yourself with more than one possible answer, reread the question. Focus on catch words that might destroy the validity of a seemingly acceptable answer. These include expressions like; *under all circumstances, at all times, never, always, under no condition, absolutely, entirely,* and *except when.* Then skim the passage once more, focusing on the underlined segments. By now you should be able to conclude which answer is *best.*

Reading-based questions may take a number of different forms. In general, some of the most common forms are as follows:

1. **Question of fact or detail.** You may have to mentally rephrase or rearrange, but you should find the answer stated in the body of the selection.

2. **Best title or main idea.** The answer may be obvious, but the incorrect choices to the "main idea" question are often half-truths that are easily confused with the main

idea. They may misstate the idea, omit part of the idea, or even offer a supporting idea quoted directly from the text. The correct answer is the one that covers the largest part of the selection.

3. **Interpretation.** This type of question asks you what the selection means, not just what it says. On firefighter exams, the questions based upon categories of building styles, for example, might fall into the realm of interpretation.

4. **Inference.** This is the most difficult type of reading-based question. It asks you to go beyond what the selection says and to predict what might happen next. You might have to choose the best course of action to take, based upon given procedures and a factual situation, or you may have to judge the actions of others. Your answer must be based upon the information in the selection and your own common sense, but not upon any other information you may have about the subject. A variation of the inference question might be stated as, "The author would expect that . . . ." To answer this question, you must understand the author's point of view and then make an inference from that viewpoint based upon the information in the selection.

5. **Vocabulary.** Some firefighter reading sections, directly or indirectly, ask the meanings of certain words as used in the selection.

Let's now work together on some typical reading comprehension selections and questions.

Selection for Questions 1 to 4

The recipient gains an impression of a typewritten letter before beginning to read the message. Factors that give a good first impression include margins and spacing that are visually pleasing, formal parts of the letter that are correctly placed according to the style of the letter, copy that is free of obvious erasures and overstrikes, and transcript that is even and clear. The problem for the typist is how to produce that first, positive impression of her work.

There are several general rules that a typist can follow when she wishes to prepare a properly spaced letter on a sheet of letterhead. The width of a letter should ordinarily not be less than four inches, nor more than six inches. The side margins should also have a desirable relation to the bottom margin, as well as the space between the letterhead and the body of the letter. Usually the most appealing arrangement is when the side margins are even, and the bottom margin is slightly wider than the side margins. In some offices, however, a standard line length is used for all business letters, and the secretary then varies the spacing between the dateline and the inside address according to the length of the letter.

1. The best title for the preceding paragraphs is   1. Ⓐ Ⓑ Ⓒ Ⓓ

    (A) "Writing Office Letters"
    (B) "Making Good First Impressions"
    (C) "Judging Well-Typed Letters"
    (D) "Good Placing and Spacing for Office Letters"

2. According to the preceding paragrpahs, which of the following might be con-   2. Ⓐ Ⓑ Ⓒ Ⓓ
   sidered the way that people quickly judge the quality of work that has been
   typed?

   (A) by measuring the margins to see if they are correct
   (B) by looking at the spacing and cleanliness of the typescript
   (C) by scanning the body of the letter for meaning
   (D) by reading the dateline and address for errors

3. According to the preceding paragraphs, what would be definitely undesirable   3. Ⓐ Ⓑ Ⓒ Ⓓ
   as the average line length of a typed letter?

   (A) 4 inches
   (B) 5 inches
   (C) 6 inches
   (D) 7 inches

4. According to the preceding paragraphs, when the line length is kept standard,   4. Ⓐ Ⓑ Ⓒ Ⓓ
   the secretary

   (A) does not have to vary the spacing at all because this also is standard
   (B) adjusts the spacing between the dateline and inside address for differ-
        ent lengths of letters
   (C) uses the longest line as a guideline for spacing between the dateline
        and inside address
   (D) varies the number of spaces between the lines

    Begin by skimming the questions and underscoring key words. Your underscored ques-
tions should look more or less like this:

    1. What is the <u>best title</u> for the preceding paragraphs?

    2. According to the preceding paragraphs, which of the following might be considered
       the way that people <u>quickly judge the quality</u> of work that has been typed?

    3. According to the preceding paragraphs, what would be definitely <u>undesirable</u> as the
       <u>average line length</u> of a typed letter?

    4. According to the preceding paragraphs, <u>when the line length is kept standard</u>, the
       secretary

    Now skim the selection. This quick reading should give you an idea of the structure of
the selection and of its overall meaning.

    Next read the selection carefully and underscore words that seem important or that you
think hold keys to the question answers. Your underscored selection should look something like this:

    The recipient gains an impression of a typewritten letter before he begins to read the
message. <u>Factors that give a good first impression</u> include <u>margins and spacing that
are visually pleasing</u>, formal parts of the letter that are <u>correctly placed</u> according to
the style of the letter, copy that is <u>free of obvious erasures and overstrikes</u>, and transcript that is <u>even and clear</u>. The problem for the typist is how to produce that first,
positive impression of her work.

    There are several general rules that a typist can follow when she wishes to prepare a
properly spaced letter on a sheet of letterhead. The width of a letter should ordinarily
<u>not be less than four inches, nor more than six inches.</u> The side margins should also

have a desirable relation to the bottom margin as well as the space between the letterhead and the body of the letter. Usually the most appealing arrangement is when the side margins are even, and the bottom margin is slightly wider than the side margins. In some offices, however, a standard line length is used for all business letters, and the secretary then varies the spacing between the dateline and the inside address according to the length of the letter.

Finally, read the questions and answer choices, and try to choose the correct answer for each question.

The correct answers are: 1. **(D)**, 2. **(B)**, 3. **(D)**, 4. **(B)**. Did you get them all right? Whether you made any errors or not, read these explanations.

1. **(D)** The best title for any selection is the one that takes in all of the ideas presented without being too broad or too narrow. Choice (D) provides the most inclusive title for this passage. A look at the other choices shows you why. Choice (A) can be eliminated because the passage discusses typing a letter, not writing one. Although the first paragraph states that a letter should make a good first impression, the passage is clearly devoted to the letter, not the first impression, so choice (B) can be eliminated. Choice (C) puts the emphasis on the wrong aspect of the typewritten letter. The passage concerns how to type a properly spaced letter, not how to judge one.

2. **(B)** Both spacing and cleanliness are mentioned in paragraph 1 as ways to judge the quality of a typed letter. The first paragraph states that the margins should be "visually pleasing" in relation to the body of the letter, but that does not imply margins of a particular measure, so choice (A) is incorrect. Meaning is not discussed in the passage, only the look of the finished letter, so choice (C) is incorrect. The passage makes no mention of errors, only the avoidance of erasures and overstrikes, so choice (D) is incorrect.

3. **(D)** This answer comes from the information provided in paragraph 2, that the width of a letter "should not be less than four inches nor more than six inches." According to this rule, seven inches is an undesirable line length.

4. **(B)** The answer to this question is stated in the last sentence of the reading passage. When a standard line length is used, the secretary "varies the spacing between the dateline and the inside address according to the length of the letter." The passage offers no support for any other choice.

Let us try another together.

Selection for Questions 5 to 9

Cotton fabrics treated with the XYZ Process have features that make them far superior to any previously known flame-retardant-treated cotton fabrics. XYZ Process treated fabrics are durable to repeated laundering and dry cleaning; are glow resistant as well as flame resistant; when exposed to flames or intense heat form tough, pliable, and protective chars; are inert physiologically to persons handling or exposed to the fabric; are only slightly heavier than untreated fabrics; and are susceptible to further wet and dry finishing treatments. In addition, the treated fabrics exhibit little or no adverse change in feel, texture, and appearance, and are shrink-, rot-, and mildew-resistant. The treatment reduces strength only slightly. Finished fabrics have "easy care" properties in that they are wrinkle-resistant and dry rapidly.

5. It is most accurate to state that the author in the preceding selection presents   5. Ⓐ Ⓑ Ⓒ Ⓓ

(A) facts but reaches no conclusion concerning the value of the process
(B) a conclusion concerning the value of the process and facts to support that conclusion
(C) a conclusion concerning the value of the process unsupported by facts
(D) neither facts nor conclusions, but merely describes the process

6. The one of the following articles for which the XYZ Process would be most suitable is   6. Ⓐ Ⓑ Ⓒ Ⓓ

(A) nylon stockings
(B) woolen shirt
(C) silk tie
(D) cotton bedsheet

7. The one of the following aspects of the XYZ Process that is not discussed in the preceding selection is its effects on   7. Ⓐ Ⓑ Ⓒ Ⓓ

(A) costs
(B) washability
(C) wearability
(D) the human body

8. The main reason for treating a fabric with the XYZ Process is to   8. Ⓐ Ⓑ Ⓒ Ⓓ

(A) prepare the fabric for other wet and dry finishing treatment
(B) render it shrink-, rot-, and mildew-resistant
(C) increase its weight and strength
(D) reduce the chance that it will catch fire

9. The one of the following that would be considered a minor drawback of the XYZ Process is that it   9. Ⓐ Ⓑ Ⓒ Ⓓ

(A) forms chars when exposed to flame
(B) makes fabrics mildew-resistant
(C) adds to the weight of fabrics
(D) is compatible with other finishing treatments

Skim the questions and underscore the words which you consider to be key. The questions should look something like this:

5. It is most accurate to state that the author, in the preceding selection, <u>presents</u>

6. The one of the following articles for which the <u>XYZ Process</u> would be <u>most suitable</u> is

7. The one of the following <u>aspects</u> of the XYZ Process which is *not* <u>discussed</u> in the preceding selection is its effect on

8. The <u>main reason for treating</u> a fabric with the XYZ Process is to

9. The one of the following which would be considered a <u>minor drawback</u> of the XYZ Process is that it

Skim the reading selection. Get an idea of the subject matter of the selection and of how it is organized.

Now read the selection carefully and underscore the words which you think are especially important. This fact-filled selection might be underlined like this:

> Cotton fabrics treated with the XYZ Process have features that make them far superior to any previously known flame-retardant-treated cotton fabrics. XYZ Process treated fabrics are durable to repeated laundering and dry cleaning; are glow resistant as well as flame resistant; when exposed to flames or intense heat form tough, pliable, and protective chars; are inert physiologically to persons handling or exposed to the fabric; are only slightly heavier than untreated fabrics; and are susceptible to further wet and dry finishing treatments. In addition, the treated fabrics exhibit little or no adverse change in feel, texture, and appearance, and are shrink-, rot-, and mildew-resistant. The treatment reduces strength only slightly. Finished fabrics have "easy care" properties in that they are wrinkle resistant and dry rapidly.

Now read each question and all its answer choices, and try to choose the correct answer for each question.

The correct answers are: 5. **(B)**, 6. **(D)**, 7. **(A)**, 8. **(D)**, 9. **(C)**. How did you do on these? Read the explanations.

5. **(B)** This is a combination main idea and interpretation question. If you cannot answer this question readily, reread the selection. The author clearly thinks that the XYZ Process is terrific and says so in the first sentence. The rest of the selection presents a wealth of facts to support the initial claim.

6. **(D)** At first glance you might think this is an inference question requiring you to make a judgment based upon the few drawbacks of the process. Closer reading, however, shows you that there is no contest for correct answer here. This is a simple question of fact. The XYZ Process is a treatment for *cotton* fabrics.

7. **(A)** Your underlinings should help you with this question of fact. Cost is not mentioned; all other aspects of the XYZ Process are. If you are having trouble finding mention of the effect of the XYZ Process on the human body, add to your vocabulary list "inert" and "physiologically."

8. **(D)** This is a main idea question. You must distinguish between the main idea and the supporting and incidental facts.

9. **(C)** Obviously a drawback is a negative feature. The selection mentions only two negative features. The treatment reduces strength slightly, and it makes fabrics slightly heavier than untreated fabrics. Only one of these negative features is offered among the answer choices.

You should be getting better at reading and at answering questions. Try this next selection on your own. Read and underline the questions. Skim the selection. Read and underline the selection. Read questions and answer choices and mark your answers. Then check your answers against the answers and explanations that follow the selection.

Selection for Questions 10 to 12

Language performs an essentially social function: It helps us to get along together, to communicate, and achieve a great measure of concerted action. Words are signs that have significance by convention, and those people who do not adopt the conventions simply fail to communicate. They do not "get along," and a social force arises that encourages them to achieve the correct associations. By "correct" is meant "as used by other members of the social group." Some of the vital points about language are brought home to an English visitor to America, and vice versa, because our vocabularies are nearly the same—but not quite.

10. As defined in the preceding selection, usage of a word is "correct" when it is   10. Ⓐ Ⓑ Ⓒ Ⓓ

   (A) defined in standard dictionaries
   (B) used by the majority of persons throughout the world who speak the same language
   (C) used by the majority of educated persons who speak the same language
   (D) used by other persons with whom we are associating

11. In the preceding selection, the author is concerned primarily with the   11. Ⓐ Ⓑ Ⓒ Ⓓ

   (A) meaning of words
   (B) pronunciation of words
   (C) structure of sentences
   (D) origin and development of language

12. According to the preceding selection, the main language problem of an English   12. Ⓐ Ⓑ Ⓒ Ⓓ
visitor to America stems from the fact that an English person

   (A) uses some words that have different meanings for Americans
   (B) has different social values than the Americans
   (C) has had more exposure to non-English-speaking persons than Americans have had
   (D) pronounces words differently than Americans do

The correct answers are: 10. **(D)**, 11. **(A)**, 12. **(A)**.

10. **(D)** The answer to this question is stated in the next to last sentence of the selection.

11. **(A)** This main idea question is an easy one to answer. You should have readily eliminated all of the wrong choices.

12. **(A)** This is a question of fact. The phrasing of the question is quite different from the phrasing of the last sentence, but the meaning is the same. You may have found this reading selection more difficult to absorb than some of the others, but you should have had no difficulty answering this question by eliminating the wrong answers.

Now try this reading selection and its questions. Once more, explanations follow the correct answers. Follow the procedure you have learned, and be sure to read the explanations even if you have a perfect score.

Selection for Questions 13 to 18

Since almost every office has some contact with data-processed records, a Senior Stenographer should have some understanding of the basic operations of data processing. Data processing systems now handle about one-third of all office paper work. On punched cards, magnetic tape, or by other means, data is recorded before being fed into the computer for processing. A machine such as the key punch is used to convert the data written on the source document into the coded symbols on punched cards or tapes. After data has been converted, it must be verified to guarantee absolute accuracy of conversion. In this manner data becomes a permanent record that can be read by electronic computers that compare, store, compute, and otherwise process data at high speeds.

One key person in a computer installation is a programmer, the man or woman who puts business and scientific problems into special symbolic languages that can be read by the computer. Jobs done by the computer range all the way from payroll operations to chemical process control, but most computer applications are directed toward management data. About half of the programmers employed by business come to their positions with college degrees; the remaining half are promoted to their positions, without regard to education, from within the organization on the basis of demonstrated ability.

13. Of the following, the best title for the preceding selection is   13. Ⓐ Ⓑ Ⓒ Ⓓ

   (A) "The Stenographer as Data Processor"
   (B) "The Relation of Key Punching to Stenography"
   (C) "Understanding Data Processing"
   (D) "Permanent Office Records"

14. According to the preceding selection, a Senior Stenographer should understand   14. Ⓐ Ⓑ Ⓒ Ⓓ
   the basic operations of data processing because

   (A) almost every office today has contact with data processed by computer
   (B) any office worker may be asked to verify the accuracy of data
   (C) most offices are involved in the production of permanent records
   (D) data may be converted into computer language by typing on a key
       punch

15. According to the preceding selection, the data that the computer understands is   15. Ⓐ Ⓑ Ⓒ Ⓓ
   most often expressed

   (A) as a scientific programming language
   (B) as records or symbols punched on tape, cards, or by other means
   (C) as records on cards
   (D) as records on tape

16. According to the preceding selection, computers are used most often to handle   16. Ⓐ Ⓑ Ⓒ Ⓓ

   (A) management data
   (B) problems of higher education
   (C) the control of chemical processes
   (D) payroll operations

17. Computer programming is taught in many colleges and business schools. The preceding selection implies that programmers in industry

    17. Ⓐ Ⓑ Ⓒ Ⓓ

   (A) must have professional training
   (B) need professional training to advance
   (C) must have at least a college education to do adequate programming tasks
   (D) do not need college education to do programming work

18. According to the preceding selection, data to be processed by computer should be

    18. Ⓐ Ⓑ Ⓒ Ⓓ

   (A) recent
   (B) complete
   (C) basic
   (D) verified

      The correct answers are: 13. **(C)**, 14. **(A)**, 15. **(B)**, 16. **(A)**, 17. **(D)**, 18. **(D)**.

13. **(C)** Choosing the best title for this selection is not easy. Although the Senior Stenographer is mentioned in the first sentence, the selection is really not concerned with stenographers or with their relationship to key punching. Eliminate choices (A) and (B). Permanent office records are mentioned in the selection, but only along with other equally important uses for data processing. Eliminate choice (D). When in doubt, the most general title is usually correct.

14. **(A)** This is a question of fact. Any one of the answer choices could be correct, but the answer is given almost verbatim in the first sentence. Take advantage of answers that are handed to you in this way.

15. **(B)** This is a question of fact, but it is a tricky one. The program language is a symbolic language, not a scientific one. Reread carefully and eliminate choice (A). (B) includes more of the information in the selection than either (C) or (D), and so is the best answer.

16. **(A)** This is a question of fact. The answer is stated in the next to the last sentence.

17. **(D)** Remember that you are answering the questions on the basis of the information given in the selection. In spite of any information you may have to the contrary, the last sentence of the selection states that half the programmers employed in business achieved their positions by moving up from the ranks without regard to education.

18. **(D)** Judicious underlining proves very helpful to you in finding the correct answer to this question buried in the middle of the selection. Since any one of the answers might be correct, the way to deal with this question is to skim the underlined words in the selection, eliminate those that are not mentioned, and choose the appropriate answer.

      On firefighter exams, many reading passages will relate to standard designations or to specified rules of procedure. When reading these passages, you must pay special attention to details relating to exceptions, special preconditions, combinations of activities, choices of actions, and prescribed time sequences. Sometimes the printed procedure specifies that certain actions are to be taken only when there is a combination of factors such as that water pressure has dropped AND staircase collapse is imminent. At other times, the procedures give choices of action under certain circumstances. You must read carefully to determine if the passage requires a combination of factors or gives a choice, then make the appropriate judgment. When a time sequence is specified, be certain to follow that sequence in the prescribed order.

The remaining reading selections in this chapter are of a style often found on firefighter exams. Beyond requiring routine reading comprehension skills, they require the special attention to sequence, combinations, and choices which we have just discussed.

Selection for Questions 19 to 22

LABEL

BREGSON'S CLEAR GLUE
HIGHLY FLAMMABLE

A clear quick-drying glue

For temporary bonding, apply glue to one surface and join immediately

For permanent bonding, apply glue to both surfaces, permit to dry and press together

Use for bonding plastic to plastic, plastic to wood, and wood to wood only

Will not bond at temperatures below 60°

PRECAUTIONS

Use with adequate ventilation
Close container after use
Keep out of reach of children
Avoid prolonged breathing of vapors and repeated contact with skin

19. Assume that you, as a member of a repair crew, have been asked to repair a wood banister in the hallway of a house. Since the heat has been turned off, the hallway is very cold except for the location where you have to make the repair. Another repair crew worker is working at that same location using a blowtorch to solder a pipe in the wall. The temperature at that location is about 67°. According to the instructions on the above label, the use of this glue to make the necessary repair is   19. Ⓐ Ⓑ Ⓒ Ⓓ

(A) advisable because the glue will bond wood to wood
(B) advisable because the heat from the soldering will cause the glue to dry quickly
(C) inadvisable because the work area temperature is too low
(D) inadvisable because the glue is highly flammable

20. According to the instructions on the label, this glue should *not* be used for which of the following applications?   20. Ⓐ Ⓑ Ⓒ Ⓓ

(A) affixing a pine table leg to a walnut table
(B) repairing leaks around pipe joints
(C) bonding a plastic knob to a cedar drawer
(D) attaching a lucite knob to a lucite drawer

21. According to the instructions on the label, using this glue to bond ceramic tile to a plaster wall by coating both surfaces with glue, letting the glue dry, and then pressing tile to the plaster wall is   21. Ⓐ Ⓑ Ⓒ Ⓓ

(A) advisable because the glue is quick-drying and clear

(B) inadvisable because the glue should be affixed to one surface of the tile only

(C) inadvisable because the glue is not suitable for bonding ceramic tile to plaster walls.

(D) inadvisable because the bonding should be a temporary one

22. The precaution described in the label, to "use with adequate ventilation" means that

22. Ⓐ Ⓑ Ⓒ Ⓓ

(A) the area you are working in should be very cold
(B) there should be sufficient fresh air where you are using the glue
(C) you should wear gloves to avoid contact with the glue
(D) you must apply a lot of glue to make a permanent bond

The correct answers are: 19. **(D)**, 20. **(B)**, 21. **(C)**, 22. **(B)**.

19. **(D)** In all caps, the label clearly states: BREGSON'S CLEAR GLUE, HIGHLY FLAMMABLE. Since another repair crew worker is working at the same location using a blowtorch, it would be most foolhardy to use the glue at this location at this time.

20. **(B)** The glue is for bonding one surface to another, not for repairing leaks. Further, the use of the glue is specific to wood and plastics.

21. **(C)** The glue is suited for bonding plastic to plastic, plastic to wood, and wood to wood only. It will not bond ceramic tile to plaster.

22. **(B)** This is really a vocabulary question. "Sufficient ventilation" means "adequate fresh air."

Selection for Questions 23 to 26

All automotive accidents, no matter how slight, are to be reported to the Safety Division by the employee involved on Accident Report Form S–23 in duplicate. When the accident is of such a nature that it requires the filling out of the State Motor Vehicle Report Form MV–104, this form is also prepared by the employee in duplicate, and sent to the Safety Divison for comparison with the Form S–23. The Safety Division forwards both copies of Form MV–104 to the Corporation Counsel, who sends one copy to the State Bureau of Motor Vehicles. When the information on the Form S–23 indicates that the employee may be at fault, an investigation is made by the Safety Division. If this investigation shows that the employee was at fault, the employee's dispatcher is asked to file a complaint on Form D–11. The foreman of mechanics prepares a damage report on Form D–8 and an estimate of the cost of repairs on Form D–9. The dispatcher's complaint, the damage report, the repair estimate, and the employee's previous accident record are sent to the Safety Division where they are studied together with the accident report. The Safety Division then recommends whether or not disciplinary action should be taken against the employee.

23. According to the preceding paragraph, the Safety Division should be notified   23. Ⓐ Ⓑ Ⓒ Ⓓ
    whenever an automotive accident has occurred by means of

    (A) Form S–23
    (B) Forms S–23 and MV–104
    (C) Forms S–23, MV–104, D–8, D–9, and D–11
    (D) Forms S–23, MV–104, D–8, D–9, D–11, and employee's accident
        record

24. According to the preceding paragraph, the forwarding of the Form MV–104 to   24. Ⓐ Ⓑ Ⓒ Ⓓ
    the State Bureau of Motor Vehicles is done by the

    (A) Corporation Counsel
    (B) dispatcher
    (C) employee involved in the accident
    (D) Safety Division

25. According to the preceding paragraph, the Safety Division investigates an auto-   25. Ⓐ Ⓑ Ⓒ Ⓓ
    motive accident if the

    (A) accident is serious enough to be reported to the State Bureau of Motor
        Vehicles
    (B) dispatcher files a complaint
    (C) employee appears to have been at fault
    (D) employee's previous accident record is poor

26. Of the forms mentioned in the preceding paragraph, the dispatcher is responsi-   26. Ⓐ Ⓑ Ⓒ Ⓓ
    ble for preparing the

    (A) accident report form   (C) damage report
    (B) complaint form   (D) estimate of cost of repairs

     The correct answers are: 23. **(A)**, 24. **(A)**, 25. **(C)**, 26. **(B)**.

23. **(A)** The first sentence makes it clear that regardless of whatever other forms might
    need to be filed, all automobile accidents should be reported to the Safety Division on
    Form S–23.

24. **(A)** Follow the steps carefully. The employee fills out Form MV–104 in duplicate and
    sends both copies to the Safety Division. The Safety Division, in turn, sends both
    copies to the Corporation Counsel who then sends one copy on to the State Bureau of
    Motor Vehicles.

25. **(C)** The Safety Division investigates if the information on the Form S–23 indicates
    that the employee may have been at fault.

26. **(B)** If the employee was indeed at fault, the dispatcher files a complaint on Form D–
    11.

    Use the skills and techniques you have just developed to practice with the traditional-
style reading comprehension questions that follow. Then apply your new expertise to the
exams in this book and to your firefighter exam.

    Before you begin the exercises, review this list of hints for scoring high on reading com-
prehension questions.

1. Read the questions and underline key words.

2. Skim the selection to get a general idea of the subject matter, the point that is being made, and the organization of the material.

3. Reread the selection giving attention to details and point of view. Underscore key words and phrases.

4. If the author has quoted material from another source, be sure that you understand the purpose of the quote. Does the author agree or disagree?

5. Carefully read each question or incomplete statement. Determine exactly what is being asked. Watch for negatives or all-inclusive words such as *always, never, all, only, absolutely, completely, none, entirely, no.*

6. Read all the answer choices. Eliminate those choices that are obviously incorrect. Reread the remaining choices and refer to the selection, if necessary, to determine the best answer.

7. Avoid inserting your own judgments into your answers. Even if you disagree with the author or even if you spot a factual error in the selection, you must answer on the basis of what is stated or implied in the selection.

8. Do not allow yourself to spend too much time on any one question. If looking back at the selection does not help you to find or figure out the answer, choose from among the answers remaining after you eliminate the obviously wrong answers, mark the question in the test booklet, and go on. If you have time at the end of the exam or exam portion, reread the selection and the question. Often a fresh look provides new insights.

READING COMPREHENSION
FOR FIREFIGHTERS

DIRECTIONS: Questions 1 to 4 relate to the following paragraph and are to be answered in accordance with the paragraph.

The unadjusted loss per $1000 valuation has only a very limited usefulness in evaluating the efficiency of a fire department, for it depends upon the assumption that other factors will remain constant from time to time and city to city. It might be expected that high fire department operation expenditures would tend to be associated with a low fire loss. A statistical study of the loss and cost data in more than 100 cities failed to reveal any such correlation. The lack of relationship, although to some extent due to failure to make the most efficacious expenditure of fire protection funds, must be attributed in part at least to the obscuring effect of variations in the natural, physical, and moral factors which affect fire risk.

1. One reason for the failure to obtain the expected relationship between fire department expenditures and fire loss data stated in the paragraph is

   (A) the changing dollar valuation of property
   (B) the unsettling effects of rapid technological innovations
   (C) the inefficiency of some fire department activities
   (D) the statistical errors made by the investigators

2. We may conclude that the "unadjusted loss per $1000" figure is useful in comparing the fire departments of two cities

   (A) only if the cities are of comparable size
   (B) only if adjustments are made for other factors that affect fire loss
   (C) under no circumstances
   (D) only if properly controlled experimental conditions can be obtained

3. The one of the following factors that affect fire risk that is most adequately reflected in the "unadjusted loss per $1000 valuation" index is

   (A) fire department operation expenditures
   (B) physical characteristics of the city
   (C) type of structures most prevalent in the city
   (D) total worth of property in the city

4. According to the paragraph, cities that spend larger sums on their fire departments

   (A) tend to have lower fire losses than cities that spend smaller sums on their fire departments
   (B) do not tend to have lower fire losses than cities that spend smaller sums on their fire departments

(C) tend to have higher fire losses than cities that spend smaller sums on their fire departments
(D) do not tend to have the same total property valuation as cities that spend smaller sums on their fire departments

DIRECTIONS: Questions 5 to 7 relate to the following paragraph and are to be answered in accordance with the paragraph.

Shafts extending into the top story, except those stair shafts where the stairs do not continue to the roof, shall be carried through and at least two feet above the roof. Every shaft extending above the roof, except open shafts and elevator shafts, shall be enclosed at the top with a roof of materials having a fire-resistant rating of one hour and a metal skylight covering at least three-quarters of the area of the shaft in the top story, except that skylights over stair shafts shall have an area not less than one-tenth the area of the shaft in the top story, but shall have an area not less than fifteen square feet in area. Any shaft terminating below the top story of a structure and those stair shafts not required to extend through the roof shall have the top enclosed with materials having the same fire-resistant rating as required for the shaft enclosure.

5. The paragraph states that the elevator shafts which extend into the top story

(A) are not required to have a skylight but are required to extend at least two feet above the roof
(B) are neither required to have a skylight nor to extend above the roof
(C) are required to have a skylight covering at least three-quarters of the area of the shaft in the top story and to extend at least two feet above the roof
(D) are required to have a skylight covering at least three-quarters of the area of the shaft in the top story but are not required to extend above the roof

6. The one of the following skylights which meets the requirements of the paragraph is a skylight measuring

(A) 4 feet by 4 feet over a stair shaft that, on the top story, measures 20 feet by 9 feet
(B) 4 1/2 feet by 3 1/2 feet over a pipe shaft that, on the top story, measures 5 feet by 4 feet
(C) 2 1/2 feet by 1 1/2 feet over a dumbwaiter shaft that, on the top story, measures 2 1/2 feet by 2 1/2 feet
(D) 4 feet by 3 feet over a stair shaft that, on the top story, measures 15 feet by 6 feet

7. Suppose that a shaft which does not go to the roof is required to have a three-hour fire-resistant rating. In regard to the material enclosing the top of this shaft, the paragraph

(A) states that a one-hour fire-resistant rating is required
(B) states that a three-hour fire-resistant rating is required
(C) implies that no fire-resistant rating is required
(D) neither states nor imples anything about the fire-resistant rating

DIRECTIONS: Questions 8 to 14 relate to the following paragraphs and are to be answered in accordance with these paragraphs.

Fire regulations require that every liquefied petroleum gas installation should be provided with the means for shutting off the supply to a building in case of an emergency. The installation of a shut-off valve immediately inside a building, which is sometimes done for the convenience of the user, does not comply with this regulation. An outside shut-off valve just outside the building seems to be the logical solution. However, the possibility of tampering illustrates the danger of such an arrangement. A shut-off valve so located might be placed in a locked box. However, this has no advantage over a valve provided within the locked cabinet containing the cylinder or an enclosure provided over the top of the cylinder. Keys may be carried by firefighters or, in an emergency, the lock may be broken. Where no valve is visible, the firefighters should not hesitate to break the lock to the cylinder enclosure. The means for shutting off the gas varies considerably in the numerous types of equipment in use. When the cover to the enclosure has been opened, the gas may be shut off as follows:

Close the tank or cylinder valves to which the supply line is connected. Such valves always turn to the right. If the valve is not provided with a handwheel, an adjustable wrench can be used. If conditions are such that shutting off the supply at once is imperative and this cannot be accomplished as above, the tubing which is commonly employed as the supply line can be flattened to the extent of closure by a hammer. If the emergency is such as to require the removal of the cylinder, the supply line should be disconnected and the cylinder removed to a safe location. A tank buried in the ground is safe against fire. When conditions indicate the need of removing a cylinder or tank, and this cannot be done due to the severity of exposure, pressures within the container can be kept within control of the safety valve by means of a hose stream played on the surface of the container. The melting of the fuse plug may also be prevented in this way.

8. According to the preceding paragraphs, in an emergency, a firefighter should break the lock of a cylinder enclosure whenever the shut-off valve

   (A) fails to operate
   (B) has no handwheel
   (C) has been tampered with
   (D) cannot be seen

9. According to the preceding paragraphs, shut-off valves for liquefied petroleum gas installations

   (A) always turn to the right
   (B) always turn to the left
   (C) sometimes turn to the right and sometimes turn to the left
   (D) are generally pulled up

10. According to the preceding paragraphs, if a cylinder should be moved but cannot be because of the severity of exposure, the pressure can be kept under control by

   (A) opening the shut-off valve
   (B) playing a hose stream on the cylinder
   (C) disconnecting the supply line into the cylinder
   (D) removing the fuse plug

11. The preceding paragraphs state that the supply line should be disconnected when the

   (A) fuse plugs melt
   (B) cylinder is removed to another location
   (C) supply line becomes defective
   (D) cylinder is damaged

12. The preceding paragraphs state that the shut-off valves for liquefied petroleum gas installations are sometimes placed inside buildings

(A) so that firefighters will be able to find the valves more easily
(B) because it is more convenient for the occupants
(C) in order to hide the valves from public view
(D) as this makes it easier to keep the valves in good working condition

13. It is suggested in the preceding paragraphs that during an emergency the supply line tubing should be flattened to the extent of closure when the

(A) supply line becomes defective
(B) shut-off valve cannot be opened
(C) shut-off valve cannot be closed
(D) supply line is near a fire

14. According to the preceding paragraphs, fire regulations require that liquefied petroleum gas installations should

(A) be made in safe places
(B) be tamperproof
(C) have shut-off valves
(D) not exceed a certain size

DIRECTIONS: Questions 15 to 24 relate to the following paragraphs and are to be answered in accordance with these paragraphs.

Air conditioning systems are complex and are made up of several processes. The circulation of air is produced by fans and ducts; the heating is produced by steam, hot water coils, coal, gas, or oil fire furnaces; the cooling is done by ice or mechanical refrigeration; and the cleaning is done by air washers or filters.

Air-conditioning systems in large buildings generally should be divided into several parts with wholly separate ducts for each part or floor. The ducts are then extended through fire partitions. As a safeguard, whenever ducts pass through fire partitions, automatic fire dampers should be installed in the ducts. Furthermore, the ducts should be lined on the inside with fire-resistant materials. In addition, a manually operated fan shut-off should be installed at a location which will be readily accessible under fire conditions.

Most air-conditioning systems recirculate a considerable portion of the air and, when this is done, an additional safeguard has to be taken to have the fan arrange to shut down automatically in case of fire. A thermostatic device in the return air duct will operate the shut-off device whenever the temperature of the air coming to the fan becomes excessive. The air filters are frequently coated with oil to help catch dust. Such oil should be of a type that does not ignite readily. Whenever a flammable or toxic refrigerant is employed for air cooling, coils containing such refrigerant should not be inserted in any air passage.

15. According to the preceding paragraphs, fan shut-offs in the air-conditioning system should be installed

(A) near the air ducts
(B) next to fire partitions

(C) near the fire dampers

(D) where they can be reached quickly

16. On the basis of the preceding paragraphs, whenever a fire breaks out in a building containing an air-conditioning system that recirculates a portion of the air,

    (A) the fan will shut down automatically
    (B) the air ducts will be opened
    (C) the thermostat will cease to operate
    (D) the fire partitions will open

17. The preceding paragraphs state that on every floor of a large building where air-conditioning systems are used there should be

    (A) an automatic damper
    (B) a thermostatic device
    (C) an air filter
    (D) a separate duct

18. From the preceding paragraphs, the conclusion can be drawn that in an air-conditioning system, flammable refrigerants

    (A) may be used if certain precautions are observed
    (B) should be used sparingly and only in air passages
    (C) should not be used under any circumstances
    (D) may be more effective than other refrigerants

19. According to the preceding paragraph, the spreading of dust by means of fans in the air-conditioning system is reduced by

    (A) shutting down the fan automatically
    (B) lining the inside of the air duct
    (C) cleaning the circulated air with filters
    (D) coating the air filters with oil

20. According to the preceding paragraphs, the purpose of a thermostatic device is to

    (A) regulate the temperature of the air-conditioning system
    (B) shut off the fan when the temperature of the air rises
    (C) operate the fan when the temperature of the air falls
    (D) assist the recirculation of the air

21. According to the preceding paragraphs, hot water coils in the air-conditioning system are limited to

    (A) cooling
    (B) heating
    (C) heating and cooling
    (D) heating and cleaning

22. The parts of an air-conditioning system which the preceding paragraphs state should be made of fire-resistant materials are the

    (A) hot water coils
    (B) automatic fire dampers
    (C) air duct linings
    (D) thermostatic devices

23. According to the preceding paragraphs, automatic fire dampers should be installed

   (A) on oil fired furnaces
   (B) on every floor of a large building
   (C) in ducts passing through fire partitions
   (D) next to the hot water coils

24. On the basis of the preceding paragraphs, the most accurate statement is that the coils containing toxic refrigerants should be

   (A) used only when necessary
   (B) lined with fire-resistant materials
   (C) coated with nonflammable oil
   (D) kept out of any air passage

DIRECTIONS: Questions 25 to 27 relate to the following paragraph and are to be answered in accordance with this paragraph.

It shall be unlawful to place, use, or to maintain in a condition intended, arranged, or designed for use, any gas-fired cooking appliance, laundry stove, heating stove, range, or water heater or combination of such appliances in any room or space used for living or sleeping in any new or existing multiple dwelling unless such room or space has a window opening to the outer air or such gas appliance is vented to the outer air. All automatically operated gas appliances shall be equipped with a device that shall shut off automatically the gas supply to the main burners when the pilot light in such appliance is extinguished. A gas range or the cooking portion of a gas appliance incorporating a room heater shall not be deemed an automatically operated gas appliance. However, burners in gas ovens and broilers that can be turned on and off or ignited by non-manual means shall be equipped with a device that shall shut off automatically the gas supply to those burners when the operation of such non-manual means fails.

25. According to this paragraph, an automatic shut-off device is not required on a gas

   (A) hot water heater   (C) space heater
   (B) laundry dryer   (D) range

26. According to this paragraph, a gas-fired water heater is permitted

   (A) only in kitchens   (C) only in living rooms
   (B) only in bathrooms   (D) in any type of room

27. An automatic shut-off device shuts off

   (A) the gas range   (C) the gas supply
   (B) the pilot light   (D) all of these

DIRECTIONS: Questions 28 to 30 relate to the following paragraph and are to be answered in accordance with this paragraph.

A utility plan is a floor plan that shows the layout of a heating, electrical, plumbing, or other utility system. Utility plans are used primarily by the persons responsible for the utilities, but they are important to the craftsman as well. Most utility installations require the leaving of openings in walls, floors, and roofs for the admission or installation of utility features. The craftsman who is, for example, pouring a concrete foundation wall must study the utility plans to determine the number, sizes, and locations of the openings he must leave for piping, electrical lines, and the like.

28. The one of the following items of information that is *least* likely to be provided by a utility plan is the

   (A) location of the joists and frame members around stairwells
   (B) location of the hot water supply and return piping
   (C) location of light fixtures
   (D) number of openings in the floor for radiators

29. According to the passage, of the following, the persons who most likely will have the greatest need for the information included in a utility plan of a building are those who

   (A) maintain and repair the heating system
   (B) clean the premises
   (C) put out the fires
   (D) advertise property for sale

30. According to the passage, a repair crew member should find it most helpful to consult a utility plan when information is needed about the

   (A) thickness of all doors in the structure
   (B) number of electrical outlets located throughout the structure
   (C) dimensions of each window in the structure
   (D) length of a roof rafter

Answer Key

1. C		7. B		13. C		19. D		25. D
2. B		8. D		14. C		20. B		26. D
3. D		9. A		15. D		21. B		27. C
4. B		10. B		16. A		22. C		28. A
5. A		11. B		17. D		23. C		29. A
6. B		12. B		18. A		24. D		30. B

Explanatory Answers

1. **(C)** The last sentence states that part of the lack of relationship between cost of running a fire department and dollar losses in fires is due to inefficiencies in the use of fire department funds.

2. **(B)** The paragraph makes it clear that many factors affect the "unadjusted loss per $1000" figure. Comparisons to the fire departments of different cities must allow for these other factors.

3. **(D)** The "loss per $1000" figure is based upon the total value of property in the city.

4. **(B)** The paragraph simply states that spending more on the fire department does not guarantee lower total property loss. It does NOT, however, state that the city that spends more has greater losses.

5. **(A)** The first sentence states that all shafts except stair shafts in which the stairs do not continue to the roof must be carried at least two feet above the roof. Elevator shafts must extend those two feet. The second sentence makes the exception that elevator shafts need not have a skylight.

6. **(B)** The 15.75-square foot skylight over the 20-square foot pipe shaft covers more than three-quarters of the area of the shaft. All other skylights are inadequate.

7. **(B)** According to the last sentence, a shaft terminating below the top story of a structure must have a top with the same fire-resistant rating as that required for the shaft enclosure, here three hours.

8. **(D)** The third from the last sentence of the first paragraph states that where no valve is visible, the firefighters should break the lock to the cylinder enclosure.

9. **(A)** The second sentence of the second paragraph states that these tank valves always turn to the right.

10. **(B)** This information is given in the next to last sentence.

11. **(B)** See the fourth sentence of the second paragraph.

12. **(B)** The second sentence tells us that the shut-off valve which is placed inside a building for the convenience of the user does not satisfy the emergency shut-off requirements.

13. **(C)** If the shut-off valve cannot be closed, the supply of gas can be shut off by flattening the supply line tubing.

14. **(C)** See the first sentence.

15. **(D)** The last sentence of the second paragraph states that manual fan shut-offs should be installed where they are readily accessible.

16. **(A)** The second sentence of the last paragraph describes a thermostatic device that automatically shuts off recirculation of air in case of fire.

17. **(D)** See the first sentence of the second paragraph.

18. **(A)** The last sentence states that when flammable refrigerants are used, the coils containing the refrigerant should not be inserted in the air passages. This constitutes a precaution that makes the use of flammable refrigerants acceptable.

19. **(D)** See the third from the last sentence.

20. **(B)** This is detailed in the second sentence of the last paragraph.

21. **(B)** See the second sentence.

22. **(C)** See the next to the last sentence of the second paragraph.

23. **(C)** See the third sentence of the second paragraph.

24. **(D)** Again, this answer is in the last sentence.

25. **(D)** An automatic shut-off device is required on all automatically operated gas appliances, however, a gas range is not considered to be an automatically operated gas appliance.

26. **(D)** A gas-fired water heater is permitted anywhere provided that the space has a window opening to the outer air or that the appliance is vented to the outer air.

27. **(C)** The automatic shut-off device shuts off the gas supply to the burners when the pilot light goes out. This prevents the room from filling with gas.

28. **(A)** The utility plan shows the layout of a utility system, not of the structure of the building itself.

29. **(A)** The heating system is one of the utility systems, so the people concerned with its maintainence and repair would find it most useful.

30. **(B)** Again, consider the definition of a utility. The electrical system is a utility system. Structural information appears on a utility plan only where it is incidental to the layout of utilities.

REASONING AND JUDGMENT

1. Of the following, the chief advantage of having firefighters under competitive civil service is that

(A) fewer fires tend to occur
(B) fire prevention becomes a reality instead of a distant goal
(C) the efficiency level of the personnel tends to be raised
(D) provision may then be made for training by the municipality

2. Of the following, it is least likely that fire will be caused by

(A) arson
(B) poor building construction
(C) carelessness
(D) inadequate supply of water

DIRECTIONS: The following three questions, numbered 3 to 5 inclusive, are to be answered on the basis of the table below. Data for certain categories have been omitted from the table. You are to calculate the missing numbers if needed to answer the questions.

	Last Year	This Year	Increase
Firefighters	9326	9744	
Lieutenants	1355	1417	
Captains		433	107
Others	_____	_____	_____
TOTAL	11469	12099	

3. The number in the Others group last year was most nearly

(A) 450   (C) 500
(B) 475   (D) 525

4. The group which had the largest percentage of increase was

(A) Firefighters   (C) Captains
(B) Lieutenants   (D) Others

5. This year the ratio between firefighters and all other ranks of the uniformed force was mostly nearly

(A) 5:1   (C) 2:1
(B) 4:1   (D) 1:1

39

6. The ideal is that in big cities a fire alarm box can be seen from any corner. This is a desirable condition mainly because

(A) several alarms may be sounded by one person running from one box to another
(B) little time is lost in sounding an alarm
(C) an alarm may be sounded from a different box if the nearest one is out of order
(D) fire apparatus can quickly reach any box

7. Suppose hydrants with a flowing capacity of less than 500 gallons per minute were to be painted red, hydrants with a flowing capacity between 500 and 1000 gallons per minute were to be painted yellow, and hydrants with a flowing capacity of 1000 gallons or greater per minute were to be painted green. The principal advantage of such a scheme is that

(A) fewer fires would occur
(B) more water would become available at a fire
(C) citizens would become more acutely aware of the importance of hydrants
(D) firefighters would save time

8. "At two o'clock in the morning, Mrs. Smart awakened her husband and said the house was on fire. Mr. Smart dressed hurriedly, ran seventeen blocks, past five fire alarm boxes, to the fire station to tell the firefighters that his house was on fire. When Mr. Smart and the firefighters had returned, the house had burned down." This is an illustration of the

(A) need for a plentiful supply of fire alarm boxes
(B) need for more fire stations
(C) necessity of preventing fires
(D) desirability of educating the public

9. "Installation of a modern fire alarm system will mean smaller fires." Of the following, the best justification for this statement is that

(A) if summoned quickly, firefighters can control a fire before it has a chance to spread
(B) if the alarm system is modern, firefighters can be given a complete picture of a fire even before they respond
(C) some fires, such as fires resulting from explosions, assume large proportions in a few seconds
(D) most industrial establishments depend on more than one method of transmitting fire alarms

10. "The firefighter must not only know how to make the proper knots, but he must know how to make them quickly." Of the following, the chief justification for this statement is that

(A) a firefighter uses many kinds of knots
(B) a slowly tied knot slips more readily than one tied quickly
(C) haste makes waste
(D) in most fire operations, speed is important

11. A survey of dip tank fires in large cleaning establishments shows that the majority of these firms started with an obvious hazard such as cutting and welding torches. This experience indicates most strongly a need for

    (A) a law prohibiting the use of welding torches
    (B) fire safety programs in this industry
    (C) a law requiring welders to be licensed
    (D) eliminating the use of dip tanks

12. In order to reduce the number of collisions between fire apparatus and private automobiles, a number of suggestions has been made regarding the training and selection of drivers. Of the following courses of action, the one which is *least* likely to lead to the reduction of accidents is to

    (A) select drivers who have the longest driving experience
    (B) require all members of a company to drive fire apparatus
    (C) select drivers who have experience in driving trucks
    (D) select an alternate driver for every piece of fire apparatus

13. "Automobile parking and double-parking on city streets is becoming daily a greater menace to effective firefighting." Of the following effects of automobile parking, the *least* serious is that

    (A) access to the fire building may be obstructed
    (B) ladder and rescue work may be retarded
    (C) traffic congestion may make it difficult for fire apparatus to get through
    (D) greater lengths of hose may be needed in fighting a fire

14. "Sometimes at a fire a firefighter endures great punishment in order to reach a desired position which could have been reached more quickly and with less hardship." Of the following, the chief implication of the above statement is that

    (A) one should take time off during a fire to think through completely problems presented
    (B) courage is an important asset in a firefighter
    (C) training is an important factor in firefighting
    (D) modern firefighting requires alert firefighters

15. The statement "Municipal inspections should be coordinated" suggests most nearly that

    (A) fire hazards are dangerous
    (B) the detection of fire hazards is connected with other problems found in a municipality
    (C) municipal and fire inspection services are of unequal importance
    (D) municipal inspections are closely tied up with noninspectional services

16. "With fireproof schools, it would appear that drills are unnecessary." The main reason for believing this statement to be false is that

    (A) panic sometimes occurs
    (B) fire extinguishers are available in every school
    (C) fire alarms are easily sounded
    (D) children are accustomed to drilling

17. "Flame is of varying heat according to the nature of the substance producing it" means most nearly that

   (A) some fires are larger than others
   (B) the best measure of the heat produced by a particular substance is its temperature
   (C) there can be no fire without flame or flame without a fire
   (D) the degree of heat evolved by the combustion of different materials is not identical

18. "No firefighter is expected to use a measuring stick to determine the exact amount of hose needed to stretch a line from the entrance of a burning building to the seat of a fire. He should be guided generally by the following rule—one length of hose for every story." This rule assumes most directly some degree of uniformity in

   (A) fire hazards
   (B) building construction
   (C) causes of fires
   (D) window areas

19. "A fire will occur when a flammable material is sufficiently heated in the presence of oxygen." It follows from this statement that a fire will occur whenever there is sufficient

   (A) heat, hydrogen, and fuel
   (B) oxygen, hydrogen, and fuel
   (C) heat, air, and fuel
   (D) gasoline, heat, and fuel

20. Suppose that the vast majority of the fires which took place in 1980 in New York City occurred in a particular type of building structure. Before a firefighter could reasonably infer that this particular type of building structure was unduly susceptible to fires in New York City in 1980, he or she would have to know

   (A) the precise incidence, in terms of percent, that is represented by the somewhat loose phrase "vast majority"
   (B) the frequency of this type of building structure as compared with that of other types of building structures
   (C) whether 1980 is validly to be taken as the fiscal year or the calendar year
   (D) whether New York City may legitimately be considered as representative of the country as a whole in regard to the statistical frequency of this type of building structure

21. "Every fire is a potential conflagration." Of the following, the most valid inference that may be drawn from this statement is that

   (A) no matter how insignificant a fire may appear to be, a first effort should be to isolate it
   (B) the method of fighting a potential fire must be adapted to the unique circumstances surrounding that fire
   (C) the apparatus and firefighters sent immediately to a fire should be sufficient to handle practically any conflagration that may be found
   (D) the full potentialities of a fire are usually realized

22. Suppose that, for a certain period of time studied, the percentage of telephone alarms which were false alarms was less than the percentage of fire box alarms which were false alarms. Then the one of the following which is most accurate, solely on the basis of the above statement, is that during the period studied

    (A) more alarms were transmitted by telephone than by fire box
    (B) more alarms were transmitted by fire box than by telephone
    (C) relatively fewer false alarms were transmitted by telephone than by fire box
    (D) relatively fewer false alarms were transmitted by fire box than by telephone

23. "There are more engine companies than hook and ladder companies. However, to conclude that the number of firefighters assigned to engine companies exceeds the number of firefighters assigned to hook and ladder companies is to make a basic assumption." Of the following, the most accurate statement of the basic assumption referred to in this quotation is that

    (A) the number of hook and ladder companies does not differ greatly from the number of engine companies
    (B) about the same number of firefighters, on the average, are assigned to each type of company
    (C) an engine company, on the average, has fewer firefighters than a hook and ladder company
    (D) the largest engine company is no larger than the largest hook and ladder company

24. If the fire company due to arrive first arrived one minute before the company second due, and the company third due arrived four minutes after the alarm was sent, then the one of the following statements which is most accurate is that the

    (A) second due company arrived three minutes before the third due company
    (B) third due company arrived three minutes after the first due company
    (C) second due company arrived two minutes after the alarm was sent
    (D) second due company arrived one minute after the first due company

25. "During June, 25 percent of all fires in a city were in buildings of type A, 40 percent were in buildings of type B, and 15 percent were in buildings of type C." Of the following, the most accurate statement is that the total number of fires during June was

    (A) equal to the sum of the percentages of fires in the three types of buildings, divided by 100
    (B) less than 100 percent
    (C) one-fourth the number of fires in buildings of type B
    (D) four times the number of fires in buildings of type A

26. "One or more public ambulances may be called to a street box by sending on the Morse key therein the following signals in the order given: the preliminary signal 777, the number of the street box, and the number to indicate how many ambulances are wanted." In accordance with these instructions, the proper signal for calling four ambulances to box 423 is

    (A) 4-777-423
    (B) 423-777-4
    (C) 7777-423
    (D) 777-423-4

27. The fire alarm box is an important element of the city's fire protection system. The one of the following factors which is of *least* value in helping a citizen to send a fire alarm quickly by means of a fire alarm box is that

   (A) the mechanism of a fire alarm box is simple to operate
   (B) fire alarm boxes are placed at very frequent intervals throughout the city
   (C) there are several different types of fire alarm boxes in use
   (D) specific directions for sending an alarm appear on each fire alarm box

28. The number of fire companies in a city is chiefly determined by all of the following conditions except

   (A) the number of hose streams likely to be required to handle such fires as may be expected
   (B) the manpower and capacity of fire department pumping apparatus and of ladder service
   (C) the type of fire alarm system in use
   (D) the accessibility of various parts of the city to fire companies

29. It is desirable that the fire department have ladders of varying lengths mainly because some

   (A) fires occur at greater distances from the ground than others
   (B) firefighters are more agile than others
   (C) firefighters are taller than others
   (D) fires are harder to extinguish than others

30. The lever on fire alarm boxes for use by citizens should be

   (A) very easy to manipulate
   (B) just a little difficult to manipulate
   (C) very difficult to manipulate
   (D) constructed without regard to ease of manipulation

31. A large fire occurs which you, as a firefighter, are helping to extinguish. An emergency arises and you believe that a certain action should be taken. Your superior officer directs you to do something else which you consider undesirable. You should

   (A) take the initiative and follow what you originally thought to be the superior line of action
   (B) think the matter over for a few minutes and weigh the virtues of the two lines of action
   (C) waste no time, but refer the problem immediately to another superior officer
   (D) obey orders despite the fact that you disagree

32. Suppose that your company is extinguishing a very small fire in a parked automobile. Your commanding officer directs you to perform some act which, as far as you can see, is not going to help in any way to put out the fire. Of the following, the best reason for obeying the order instantly and without question is that

   (A) the fire department is a civil organization
   (B) your officer, after all, has been in the service for a much longer period than you have been
   (C) without discipline, the efficiency of your company would be greatly reduced
   (D) the first duty of the commanding officer is to command

33. Assume that you are a firefighter. While you are walking along in a quiet residential neighborhood about 3 A.M. on a Sunday morning, another pedestrian calls your attention to smoke coming from several windows on the top floor of a three-story apartment house. Of the following, the best action for you to take immediately is to

    (A) race through the house, wake the tenants, and lead them to safety
    (B) run to the nearest fire alarm box, turn in an alarm, and then run back into the house to arouse the tenants
    (C) run to the nearest fire alarm box, turn in an alarm, and stay there to direct the fire apparatus when it arrives
    (D) direct the other pedestrian to the nearest alarm box with directions to stay there after sending the alarm while you go to the apartment from which the smoke is issuing

34. Assume that you are a firefighter, off duty and in uniform in the basement of a department store. A large crowd is present. There are two stairways, 50 feet apart. A person whom you cannot see screams, "Fire!" You should first

    (A) rush inside and sound the alarm at a fire alarm box
    (B) find the person who shouted "Fire!", ascertain where the fire has occurred, and proceed to extinguish the fire
    (C) jump on top of a nearby counter, order everyone to be quiet and not to move; find out who screamed "Fire!" and reprimand that person publicly
    (D) jump on top of a counter, obtain the attention of the crowd, direct the crowd to walk to the nearest stairway, and announce that there is no immediate danger

35. A number of prominent municipal officials are extremely interested in the operation of the local fire department. These officials frequently arrive at the scene of a fire almost as quickly as the firefighters. It is desirable that the fire department welcome the interest of these municipal officials mainly because of the probability that

    (A) the officials will acquire considerable technical information
    (B) the organization of the fire department will be modified
    (C) fire administration will remain unaffected
    (D) the work and problems of the fire department will receive due recognition

36. The owner of a building at which you helped put out a fire complains bitterly to you that the firefighters broke a number of cellar windows even before setting out to extinguish the fire in the cellar. Of the following, the best action for you to take is to

    (A) question the validity of the data as described by the owner
    (B) request the owner to put the statement in writing
    (C) explain the reason for breaking the windows
    (D) suggest that the owner have the cellar windows replaced with unbreakable glass

37. An elderly man approaches you, as officer on duty, to complain that the noise of apparatus responding to alarms wakes him in the middle of the night. Of the following, the best way for you to handle this situation is to

    (A) quickly change the subject since the man obviously is a crank
    (B) ask the man for specific instances of apparatus making noise at night
    (C) explain the need for speed in response of apparatus
    (D) promise to avoid making unnecessary noise at night

38. A firefighter who was assigned to deliver a talk before a civic organization first made a detailed outline of his talk. Then, with the help of two members of his company, he prepared several demonstrations and charts. The firefighter's procedure was

(A) good, chiefly because preparation increases the effectiveness of a talk
(B) poor, chiefly because a talk should be guided by the questions and comments of the audience and not follow a predetermined outline
(C) good, chiefly because the audience will be impressed by the care that the firefighter has taken in preparing the talk
(D) poor, chiefly because the talk will appear rehearsed rather than spontaneous and natural

39. Suppose that you were assigned to be in charge of a new headquarters bureau that will have extensive correspondence with the public and very frequent mail contact with the other city departments. It is decided at the beginning that all communications from the headquarters officer are to go out over your signature. Of the following, the most likely result of this procedure is that

(A) the administrative head of the bureau will spend too much time preparing correspondence
(B) execution of bureau policy will be unduly delayed
(C) subordinate officers will tend to avoid responsibility for decisions based on bureau policy
(D) uniformity of bureau policy as expressed in such communications will tend to be established

40. Suppose that you have been asked to answer a letter from a local board of trade requesting certain information. You find that you cannot grant this request. The best way to begin your answering letter is by

(A) quoting the laws or regulations which forbid the release of this information
(B) stating that you are sorry that the request cannot be granted
(C) explaining in detail the reasons for your decision
(D) commending the organization for its service to the community

41. The relationship between the fire department and the press is like a two-way street because the press is not only a medium through which the fire department releases information to the public, but the press also

(A) is interested in the promotion of the department's program
(B) can teach the department good public relations
(C) makes the department aware of public opinion
(D) provides the basis for community cooperation with the department

42. Assume that you have been asked to prepare an answer to a request from a citizen of the city addressed to the chief of the department. If this request cannot be granted, it is most desirable that the answering letter begin by

(A) telling the person that you were glad to receive the request
(B) listing the laws which make granting the request impossible
(C) saying that the request cannot be granted
(D) discussing the problem presented and showing why the fire department cannot be expected to grant the request

43. Assume that you have been requested by the chief of the department to prepare for the public distribution a statement dealing with a controversial matter. For you to present the department's point of view in a terse statement, making no reference to any other matter, is, in general

    (A) undesirable; you should show all the statistical data you used, how you obtained the data, and how you arrived at the conclusions presented
    (B) desirable; people will not read long statements
    (C) undesirable; the statement should be developed from ideas and facts familiar to most readers
    (D) desirable; the department's viewpoint should be made known in all controversial matters

44. The most important function of good public relations in a fire department should be in

    (A) training personnel
    (B) developing an understanding of fire dangers on the part of the public
    (C) enacting new fire laws
    (D) recruiting technically trained firefighters

45. The primary purpose of a public relations program of an administrative organization should be to develop mutual understanding between

    (A) the public and those who benefit by organized service
    (B) the public and the organization
    (C) the organization and its affiliate organizations
    (D) the personnel of the organization and the management

46. You arrive in your office at 11 A.M., having been on an inspection tour since 8 A.M. A man has been waiting in your office for two hours. He is abusive because of his long wait, and he accuses you of sleeping off a hangover at the taxpayers' expense. You should

    (A) say that you have been working all morning, and let him sit in the outer office a little longer until he cools off
    (B) tell him you are too busy to see him and make an appointment for later in the day
    (C) ignore his comments, courteously find out what his business is, and take care of him in a perfunctory manner
    (D) explain briefly that your duties sometimes take you out of your office and that an appointment would have prevented inconvenience

47. During the routine inspection of a building, you are told very unfavorable personal comments concerning several top officials of the department by a citizen. Of the following, it would usually be best for you to

    (A) try to change the subject as soon as possible
    (B) attempt to convince the citizen that he is in error
    (C) advise the citizen that your opinion might be like his, but that you can't discuss it
    (D) tell the citizen that it would be more proper for him to put his comments in writing

48. "The fire officer should, as far as operations permit, answer any reasonable questions asked by the occupant of the premises involved in a fire." However, of the following, the subject about which he should be most guarded in his comments is the

    (A) time of the arrival of companies
    (B) methods used to control and extinguish the fire

(C) specific cause of the fire

(D) probable extent of the damage

49. The public is most likely to judge personnel largely on the basis of their

(A) experience

(B) training and education

(C) civic-mindedness

(D) manner and appearance while on duty

50. When employees consistently engage in poor public relations practices, of the following, the cause is most often

(A) disobedience of orders

(B) lack of emotional control

(C) bullheadedness

(D) poor supervision

51. A citizen's support of a fire department program can best be enlisted by

(A) minimizing the cost to the community

(B) telling him how backward the community is in its practices and why such a situation exists

(C) telling him that it is his civic duty to do all that he can to support the program

(D) informing him how the program will benefit him and his family

52. The term "public relations" has been defined as the aggregate of every effort made to create and maintain good will and to prevent the growth of ill will. This concept assumes particular importance with regard to public agencies because

(A) public relations become satisfactory in inverse ratio to the number of personnel employed in the agency

(B) legislators may react unfavorably to the public agency no matter how its public relations policy is developed

(C) they are much more dependent upon public good will than are commercial organizations

(D) they are tax-supported and depend on the active and intelligent support of an informed public

53. Of the following, the proper attitude for an administrative officer to adopt toward complaints from the public is that he or she should

(A) not only accept complaints but should establish a regular procedure whereby they may be handled

(B) avoid encouraging correspondence with the public on the subject of complaints

(C) remember that it is his or her duty to get a job done, not to act as a public relations officer

(D) recognize that complaints are rarely the basis for significant administrative action

54. The method of reducing the number of false alarms by education alone is most effective for which one of the following types of individuals who pull false alarms?

(A) mental patients

(B) drunks

(C) children of school age

(D) former firefighters

55. Excluding direct enforcement measures, the most promising approach to relief of the municipal false alarm problem seems to be concerned with

    (A) better location and distribution of alarm boxes
    (B) replacement of telegraph systems with telephone systems
    (C) improved electrical circuitry for verification
    (D) public relations and education

56. An aspect of public relations which has been severely neglected by government agencies, but which has been stressed in business and industry, is

    (A) use of radio
    (B) public speaking
    (C) advertising
    (D) training employees in personal contacts with the public

57. In dealing with the public, it is helpful to know that generally most people are more willing to do something when they

    (A) are not responsible
    (B) understand the reasons
    (C) will be given a little assistance
    (D) must learn a new skill

58. A study shows that false alarms occur mostly between noon and 1 P.M. and between 3 and 10 P.M. The most likely explanation of these results is that many false alarms are sent by

    (A) school children
    (B) drunks
    (C) mental patients
    (D) arsonists

59. While visiting the lounge of a hotel, a firefighter discovers a fire which apparently has been burning for some time and is rapidly spreading. Of the following, the first action for him to take is to

    (A) find the nearest fire extinguisher and attempt to put out the fire
    (B) notify the desk clerk of the fire and send an alarm from the nearest street alarm box
    (C) send an alarm from the nearest street alarm box only
    (D) run throughout the hotel and warn all occupants to evacuate the building

60. When a fire occurs in the city of Merritt, the alarm transmitted over the fire alarm telegraph system sounds in every firehouse in Merritt. Of the following, the best justification for this practice is that

    (A) some companies are more efficient than others
    (B) to listen to a large number of alarms keeps the firefighters alert
    (C) for maximally effective disposition of fire apparatus, the staff at the central headquarters should be fully aware of the entire situation at any moment
    (D) while certain companies are responding to one alarm, other companies must be prepared to respond to new alarms in the same or nearby districts

61. Local firefighters have been called to the assistance of airport firefighters at the site of a major airplane crash. The primary concern of the firefighters must be

(A) extinguishment of the fire
(B) protection of life
(C) protection of exposures
(D) overhauling

62. Your fire company has just responded to a fire in the hold of a ship docked at a pier. Prior to your arrival, the ship's crew began to take certain actions to control the fire. Which of their following recent actions would prove *least* effective in controlling the fire?

(A) They sealed the hatch covers on the hold with waterproof tape.
(B) They shut down the ventilators from the hold and sealed them.
(C) They flooded the hold with carbon dioxide.
(D) They placed thermostats throughout the vessel to detect any increase in temperature.

63. Experience in fighting fires in unsprinklered high-rise structures has created a new body of knowledge for firefighters. Which of the following statements is most correct with regard to lessons learned from such fires?

(A) Central air conditioning, if allowed to continue to supply fresh air in the vicinity of the fire will cool the fire down.
(B) New lightweight construction helps to restrict vertical extension of the fire to higher floors.
(C) Extreme temperatures have no significant impact on a firefighter's effective work-time.
(D) Synthetic materials used in furnishings increase the threat to life and safety.

64. Your fire company responds to a fire involving drums of hazardous waste materials. Of the following actions, which would be the *least* correct for you to take?

(A) Treat empty drums carefully, since they may be more hazardous than full drums.
(B) Avoid using water on a fuming substance until you have identified the substance.
(C) Smell or touch the substance involved so as to get an indication of just what it is that is burning.
(D) Treat small leaks in small containers very carefully.

65. Your ladder company is operating at a serious fire in an old loft building. During operations, you become aware of a large accumulation of water on the third floor. The depth of the water is greater in the center of the room than on the sides. In addition, large slabs of plaster are falling from the walls. It would be most reasonable for you to

(A) continue with your assigned firefighting duties. Large amounts of water are always necessary to extinguish a large fire.
(B) take care to avoid being hit by the falling plaster, a normal occurrence in building fires
(C) skirt the deep water in the middle of the room. Old buildings generally sag in the middle.
(D) give this information to your superior officer so that he or she may assess the possibility of building collapse

66. As a firefighter you are summoned to a hazardous waste emergency involving radioactive isotopes. Of the following, the factor *least* likely to present a health hazard is

(A) atmospheric conditions
(B) length of exposure
(C) nature of the radiation
(D) intensity of the radiation

67. Your fire company responds to the site of a serious auto accident and finds the driver unconscious in the car. The victim's position severly limits your survey of the injuries involved. However, the vehicle itself may give some clues. Of the following, which would be the *least* correct assumption?

(A) A cracked windshield or rear view mirror may indicate injury to the head and spine.
(B) A damaged steering wheel or column may indicate chest, head, or abdominal injury.
(C) Dents on the dashboard may indicate leg, knee, or pelvis trauma.
(D) Since the driver's seat belt is still in position, injuries to spine or abdomen are not likely.

68. The firefighters in a ladder company are required to search for victims in apartments above the site of a store fire. Of the following, the *least* reasonable procedure is to

(A) have a plan and move toward windows and secondary means of egress
(B) in each bedroom, flip a mattress into a "U" position to indicate that the area has been searched
(C) pause from time to time to listen for crying, moaning, or coughing
(D) have the firefighters, after completing their first search for victims, repeat their search a second time to make sure all victims have been located

69. In case of an early morning major explosion in which a number of blocks are threatened with being engulfed by a fire, the structure which should demand the *least* attention from firefighters is

(A) a hotel
(B) a theater
(C) a single-family dwelling
(D) an apartment house

70. The fire department has criticized the management of several hotels for failure to call the fire department promptly when fires are discovered. The most probable reason for this delay by the management is

(A) fire insurance rates are affected by the number of fires reported
(B) most fires are extinguished by the hotel's staff before the fire department arrives
(C) hotel guests frequently report fires erroneously
(D) it is feared that hotel guests will be alarmed by the arrival of fire apparatus

Answer Key

1. C	11. B	21. A	31. D	41. C	51. D	61. B
2. D	12. B	22. C	32. C	42. C	52. D	62. D
3. A	13. D	23. B	33. D	43. C	53. A	63. D
4. C	14. C	24. D	34. D	44. B	54. C	64. C
5. B	15. B	25. D	35. D	45. B	55. D	65. D
6. B	16. A	26. D	36. C	46. D	56. D	66. A
7. D	17. D	27. C	37. C	47. A	57. B	67. D
8. D	18. B	28. C	38. A	48. C	58. A	68. D
9. A	19. C	29. A	39. D	49. D	59. B	69. B
10. D	20. B	30. B	40. B	50. D	60. D	70. D

Explanatory Answers

1. **(C)** The rationale behind the civil service testing program is that the person with the higher score will become a better firefighter. If this is correct, then a firefighting force made up of high scorers on civil service exams will be more efficient than one made up of people appointed on the basis of friendships or political connections.

2. **(D)** A fire might get out of control because of an inadequate supply of water, but the lack of water could not be its cause.

Completed Table
(Questions 3 to 5)

	Last Year	This Year	Increase
Firefighters	9326	9744	418
Lieutenants	1355	1417	62
Captains	326	433	107
Others	462	505	43
TOTAL	11469	12099	630

3. **(A)** To calculate the number of people in the Others group last year, you must first find the number of Captains last year. Since the increase from last year to this year among the Captains was 107, subtract 107 from 433 to find out the number of Captains last year. By adding together the number of Firefighters, Lieutenants, and Captains and subtracting that sum from last year's Total, we find that the number of Others last year was 462. 462 is closest to 450, so the answer to the question is (A).

4. **(C)** To find percent of change, find the difference between the numbers and divide that difference by the original number: $107 \div 326 = 33\%$. Try the others and you will find that the Captains had the greatest percentage of increase by far.

5. **(B)** There are 9744 Firefighters and 2355 people of all other ranks. This constitutes a ratio of about 4:1.

6. **(B)** In firefighting, time is crucial. If a fire alarm box can be found at once, no time is lost in sounding the alarm.

7. **(D)** Obviously, the more information available to firefighters, the less time they need to spend finding things out for themselves and the more quickly they can attack the fire.

8. **(D)** If Mr. Smart had known how to use a fire alarm box, he might have alerted the firefighters far more quickly, and his house might have been saved. The public must be educated to use all resources.

9. **(A)** Presumably a modern fire alarm system allows for alarms to be transmitted and received most rapidly and accurately. The more quickly the firefighters can respond, the more quickly they can control the fire and the smaller it will be.

10. **(D)** It is impossible to overemphasize the need for speed in all aspects of firefighting.

11. **(B)** Obviously, the number of industrial fires can be reduced by proper fire safety education of workers in fire-prone industries.

12. **(B)** Experience is the best teacher. If a few firefighters drive frequently, they will become expert. If firefighters rotate the driving responsibility, each will drive less often, and no one will have the opportunity to develop expertise.

13. **(D)** The need for greater lengths of hose is less serious than problems in reaching the site of the fire or in gaining access to the building.

14. **(C)** A better trained firefighter is more likely to recognize the best way to approach a problem.

15. **(B)** If fire inspections are made in conjunction with other safety inspections, remedial actions and follow-up can also be coordinated for efficiency and guarantees of compliance.

16. **(A)** Fireproof schools do not have fireproof contents. In case of fire, evacuation of students is a must. Fire drills are necessary so that teachers and students may become familiar with procedures and will not panic in case of fire.

17. **(D)** The heat produced by all fires is not alike.

18. **(B)** If one length of hose is required for each story of a building, it must be assumed that the height of a story is similar in most buildings.

19. **(C)** Heat is always necessary for fire; flammable material is the fuel; oxygen is the component of air that promotes combustion.

20. **(B)** If most of the buildings are of a particular construction, then it stands to reason that most of the fires would occur in that kind of building. The answer to this question is entirely based upon the frequency of that type of building as compared to all others.

21. **(A)** In other words, even a tiny fire can grow; take all fires seriously.

22. **(C)** This is a combination reading comprehension and reasoning question. The correct answer is simply a restatement of the first sentence.

23. **(B)** If the same number of firefighters were assigned to engine companies as to hook and ladder companies, then if there were more engine companies, there would be more firefighters assigned to engine companies.

24. **(D)** Since the first company arrived one minute before the second, the second company arrived one minute after the first. This is the only certain answer.

25. **(D)** If 25 percent of all fires were in buildings of Type A, the total number of fires was four times the number of fires in Type A buildings.

26. **(D)** The preliminary signal is the first signal, here 777, followed by box number 423 and the four ambulances wanted.

27. **(C)** Quite to the contrary, if directions for using alarm boxes vary from one to the next, sending of alarms will be delayed while citizens study the directions. It is important that the directions be uniform throughout the city.

28. **(C)** The type of alarm system has little bearing upon the number of fire companies. What is important is that there should be sufficient manpower and sufficient equipment to reach all parts of the city efficiently.

29. **(A)** Longer ladders are more difficult to maneuver, so it is good to have short ladders for low fires. However, long ladders must be available to fight fires on higher floors.

30. **(B)** The logic of this answer is that slight difficulty will discourage accidental false alarms.

31. **(D)** There would be chaos if each firefighter were to create his or her own plan of action and follow it. You MUST obey orders. Your superior officer has had greater experience in fighting fires. After the fire, at the station house, you might discuss the orders with your superior officer and learn of the reasoning which lay behind them.

32. **(C)** Teamwork is vital in firefighting. Discipline coordinates teamwork.

33. **(D)** Sending in the alarm is the first priority, but the other pedestrian is perfectly qualified to do this chore. Assign the pedestrian to send the alarm and direct the firefighters while you, as a trained firefighter, begin to rouse and rescue.

34. **(D)** The fire is not visible and is not yet a clear and present danger. Of far greater concern is panic and stampede. First deal with the crowd, if possible directing someone to turn in an alarm at the same time.

35. **(D)** So long as municipal officials do not get into the way of firefighters, their interest and appreciation should be most welcome. Recognition is good for morale, and appreciation of the firefighters' work can do no harm at budget time.

36. **(C)** The best course in dealing with irate citizens is patient explanation.

37. **(C)** Again, explanation is the best policy.

38. **(A)** Obviously, preparation makes for an organized, useful talk. Although the audience may be impressed with the firefighter's careful presentation, impressing the public must not be the goal of the firefighter.

39. **(D)** If one person oversees and signs all correspondence, that person has control over the content of that correspondence and can assure consistency of policy and tactfulness of expression.

40. **(B)** This is a matter of tact and public relations. Begin with your regrets, then explain the reasons for your refusal.

41. **(C)** A free press reports that which it hears. The press reflects public opinion; the fire department must listen.

42. **(C)** A business letter, and that is what this is, should come straight to the point, then explain in greater detail.

43. **(C)** You must not appear to be arbitrary in dealing with controversial matters with which the public is genuinely concerned. Explain in a way that is clear and reasonable.

44. **(B)** Public relations entails relating to the public. Of the choices, only explaining fire dangers involves dealing with the public.

45. **(B)** The public relations program of an organization has to do with the relations between that organization and the public.

46. **(D)** There is no point to further antagonizing the man. On the other hand, by all means, let him know that you were involved in fire department business. Suggesting that he might have made an appointment is constructive criticism.

47. **(A)** This may be a hard spot to get out of. No professional should discuss his or her superiors with the public. Do your best to find another topic of conversation. Avoid expressing any opinion.

48. **(C)** The firefighter is not qualified to determine the cause of a fire. That is the role of investigators. It is best to refer questions about the cause of a fire to the experts.

49. **(D)** The public bases judgments first upon results (not offered here as an answer choice), then upon appearances.

50. **(D)** Occasional poor public relations practices may stem from lack of self-control or disobedience of orders. If this is a consistent problem, it more likely stems from poor supervision and lack of instruction in more desirable practices.

51. **(D)** Self-interest is a powerful stimulus.

52. **(D)** By definition, public agencies are supported by the public in terms of finances and cooperation. Good will is vital.

53. **(A)** A regular procedure for handling complaints leads to quick disposition of those complaints and may even minimize their number. A public that knows that its concerns are taken seriously is less likely to complain.

54. **(C)** It is impossible to change the behavior of mentally ill persons or drunks through education alone. And if former firefighters are turning in false alarms, they have severe psychological problems; they are already fully aware of the dangers of false alarms.

Schoolchildren, on the other hand, can be directed away from undesirable behavior through education and explanation.

55. **(D)** The public must be made aware of the dangers of false alarms and of the expense to the taxpayers.

56. **(D)** Government agencies tend to neglect employee training in nonperformance areas. Some training in dealing with the public might lead to a better image and greater cooperation for the government agencies.

57. **(B)** People are always more willing to cooperate with requests when they understand the reasons. Education is the key.

58. **(A)** From noon to 1 P.M. is the lunch hour; from 3 P.M. to 10 P.M., school is out; the culprits are schoolchildren.

59. **(B)** The alarm must be sounded immediately for a well-developed fire. The desk clerk is best equipped to notify hotel occupants of the fire and should do that simultaneously with the firefighter's sending in the alarm. The firefighter should then return to assist with evacuation.

60. **(D)** All fire companies must be aware of the location and involvement of all other companies at all times. In this way, they can know which companies are available to respond to subsequent alarms and the entire city can be certain to be protected at all times.

61. **(B)** The main objective of all firefighting at all times is the saving of lives of civilians and of other firefighters.

62. **(D)** Thermostats, strategically placed long before a fire, might have given advance warning and have been useful in preventing the development of the fire. However, once the ship is actually on fire, placing thermostats is akin to closing the barn door after the horses have escaped. All the other actions should be useful in controlling the fire.

63. **(D)** Synthetic materials produce highly toxic products of combustion that seriously increase the threat to life and safety. All the other statements are untrue. Introduction of fresh air contributes to the intensity of a fire. Fire spreads upward rapidly in the new lightweight high-rise construction. At extreme temperatures, a firefighter's effective worktime is approximately five minutes.

64. **(C)** Under no circumstances should you touch, smell, or taste any suspected hazardous substance, nor, for that matter, any unknown substance involved in a fire. All the other actions would be correct.

65. **(D)** Collapse of the building may be imminent. The weight of water causes beams to sag in the middle, which, in turn, pulls them from their supports on the sides of the building. The large slabs of falling plaster are futher indication of the shifting of floors and walls. The situation in the old loft building is by no means normal or routine.

66. **(A)** Dangers of radiation come from length of exposure, the nature of the radiation, and the intensity of the radiation. Atmospheric conditions have little influence on the hazards from radiation exposure.

67. **(D)** While the use of seat belts does save lives and limit many injuries, it is not a 100 percent effective measure. Use of seat belts, especially without the shoulder harness in place, does not always prevent injury to abdomen, spine, and pelvis. The other assumptions all deserve consideration.

68. **(D)** A second search is very much in order, but it should be made by a different group of firefighters. If the first group has not found victims, it makes sense to send another group which may have different, more effective methods. The other procedures are all proper and helpful.

69. **(B)** In the early morning, a theater is unlikely to be occupied. Since lifesaving is the paramount concern, an unoccupied structure should not divert attention from buildings where lives are in danger.

70. **(D)** Your knowledge of human nature should tell you that the hotel management is concerned with public relations. Fear of panic among guests and of publicity that the hotel is fire-prone combine to discourage hotel management from reporting fires promptly.

SPATIAL ORIENTATION

Spatial orientation questions measure your ability to keep a clear idea of where you are in relation to the space in which you happen to be. The ability to orient yourself in space is very important to the firefighter who must reach a fire in a minimum of time and in the safest and most effective manner. Spatial orientation is also vital to the firefighter in a smoky environment. Knowing the direction in which you are facing after a number of turns with no visual clues may be a skill that will save your own life.

Your firefighting training will give you clues and instruction to orient yourself in space. You will have lots of practice before you are put into a position where you must rely on this skill. The spatial orientation questions on the exam are meant to test for your native aptitude in this regard and to measure how carefully you read and how logically you follow through.

Firefighter exam spatial orientation questions tend to emphasize either where you are in a diagram or how to go from one spot to another on the diagram or on a map. Since these are not memorization questions, you will probably be allowed to use your pencil to write on the diagrams or maps as a way of testing your answer choices. When using your pencil to write on a diagram or map, be sure to write lightly. Erase any of your jottings that do not work out or that are no longer needed. If several questions are based on the same diagram or map and you have made pencil markings for them, the diagram can get quite confusing with markings for prior questions. Hence, if there is another question to be answered on the basis of the same diagram, you should erase your markings as soon as you are done with them.

Many diagrams or maps use symbols, also called "legends." Look at the whole page to see if there is a key to the legend. A dotted line may indicate movement. An arrow may indicate direction of movement or of permitted movement. An important feature of many diagrams and maps is the symbol indicating the directions of North, South, East, and West.

Often questions are based on phrases like "turn left," "to the right," or "to the left of the rear entrance." The test-maker often approaches a diagram or map from the side or from the top so that "left" and "right" do not correspond to where you are sitting in relation to the diagram. Just turn the test booklet sideways or upside down to enter the map or diagram from the standpoint of the question. Turn it so that "left" or "right" on the diagram or map is in the same direction as your left or right hand.

Occasionally, a question will give you information in verbal form and ask you to relate that information to a map or diagram. For example, the question may say that a fire engine went north two blocks, then turned east for two blocks, turned south for one block, and then went east one more block. The answer choices may consist of just lines showing the directions of movement. The correct line will go up, go left, come down halfway, then go left again. In a question of this sort, you may find it helpful to make a diagram of your own in the margin of the test booklet. If you diagram the problem yourself, you are less likely to be confused by wrong answers.

DIRECTIONS: Answer question 1 on the basis of the street map and the information below. Circle the letter of your answer choice.

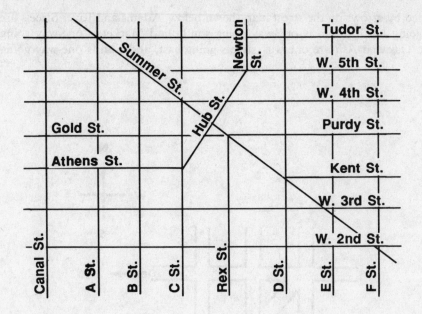

The portion of a city street map showing an area that is divided into four fire control sectors as follows:

—Sector Adam: Bounded by Tudor Street, Newton Street, Hub Street, Athens Street, and Canal Street.
—Sector Boy: Bounded by Tudor Street, F Street, West 4th Street, Hub Street, and Newton Street.
—Sector Charles: Bounded by West 4th Street, F Street, West 2nd Street, C Street, and Hub Street.
—Sector David: Bounded by Athens Street, C Street, West 2nd Street, and Canal Street.

1. There is a fire in the block bounded by West 4th Street, Summer Street, and Hub Street. The fire is in Sector

(A) David
(B) Charles
(C) Boy
(D) Adam

DIRECTIONS: Answer question 2 on the basis of the passage and diagram below.

As indicated by arrows on the street map shown below, Adams and River Streets are one-way going north. Main is one-way going south, and Market is one-way going northwest. Oak and Ash are one-way streets going east, and Elm is one-way going west.

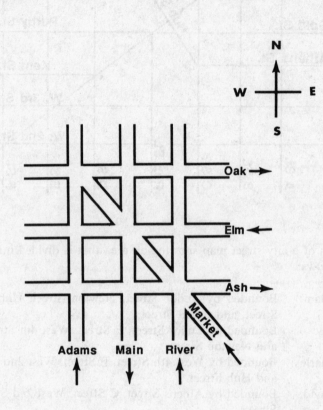

2. A fire engine heading north on River Street between Ash and Elm Streets receives a call to proceed to the intersection of Adams and Oak. In order to travel the shortest distance and not break any traffic regulations, the engine should turn

(A) left on Elm and right on Market
(B) left on Market and proceed directly to Oak
(C) left on Elm and right on Adams
(D) left on Oak and proceed directly to Adams

DIRECTIONS: Answer questions 3 to 5 on the basis of the map below.

The flow of traffic is indicated by the arrows. You must follow the flow of the traffic.

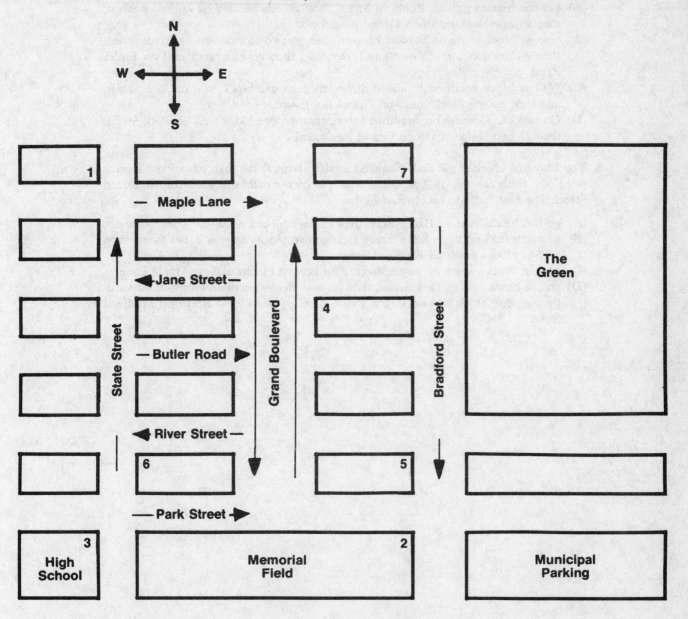

3. If you are located at point (1) and travel east one block, then turn right and travel for four blocks, and then turn left and travel one block, you will be closest to point

(A) 6
(B) 5
(C) 3
(D) 2

4. You have responded to an alarm sent from the box at the corner of Butler Road and Bradford Street and learn that the fire is at Butler Road and Grand Boulevard. Which is the best route to take to the scene?

(A) Go one block south on Bradford Street, then go one block west on River Street, then go one block north on Grand Boulevard.

(B) Go one block south on Bradford Street, then go two blocks west on River Street, then go one block north on State Street, and then go one block east on Butler Road.

(C) Go one block north on Bradford Street, then go one block west on Jane Street, and then go one block south on Grand Boulevard.

(D) Go two blocks south on Bradford Street, then go one block west on Park Street, then go two blocks north on Grand Boulevard.

5. You have just checked out and dismissed a false alarm at the high school and receive word of a trash can fire on The Green near the corner of Maple Lane and Bradford Street. The best route to take would be to

(A) go two blocks east on Park Street, then go four blocks north on Bradford Street

(B) go two blocks north on State Street, then go two blocks east on Butler Road, then go two blocks north on Bradford Street

(C) go four blocks north on State Street, then go two blocks east on Maple Lane

(D) go one block east on Park Street, then go two blocks north on Grand Boulevard, then go one block east on Butler Road, then go two blocks north on Bradford Street

DIRECTIONS: Answer questions 6 to 12 on the basis of the diagram below.

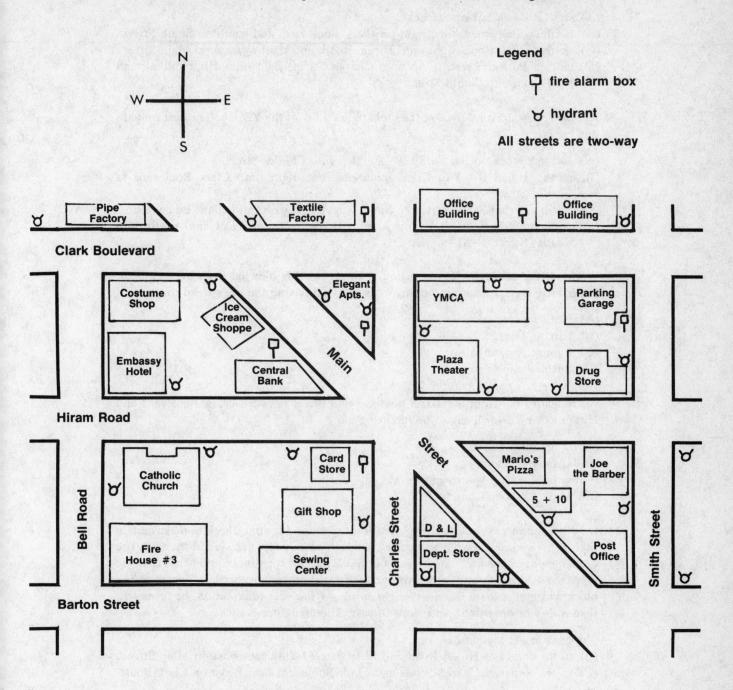

6. The parade down Main Street has cleared Clark Boulevard and is crossing Hiram Road when an alarm is received of a fire at the Post Office. The best hydrant for the firefighters to attach their first hose to should be the hydrant

(A) at the corner of Barton Street and Main Street
(B) mid-block on the northeast side of Main Street
(C) mid-block on Smith Street between Hiram Road and Barton Street
(D) at the corner of Hiram Road and Smith Street

7. The best route for the firefighters to take to the Post Office fire in Question 6 is

    (A) straight down Barton Street
    (B) north on Bell Road, then right on Clark Boulevard, and south on Smith Street
    (C) north on Bell Road, right onto Hiram Road, and right again onto Main Street
    (D) east on Barton Street, north onto Charles Street, right onto Hiram Road, and right again onto Smith Street

8. At 5:00 P.M. word is received at the firehouse of a fire at the YMCA. The best route to this fire is

    (A) east one block on Barton Street, then left into Charles Street
    (B) north on Bell Road to Clark Boulevard, and right onto Clark Boulevard to Charles Street
    (C) east on Barton Street, left onto Smith Street, and west onto Clark Boulevard
    (D) north on Bell Road to Hiram Road, east onto Hiram Road, and north onto Charles Street

9. While the fire at the YMCA is raging, a strong wind is blowing from the southeast. Firefighters should train their hoses so as to avoid having the fire spread to the

    (A) Plaza Theater
    (B) Parking Garage
    (C) Elegant Apartments
    (D) Textile factory

10. A fire alarm sent from the alarm box on Main Street between Clark Boulevard and Hiram Road is *least* likely to be reporting a fire

    (A) in an office building
    (B) at the pipe factory
    (C) at the Elegant Apartments
    (D) at the Catholic Church

11. On a hot summer evening, a firefighter is sent to make a routine check to make certain that no fire hydrants have been opened so as to reduce water pressure in the area. The firefighter leaves the firehouse by way of the Barton Street entrance and heads east. He takes the second left turn, proceeds one block, and makes another left, then goes one block and makes three successive right turns. At the next intersection, he turns left, then makes his first right, and stops the car. The firefighter is now

    (A) back at the firehouse
    (B) at the corner of Hiram Road and Main Street facing southeast on Main Street
    (C) at the corner of Charles Street and Clark Boulevard facing east on Clark Boulevard
    (D) on Bell Road at the side entrance of the Embassy Hotel

12. The most poorly protected block in terms of ease of a pedestrian's reporting a fire is the block bounded by

    (A) Charles Street, Main Street, and Barton Street
    (B) Hiram Road, Main Street, Barton Street, and Smith Street
    (C) Clark Boulevard, Charles Street, Hiram Road, and Smith Street
    (D) Hiram Road, Bell Road, Barton Street, and Charles Street

DIRECTIONS: Answer questions 13 to 16 on the basis of the floor plan below.

CENTRAL HOTEL

Central Hotel is an elegant old five-story structure. The plan of each residential floor, four floors above the lobby, is identical. ⌇ represents a standpipe connection.

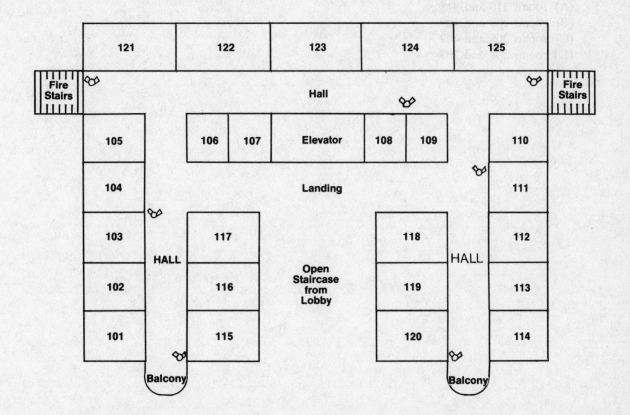

13. In the event of a fire in the hotel lobby, the best means by which to evacuate the tenant in room 124 is

(A) via the elevator
(B) down the fire stairs next to room 125
(C) across the hall and down the wide open staircase
(D) across the hall and down the corridor to the balcony

14. In the same lobby fire, the tenant in room 217 should be evacuated by

(A) running down the hall to the balcony
(B) taking the elevator
(C) whipping around the corner and down the open staircase
(D) the fire stairs between rooms 105 and 121

15. In case of fire in room 406, firefighters should

    (A) take the elevator to the roof, enter the fifth floor from the fire stairs, and connect to the standpipe at room 421

    (B) run up the open staircase and connect hoses to the standpipe between rooms 403 and 404

    (C) enter the building by way of a ladder placed in front and up to the balcony, then connect to the standpipe at rooms 403 and 404

    (D) run up the fire stairs and connect to the standpipe at room 421

16. Aside from danger to the tenant of the involved room, the greatest danger from a fire in room 309 is to the tenants of

    (A) rooms 310 and 318
    (B) rooms 209 and 409
    (C) rooms 308 and 409
    (D) rooms 308 and 209

Answer Key

1. **D**		5. **C**		9. **D**		13. **B**
2. **A**		6. **C**		10. **A**		14. **A**
3. **D**		7. **B**		11. **C**		15. **D**
4. **A**		8. **A**		12. **B**		16. **C**

Explanatory Answers

1. **(D)** Sector Adam is bounded by heavy black lines. The block with the fire is marked with an X.

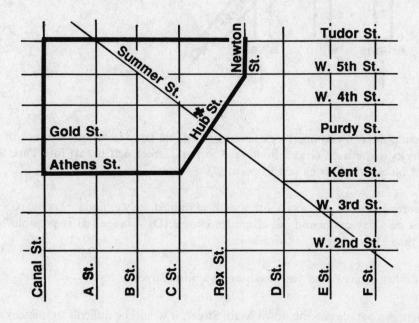

2. **(A)** The route is marked on the diagram. Choice (C) would be slightly longer. Choices (B) and (D) are impossible from the given starting point.

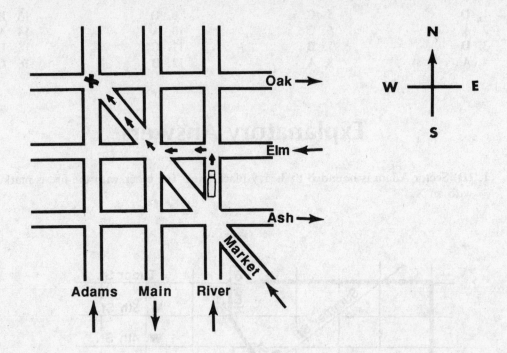

3. **(D)** From point (1), east one block is right on Maple Lane to Grand Boulevard; right four blocks is south on Grand Boulevard to Park Street; left is east into Park Street and one block brings us closest to point (2).

4. **(A)** Bradford Road is one-way southbound, so immediately eliminate choice (C). Park Street is one-way eastbound, so eliminate choice (D). Choice (B) is possible but is longer than choice (A) to no advantage.

5. **(C)** All other choices send you north on Bradford Street.

6. **(C)** There is a parade coming down Main Street; it would be difficult to maneuver fire trucks and hoses on either side of the parade route. The hydrant at choice (C) is closer than that at (D).

7. **(B)** Certainly the route straight down Barton Street is the most direct, but remember that parade. Under the special circumstances of the day, the best route is to entirely skirt the parade. The parade has cleared Clark Boulevard and the detour around the parade route is not really far.

8. **(A)** Look at the time of day. Clark Boulevard is clogged with workers on their way home and should be avoided if at all possible. Choice (D) is acceptable, but (A) is faster.

9. **(D)** Wind from the southeast is blowing toward the northwest. The Textile Factory is directly northwest from the YMCA.

10. **(A)** The fire box on Main Street between Clark Boulevard and Hiram Road is between the Ice Cream Shoppe and the Central Bank. Direct vision from this box to the office buildings is completely blocked by the Elegant Apartments. A person at this location is unlikely to be aware of a fire at either office building.

11. **(C)** The firefighter headed east on Barton, making the second left onto Main Street. He went one block, then turned left (west) onto Hiram Road. After one block, he turned right (north) onto Bell Road, right again (east) onto Clark Boulevard, and right once more (southeast) onto Main Street. At the next intersection, he turned left (north) onto Charles Street, then right placing him on Clark Boulevard, facing east. The arrows mark his route:

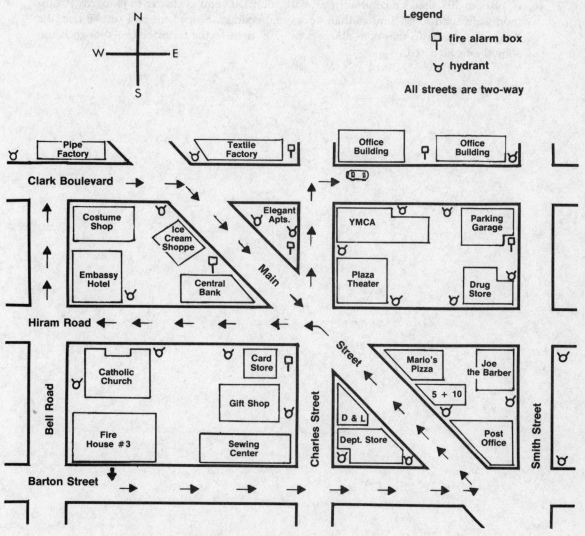

12. **(B)** There is no fire alarm box on this block; it is quite far from the nearest box on the next block and not close to the firehouse. The block in choice (A) also has no box but is closer to the nearest box and to the firehouse.

13. **(B)** Fire, heat, and smoke will rise rapidly up the open staircase from the lobby and will fill the halls of the floors above. Tenants being evacuated should travel the shortest possible distances in the halls. The fire stairs are close and are safe because they are

enclosed. Elevators are never safe in a fire. Further, since the fire is in the lobby, patrons should be kept away from the lobby.

14. **(A)** The reasoning is the same as in Question 13. This tenant will reach fresh air and safety most quickly and with least exposure to smoke by running to the balcony for rescue from outside.

15. **(D)** As we said, the elevator and open staircase present hazards to all people, firefighters as well as civilians. The safely enclosed fire stair is a much quicker and safer route by which to carry hoses to the upper floor.

16. **(C)** Room 308 shares a common wall with room 309 and is clearly in jeopardy. Common walls tend to be flimsier than floors and ceilings. Since heat and smoke rise, the danger is greater to the room above room 309 than to the room below, though it too should be evacuated.

OBSERVATION AND MEMORY

Firefighters do not work alone. Whatever it is that they are doing—fire inspections, equipment maintenance, firefighting—they always do their work in pairs, groups, or teams. Much of the firefighter's work is done under direction and supervision. However, it is important that the firefighter keep alert at all times and not rely solely on the observations and judgments of others.

Fire inspections provide the ideal opportunity for firefighters to learn neighborhood and building layouts, location of firefighting aids, and possible hazardous situations. It is vital for the firefighter out on inspection to observe everything about the premises. It is to be hoped that observation of problem situations during inspections will lead to corrective and preventive measures so that fires never occur at the premises.

But, fires do occur. In the event of fire, the firefighter who has noticed and remembered that chemicals stored near the site of the fire might give off toxic fumes when burned will be the better prepared and more effective firefighter. In event of fire, the firefighter who has noticed and remembered the locations of fire stairs and firehose connectors will be on the scene more quickly and will attack the fire with the best efficiency. Should a firefighter find him or herself inside a premises engulfed in dense smoke, memory of layout and points of exit may save his or her own life.

Since observation and memory are so important to the firefighter, many firefighter exams try to measure this ability in their applicants. The usual method of measuring this skill is to distribute a diagram or picture (or perhaps two diagrams or pictures) to applicants to study for a given period of time. After the study period, most usually five minutes per picture or diagram, the pictorial materials are collected. Then the applicant must answer a series of multiple-choice questions based upon information in the pictures or diagrams.

The pictures and diagrams on firefighter exams are usually not too complex. Five minutes is adequate for absorbing the information provided that the five minutes is used systematically. Your problem is that you cannot write anything down. All notes must be mental notes. It is therefore, worthwhile to establish in advance the categories into which you will want to fit information and then to approach the pictorial material to fill in the blanks of each category.

1. *People.* Are there any people in the picture or diagram? If not, proceed to the next category. If so, where are they located? What landmark objects (windows, doors, fire escapes, fire sources) are they near? Are they in danger?

2. *Fire activity.* Is there fire? Is there smoke? Where is the activity located? Again, how is the fire activity located with respect to the landmarks? How does it endanger life and property? Is there a predictable direction in which it might spread? (In connection with this last point, note whether or not there is a directional indicator—N–S–E–W—anywhere on the page.)

3. *Layout.* If the picture is an outdoor scene, look for fire equipment such as alarm boxes and hydrants. Note what these items are nearest to and farthest from. Study heights of buildings, natures of buildings, fire escapes. If you have in hand an indoor scene or floor plan, focus on which rooms are adjacent, means of entering one room from the next, locations of doors, windows, and fire escapes. Count and try to remember numbers of doors and doorways and numbers of windows.

4. *Special details.* Is there a smoke alarm? Where? Is there an obstruction noted? Is there an indication that doors are open or closed? If so, which ones? What is unique to this particular scene or diagram? What feature would lead to a likely question on an exam?

Give yourself five minutes to study each of the following drawings and diagrams, doing one exercise at a time. That is, study one diagram for five minutes, set it aside, answer the multiple-choice questions, then study the answer discussion before going on to the next exercise.

Exercise One–Diagram

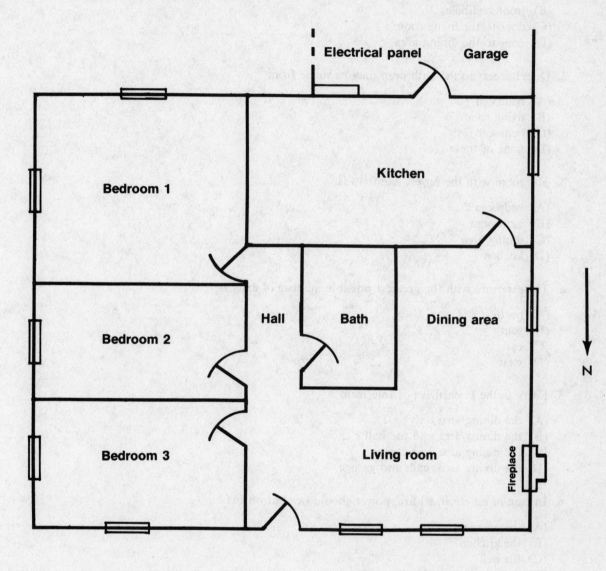

Exercise One—Questions

1. The main entry to these premises is made by

    (A) door to the kitchen
    (B) front vestibule
    (C) door to the living room
    (D) door to the dining area

2. Direct access to the bathroom may be made from

    (A) bedroom 1
    (B) living room
    (C) bedroom 2
    (D) none of these

3. The room with the fewest windows is

    (A) bedroom 2
    (B) bathroom
    (C) dining area
    (D) kitchen

4. The exposure with the greatest possible number of exits is

    (A) north
    (B) south
    (C) east
    (D) west

5. Entry to the kitchen is possible from

    (A) the dining area only
    (B) the dining area and the hall
    (C) the dining area and garage
    (D) the dining area, hall, and garage

6. In case of an electrical fire, power should be shut off in

    (A) the garage
    (B) the kitchen
    (C) the hall
    (D) there is no way to tell

7. The stairway goes upstairs by way of

    (A) the garage
    (B) the hall
    (C) bedroom 2
    (D) there is no stairway

8. If there were an out-of-control fire spreading from the fireplace, a person's best escape from the bathroom would be

    (A) out the bathroom door to the hall, into bedroom 1, and out the south window

(B) out the bathroom door into the hall, through the living room and dining area, through the kitchen and out by way of the garage

(C) out the bathroom door, down the hall and out the front door

(D) across the hall into bedroom 2 and out the window

9. Connecting rooms are

(A) bedroom 1 and bedroom 2
(B) bedroom 2 and bedroom 3
(C) bedroom 3 and dining area
(D) dining area and kitchen

10. The premises here diagrammed are most probably

(A) a single-family detached house
(B) a semi-detached house
(C) an apartment in a low-rise building
(D) an apartment in a high-rise building

Answer Key

1. **C**	3. **B**	5. **C**	7. **D**	9. **D**
2. **D**	4. **A**	6. **A**	8. **C**	10. **A**

Explanatory Answers

1. **(C)** The front door enters directly into the living room. The door to the kitchen from the garage is a secondary entrance. There is no anteroom indicated in front of the living room.

2. **(D)** The bathroom door opens only onto the hallway.

3. **(B)** The bathroom has no windows at all. Each of the other areas has one window.

4. **(A)** Most directional markers indicate North as up. This marker is unusual in that North is pointing down. Reverse the diagram so as to make North point up and thereby get your bearings. The front of the house is the northern exposure. There are three windows and a door along the front, (north) of the house.

5. **(C)** The hall does not go through to the kitchen; however, entry to the kitchen may be made through either the dining area or the garage.

6. **(A)** The electrical panel is in the garage.

7. **(D)** There is no stairway.

8. **(C)** This question requires both memory of the layout and common sense. Exit is always faster and safer by way of a door than by way of a window, if it is possible to

reach a door. The fire is at the end of the living room farthest from the door. The short run out the door makes most sense.

9. **(D)** Of these, only the dining area and kitchen connect.

10. **(A)** This takes some inference in conjunction with memory. The presence of a garage entry directly to the kitchen rules out an apartment house of any height. Windows on all four sides indicate that the house is free-standing and fully detached. There are no windows in a party wall.

In studying this floor plan, you should have noticed:

1. *People.* None.

2. *Fire activity.* None.

3. *Layout.* Northern orientation of house. Front entry to living room. L-shaped living/dining area. Access to bathroom from hall only. Three nonconnecting bedrooms in a row. Corner bedrooms with two windows each. Hallway does not go through to kitchen. Door from dining area to kitchen (as opposed to a possible open doorway without door). Access to kitchen through garage. Fully interior bathroom.

4. *Special details.* Electrical panel in garage. Fireplace in living room. No fire-protective devices.

Exercise Two—Diagram

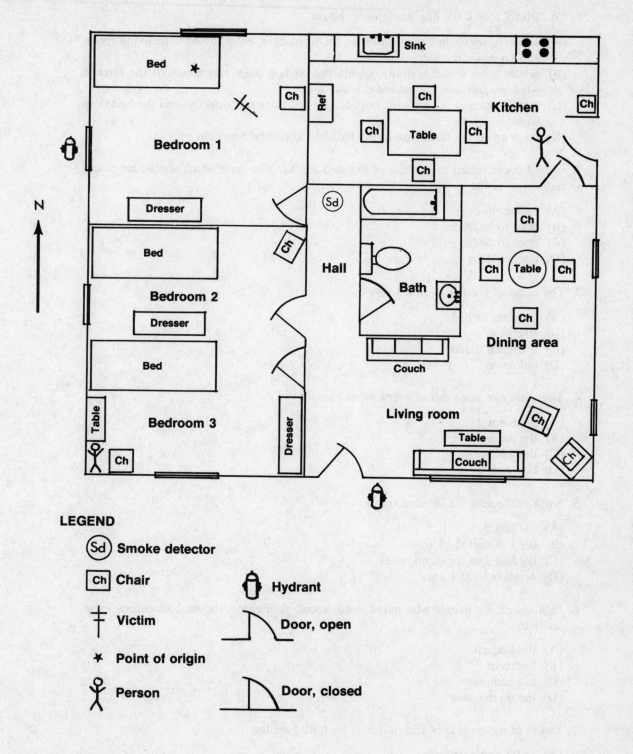

LEGEND

(Sd) **Smoke detector**

[Ch] **Chair**

✝ **Victim**

✻ **Point of origin**

👤 **Person**

🔥 **Hydrant**

Door, open

Door, closed

Exercise Two—Questions

1. To directly attack the fire, firefighters should

    (A) attach hoses to the hydrant next to the front door, enter through the living room, and go down the hall
    (B) attach hoses to the hydrant outside the kitchen door, run them out the kitchen window, and into the bedroom window
    (C) attach hoses to the hydrant outside the bedroom and directly into the bedroom window
    (D) hook up to the standpipe in the hall and enter the bedroom

2. Closed doors retard the spread of fire and smoke. The door which should be closed right away is the

    (A) bathroom door
    (B) door to bedroom 2
    (C) door to bedroom 3
    (D) kitchen door

3. The cause of the fire is most likely

    (A) smoking in bed
    (B) electrical
    (C) a kitchen flareup
    (D) unknown

4. There are two ways out of every room *except*

    (A) bedroom 2
    (B) the dining area
    (C) the bathroom
    (D) bedroom 3

5. Smoke detectors are in place in

    (A) the hall
    (B) the hall and the kitchen
    (C) the hall and the dining area
    (D) nowhere in this unit

6. In a search for people who might need rescue, firefighters should concentrate especially in

    (A) the kitchen
    (B) bedroom 3
    (C) the bathroom
    (D) the dining area

7. Doors to the outside of this unit may be found on the

    (A) south and east walls
    (B) north and east walls
    (C) south and west walls
    (D) north, south, and east walls

8. The total number of people in the unit is

    (A) 0
    (B) 1
    (C) 2
    (D) 3

9. A room with one window is

    (A) bedroom 1
    (B) bedroom 3
    (C) the living room
    (D) the bathroom

10. If the fire were to extend throughout the unit, the risk of involving furnishings would be greatest in

    (A) bedroom 3
    (B) the living room
    (C) the dining area
    (D) the kitchen

Answer Key

1. **C**	3. **A**	5. **A**	7. **A**	9. **B**
2. **B**	4. **C**	6. **B**	8. **D**	10. **B**

Explanatory Answers

1. **(C)** For the most direct attack on the fire, attach to the hydrant right outside the involved room. The hydrant outside the front door should be utilized immediately thereafter for protection of the other rooms in the unit. With careful observation and memory, you should be aware that there is no hydrant outside the kitchen door and that the device in the hall is a smoke detector, not a standpipe.

2. **(B)** The bathroom door and the door to bedroom 3 are already closed. Danger to the immediately adjacent bedroom 2 is far greater than the danger to the remote kitchen.

3. **(A)** The point of origin of the fire is a bed, and the victim is in the bedroom. Suspicion of smoking in bed as the cause of the fire is clearly legitimate.

4. **(C)** Only the bathroom has only one exit. Bedrooms 2 and 3 each have both door and window, and the dining area is wide open to the living room and kitchen and has window access as well.

5. **(A)** There is one smoke detector at the end of the hall.

6. **(B)** The victim on the floor in bedroom 1 and the person in the kitchen are quite obvious. Note the person hiding behind a chair in the corner of bedroom 3. Of course,

the bathroom and dining area must be checked too, but they offer fewer hiding places and require less concentrated attention.

7. **(A)** Look at the directional marker. The north wall of the unit is at the top of the page, so the doors are on the south and east walls.

8. **(D)** The three people are the victim in bedroom 1, the person standing in the kitchen, and the person hiding in bedroom 3.

9. **(B)** Even though it is a corner bedroom, bedroom 3 has only one window. Bedroom 1 and the living room both have two windows, and the bathroom has none at all.

10. **(B)** The living room is full of upholstered furniture. The other rooms are more sparsely furnished.

In studying this diagram, you should have noticed:

1. *People.* Three, one each in bedroom 1, bedroom 3, and the kitchen.

2. *Fire activity.* Point of origin in the bedroom.

3. *Layout.* Exit doors from living room and kitchen. Three nonconnecting bedrooms in a row. Fully interior bathroom with exit to hallway only. L-shaped living/dining area. Hallway does not go through to kitchen. Locations of windows. General volume and placement of furniture. Open and closed doors.

4. *Special details.* Hydrants at front door and outside window of bedroom 1. Smoke detector in hall. Position of victim.

Exercise Three—Sketch

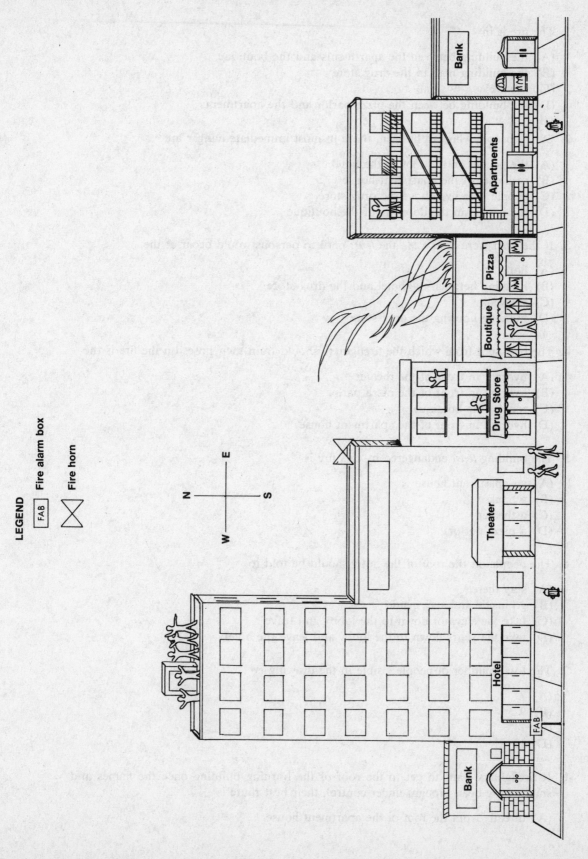

Exercise Three—Questions

1. The fire is in

   (A) the building between the apartments and the boutique
   (B) the building next to the drug store
   (C) a bank
   (D) the building between the pizza parlor and the apartments

2. Of the people who can be seen, those in most immediate danger are

   (A) the people on the roof of the hotel
   (B) residents of the apartment house
   (C) people who live above the drug store
   (D) the person in the doorway of the boutique

3. If the hour were 8:00 A.M., the *least* peril to persons would occur at the

   (A) hotel
   (B) building between the hotel and the drug store
   (C) sidewalk
   (D) building housing the drug store

4. The best spot from which the firefighters should train their hoses on the fire is the

   (A) hydrant in front of the theater
   (B) hydrant in front of the pizza parlor
   (C) roof of the boutique
   (D) hydrant in front of the apartment house

5. The building *least* endangered by this fire is

   (A) an apartment house
   (B) a bank
   (C) a theater
   (D) a pizza parlor

6. The people on the roof of the hotel should be told to

   (A) stay there
   (B) go inside and stay inside
   (C) take the elevator down to the lobby and leave
   (D) take the stairs down to the lobby and leave the hotel

7. The total number of people visible in the drawing is

   (A) 7
   (B) 9
   (C) 11
   (D) 12

8. If firefighters want to get to the roof of the burning building once the flames and smoke have been brought under control, their best route is

   (A) directly from the roof of the apartment house

(B) from the roof of the hotel to the roof of the drug store to the roof of the burning building
(C) from the roof of the building housing the drug store to the roof of the building next door to the roof of the burning building
(D) from the roof of the bank to the roof of the theater to the roof of the burning building

9. The fire was probably reported from the fire alarm box

   (A) in front of the hotel
   (B) on the wall of the theater
   (C) in front of the pizza parlor
   (D) on the fire escape of the apartment house

10. The wind is blowing from the

    (A) west
    (B) east
    (C) north
    (D) south

Answer Key

1. **A**	3. **B**	5. **B**	7. **C**	9. **A**
2. **C**	4. **D**	6. **D**	8. **C**	10. **B**

Explanatory Answers

1. **(A)** The fire is in the pizza parlor, which is between the apartments and the boutique.

2. **(C)** The people who live in the apartments over the drug store are separated from the fire by only one building. With the wind blowing in their direction, they are clearly in greater danger than people in the apartment house next to the fire but upwind of it. The person at street level can just walk away, while the people on the hotel roof are quite a bit removed.

3. **(B)** The building between the hotel and the drug store is a theater. At 8:00 A.M. a theater is likely to be deserted, creating peril to property but not to life.

4. **(D)** From the hydrant in front of the apartment house, the firefighters can move in quite close behind the fire. This is an ideal location. They can approach the fire directly without having to force their way through heat, flame, and smoke. The hydrant in front of the theater is farther away and the approach is from a less favorable direction. There is no hydrant in front of the pizza parlor itself. The roof of the boutique is too dangerous a spot from which to fight the fire at this active stage.

5. **(B)** The two banks are the two least endangered buildings. The bank beside the apartment house is separated from the fire by the apartment house and is away from the

path of the fire. The bank next door to the hotel is the building farthest from the fire and is separated from it by a couple of large, sturdy buildings.

6. **(D)** This judgment question must be based upon your recall of the whole scene. The people on the hotel roof (and the people inside the hotel) are in no immediate danger, but smoke and fire can spread and precautions must be taken. With the emergency in the area, electrical interruption could occur at any time. People should avoid the elevator and evacuate the hotel by way of the stairs. The roof of the hotel offers a wonderful vantage point for watching the fire, but a fire is not a show, and the smoke will eventually reach them. They must go.

7. **(C)** 5 on the hotel roof + 2 pedestrians + 2 in the apartments over the drug store + 1 in the doorway of the boutique + 1 on the apartment house fire escape = 11.

8. **(C)** Consider relative heights of adjacent buildings in answering this question. The drop from the apartment house is too steep, especially onto a roof weakened by fire. The route from the drug store allows for a gradual, cautious approach.

9. **(A)** Did you notice the legend on the page? The fire alarm box in front of the hotel is the only one in the picture. The device on the theater is a fire horn.

10. **(B)** The flame and smoke, and therefore the wind, are blowing from the east towards the west.

In studying this sketch, you should have noticed:

1. *People.* Numbers, locations, and locations with reference to the fire.

2. *Fire activity.* Location and location with reference to landmarks. Direction. Extent of involvement.

3. *Layout.* Types of buildings. Sizes of buildings. Arrangement of buildings. Relationships between buildings in terms of size and types of occupancy.

4. *Special details.* Legend on page. Number and location of hydrants, fire alarm box, and fire horn.

Exercise Four—Diagram

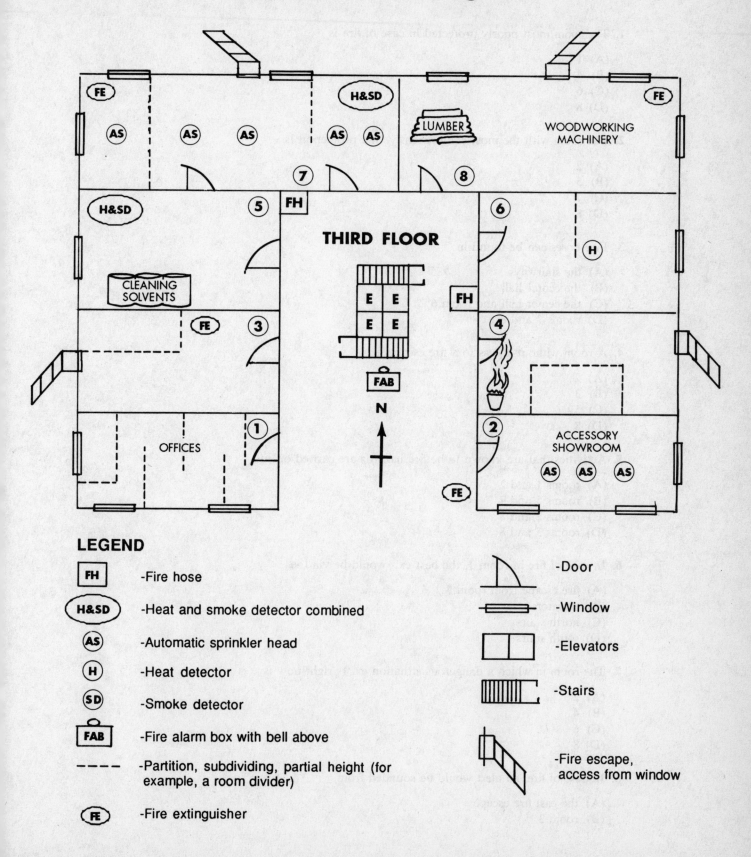

THIRD FLOOR

WOODWORKING MACHINERY

LUMBER

CLEANING SOLVENTS

OFFICES

ACCESSORY SHOWROOM

N

LEGEND

FH	-Fire hose
H&SD	-Heat and smoke detector combined
AS	-Automatic sprinkler head
H	-Heat detector
SD	-Smoke detector
FAB	-Fire alarm box with bell above
– – – –	-Partition, subdividing, partial height (for example, a room divider)
FE	-Fire extinguisher
	-Door
	-Window
	-Elevators
	-Stairs
	-Fire escape, access from window

Exercise Four—Questions

1. The room most poorly protected in case of fire is

(A) 1
(B) 4
(C) 6
(D) 8

2. The room with the most comprehensive fire protection is

(A) 2
(B) 5
(C) 7
(D) 8

3. Fire hoses can be found in

(A) the stairways
(B) the center hall
(C) the center hall and room 6
(D) rooms 7 and 8

4. A room without access to a fire escape is

(A) 2
(B) 3
(C) 5
(D) 8

5. Activities that are known to be fire hazards are carried on in

(A) rooms 1 and 2
(B) rooms 5 and 7
(C) rooms 2 and 8
(D) rooms 5 and 8

6. In case of fire in room 1, the best exit would be via the

(A) fire escape from room 3
(B) elevators
(C) north stairs
(D) south stairs

7. The room in which a dangerous situation exists right now is

(A) 2
(B) 4
(C) 6
(D) 8

8. In case of fire, an alert would be sounded from

(A) the east fire escape
(B) room 3

(C) the hall
(D) the elevators

9. The intricate pattern of partitions would make quick exit most difficult from room

(A) 1
(B) 2
(C) 6
(D) 7

10. The total number of windows on this floor is

(A) 14
(B) 16
(C) 18
(D) 20

Answer Key

1. **A**	3. **B**	5. **D**	7. **B**	9. **A**
2. **C**	4. **A**	6. **D**	8. **C**	10. **C**

Explanatory Answers

1. **(A)** Room 1 has no fire protective device of any kind and has no independent exit to the outside. Room 6 is not much better off but at least has a heat detector. Room 4 has a fire escape, though no internal protective device. Room 8 is well protected.

2. **(C)** Room 7 has automatic sprinklers, heat and smoke detector, a fire extinguisher, and fire escape. Room 2 has only automatic sprinklers and room 5 only heat and smoke detectors. Room 8 does have a smoke detector, fire extinguisher, and fire escape but lacks a sprinkler system.

3. **(B)** Fire hoses are only in the hallway.

4. **(A)** Room 2 has no access to a fire escape. Rooms 3 and 8 have fire escapes while room 5 has a door connecting it with room 7, which has a fire escape.

5. **(D)** Use and storage of cleaning solvents and woodworking are activities that can pose fire hazards. The usual activities in offices and accessory showrooms do not present fire hazards. The nature of the business in the other areas is unspecified.

6. **(D)** Room 1 is not connected with a fire escape, so its occupants would have to go into the hall to escape the fire. Once in the hall, the quickest way out would be via the nearest, the south, stairway. Entering room 3 to go out the window and down the fire escape would be slower and more cumbersome. Also, room 3 is the next room to become involved in the fire. Elevators should be avoided in a fire situation.

7. **(B)** There is a wastebasket afire in room 4.

8. **(C)** There is a fire alarm box topped by a bell in the hall next to the south staircase. The device in room 3 is a fire extinguisher, not an alarm.

9. **(A)** Room 1 is full of half partitions.

10. **(C)** Be sure to count the windows that are connected to fire escapes. There are 18 windows in all.

In studying this diagram, you should have noticed:

1. *People*. None.

2. *Fire activity*. Wastebasket fire in room 4.

3. *Layout*. Center core building with four elevators and two sets of stairs. All rooms have doors to hallway and at least one window. Corner rooms have multiple windows. Rooms 5 and 7 connect. Fire escapes from rooms 3, 4, 7, and 8.

4. *Special details*. Two fire hoses, fire extinguisher, and fire alarm box with bell in hall. Fire hazardous activities in rooms 5 and 8. Fire extinguishers in rooms 3, 7, and 8 and the hall. Sprinkler systems in rooms 2 and 7. Heat detector in room 6; smoke detector in room 8; heat and smoke detectors in rooms 5 and 7. Diagram is of the third floor of building. Extensive legend.

Exercise Five—Sketch

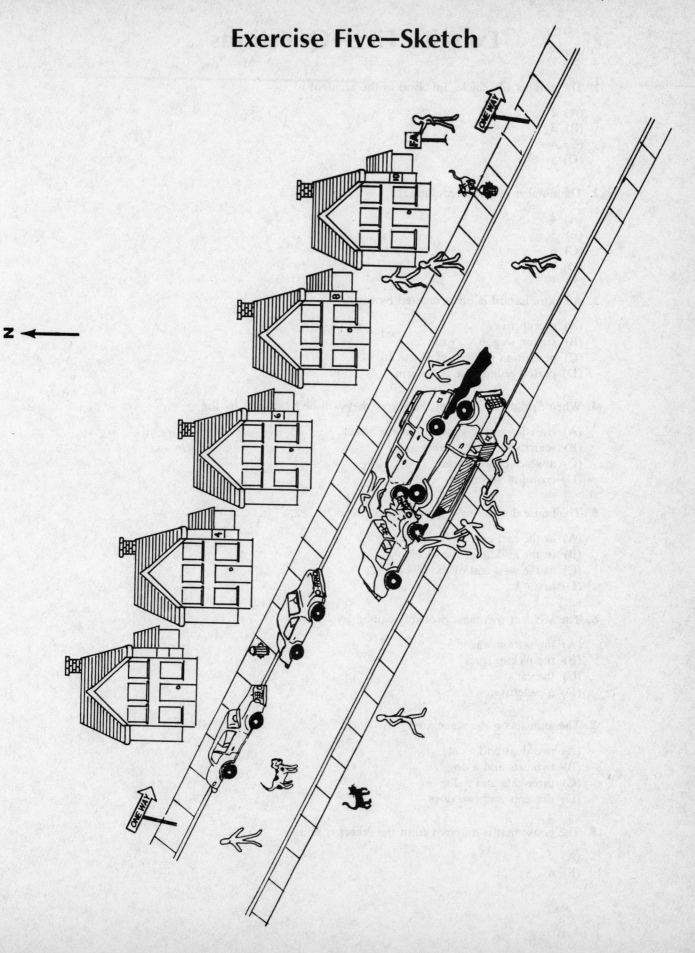

Exercise Five—Questions

1. The number of vehicles involved in the accident is

   (A) 2
   (B) 3
   (C) 4
   (D) 5

2. The number of known victims is

   (A) 4
   (B) 5
   (C) 6
   (D) 7

3. An extra hazard is being created by the

   (A) pickup truck
   (B) station wagon
   (C) pedestrian in front of house 10
   (D) person pulling the fire alarm

4. When firefighters arrive at the scene, they will be hampered by the

   (A) situation at the west end of the street
   (B) scarcity of fire hydrants
   (C) absence of witnesses
   (D) crowd of spectators

5. The house that is missing its house number is

   (A) at the east end of the block
   (B) in the middle of the block
   (C) at the west end of the block
   (D) house 8

6. The accident was most probably caused by

   (A) the station wagon
   (B) the pickup truck
   (C) the car
   (D) a pedestrian

7. The animals on the scene are

   (A) two dogs and a cat
   (B) two cats and a dog
   (C) three cats and a dog
   (D) two cats and two dogs

8. The house that is different from the others is house

   (A) 4
   (B) 6

(C) 10
(D) all are exactly alike

9. The vehicle that is totally overturned is the

(A) station wagon
(B) sedan
(C) pickup truck
(D) convertible

10. When firefighters arrive on the scene, they must first direct their attention to the

(A) victims
(B) houses in danger of catching fire
(C) cars on fire
(D) bystanders

Answer Key

1. **B**	3. **B**	5. **C**	7. **B**	9. **C**
2. **C**	4. **A**	6. **A**	8. **D**	10. **A**

Explanatory Answers

1. **(B)** There are three vehicles piled up in the middle of the street. The other two cars are parked.

2. **(C)** Six people are lying in the street. We have no way of knowing how many are in the cars.

3. **(B)** The puddle being formed from the back end of the station wagon is most likely gasoline. With the three vehicles actively on fire, a gasoline puddle creates a serious explosion threat.

4. **(A)** At the west end of the street, two cars are parked directly in front of and completely blocking access to the fire hydrant. This is both dangerous and illegal.

5. **(C)** House 2 at the far west end of the block is missing its house number.

6. **(A)** Note the one-way signs at each end of the street. It is possible that one or more of the vehicles spun around at impact, but on the face of it, it appears that the station wagon caused the accident by driving the wrong way on a one-way street.

7. **(B)** There are two cats and a dog in the scene. The animals have no bearing on the fire problem, but they are there and an observant test-taker must notice them.

8. **(D)** All the houses are exactly alike.

9. **(C)** The pickup truck has completely overturned and is resting on the top of its cab. The other two vehicles are more or less upright.

10. **(A)** Saving life is more important than saving property. The bystanders must get out of the way by themselves; the victims need assistance.

In studying this sketch, you should have noticed:

1. *People.* Six lying in the street, one turning in a fire alarm at the street box, five approaching the scene.

2. *Fire activity.* Three vehicles apparently on fire after a collision.

3. *Layout.* Three cars in pileup in middle of street, one upside-down. Five identical houses in a row. Two parked cars blocking one hydrant. Second hydrant at the end of the street. Fire alarm box on corner.

4. *Special details.* Puddle coming from rear end of station wagon. Overturned pickup truck. One-way street, eastbound. Two cats; one dog.

MECHANICAL SENSE

A firefighter is not expected to be a mechanic, an electrician, a plumber, nor an engineer. However, the daily work of firefighters does bring them into contact with tools and machinery, does require an understanding of how things fit together, and of how mechanical things work. It is important for the firefighter to have some mechanical insight, some sense of "what makes it go," and some idea of how to fix it or how to improvise.

Your firefighter exam will not presuppose that you have had any training with specific tools or in mechanical principles or firefighting concepts. However, it will assume that you are not a mechanical "illiterate." It will assume that you have shown an interest in things mechanical and have picked up information along the way from school subjects and from experience and experimentation. And it may seek to discover whether you are able to reason along mechanical lines.

The following selection of questions will give you a fair sampling of the kinds of questions you may encounter. These questions are designed to gauge your interest, insight, and basic understanding. A full set of explanations follows the question set.

DIRECTIONS: Choose the best answer and circle the letter of its answer.

1.

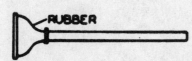

The tool above is most likely to be used by a

(A) stonemason
(B) plumber
(C) carpenter
(D) machinist

2. The main advantage of four-wheel drive is

(A) higher speed
(B) better traction
(C) better gas mileage
(D) greater durability of the vehicle

93

3.

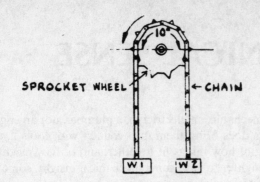

SPROCKET WHEEL → ← CHAIN

One complete revolution of the sprocket wheel will bring weight W2 higher than weight W1 by

(A) 20 inches
(B) 40 inches
(C) 30 inches
(D) 50 inches

4. If a co-worker is not breathing after receiving an electric shock but is no longer in contact with the electricity, it is most important for you to

(A) wrap the victim in a blanket
(B) force him to take hot liquids
(C) start artificial respiration promptly
(D) avoid moving him

5.

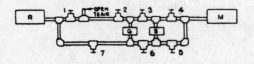

In the figure above, assume that all valves are closed. For air to flow from R, through G, then through S to M open

(A) valves 1, 2, 6, and 4
(B) valves 7, 3, and 4
(C) valves 7, 6, and 4
(D) valves 7, 3, and 5

6. The material used with solder to make it stick better is

(A) oakum
(B) lye
(C) oil
(D) flux

7. The function of the generator or alternator is to

(A) start the engine
(B) carry electricity from the battery to the engine

(C) keep the battery charged
(D) control the production of hydrocarbons

8. The principal objection to using water from a hose to put out a fire involving electrical equipment is that

(A) serious shock may result
(B) it may spread the fire
(C) metal parts may rust
(D) fuses may blow out

9.

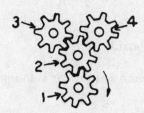

The figure above shows four gears. If gear 1 turns as shown, then the gears turning in the same direction are

(A) 2, 3, and 4   (C) 2 and 3
(B) 2 and 4   (D) 3 and 4

10. Light fixtures suspended from chains should be wired so that the

(A) wires do not support the fixture
(B) wires help support the fixture
(C) chains have an insulated link
(D) chain is not grounded to prevent short circuits

11. Asbestos is commonly used as the covering of electric wires in locations where there is likely to be high

(A) voltage   (C) temperature
(B) humidity   (D) current

12.

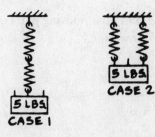

In the figure above, all four springs are identical. In Case 1 with the springs end to end, the stretch of each spring caused by the five-pound weight is

(A) one-half as much as in Case 2   (C) twice as much as in Case 2
(B) the same as in Case 2   (D) four times as much as in Case 2

13.

You might use this instrument if you wanted to

(A) pitch a tent
(B) poke holes in a fabric
(C) locate studs in a wall
(D) drill holes at equal short distances along a board

14. A green puddle under the front end of a car means that the car is losing

(A) power steering fluid
(B) antifreeze
(C) transmission fluid
(D) crankcase oil

15.

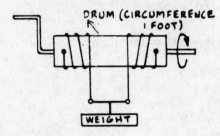

One complete revolution of the windlass drum shown above will move the weight up

(A) ½ foot
(B) 1½ feet
(C) 1 foot
(D) 2 feet

16. The frequency of oiling and greasing of bearings and other moving parts of machinery depends mainly on the

(A) size of the parts requiring lubrication
(B) speed at which the parts move
(C) ability of the operator
(D) amount of use of the equipment

17.

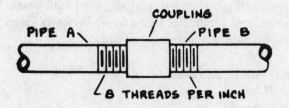

The tool above is a

(A) screw
(B) screwdriver
(C) drill bit
(D) corkscrew

18.

If both pipes A and B are free to move back and forth but are held so they cannot turn, and the coupling is turned four revolutions with a wrench, the overall length of the pipes and coupling will:

(A) decrease 1/2 inch
(B) remain the same
(C) increase or decrease 1 inch depending upon the direction of turning
(D) increase 1/2 inch

19. The plug of a portable tool should be removed from the convenience outlet by grasping the plug and not by pulling on the cord because

(A) the plug is easier to grip than the cord
(B) pulling on the cord may allow the plug to fall on the floor and break
(C) pulling on the cord may break the wires off the plug terminals
(D) the plug is generally better insulated than the cord

20. The muffler on a car serves to

(A) filter the exhaust fumes
(B) keep the car warm
(C) reduce exhaust sounds
(D) protect against body damage

21.

The figure above represents an enclosed water chamber, partially filled with water. The number 1 indicates air in the chamber and 2 indicates a pipe by which water enters the chamber. If the water pressure in the pipe, 2, increases then the

(A) water pressure in the chamber will be decreased
(B) water level in the chamber will fall
(C) air in the chamber will be compressed
(D) air in the chamber will expand

22. Black smoke coming from the muffler means that

(A) there is too much lubricating oil
(B) the car needs an oil change
(C) the carburetor is delivering too rich a mixture
(D) the carburetor is delivering too weak a mixture

23. When removing the insulation from a wire before making a splice, care should be taken to avoid nicking the wire mainly because the

(A) current carrying capacity will be reduced
(B) resistance will be increased
(C) wire tinning will be injured
(D) wire is more likely to break

24. Wood screws properly used as compared to nails properly used

(A) are easier to install
(B) generally hold better
(C) are easier to drive flush with surface
(D) are more likely to split the wood

25. Worn universal joints make themselves known by

(A) a "clunk" when the car is first started and put into driving gear
(B) steam rising from the hood of the car
(C) difficulty in starting the engine in cold weather
(D) sticking doors

26.

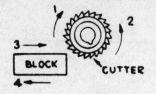

The figure above shows a cutter and a steel block. For proper cutting, they should move respectively in directions

(A) 1 and 4
(B) 2 and 3
(C) 1 and 3
(D) 2 and 4

27.

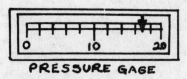

The reading shown on the gage is

(A) 10.35
(B) 13.5
(C) 10.7
(D) 17.0

28. An electric light bulb operated at *more* than its rated voltage will result in a

(A) longer life and dimmer light
(B) longer life and brighter light
(C) shorter life and brighter light
(D) shorter life and dimmer light

29. The wrench that is used principally for pipe work is

(A)

(B)

(C)

(D)

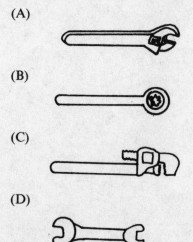

30. A slight coating of rust on small tools is best removed by

(A) applying a heavy coat of vaseline
(B) rubbing with kerosene and fine steel wool
(C) scraping with a sharp knife
(D) rubbing with a dry cloth

31. Shock absorbers are part of the

(A) engine
(B) upholstery
(C) suspension
(D) exhaust system

32.

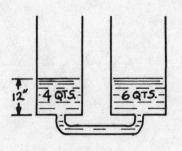

To bring the level of the water in the tanks to a height of 2¹/₂ feet, the quantity of water to be added is

(A) 10 quarts
(B) 20 quarts
(C) 15 quarts
(D) 25 quarts

33.

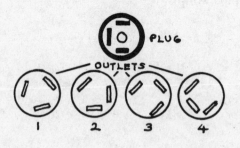

The outlet that will accept the plug is

(A) 1
(B) 2
(C) 3
(D) 4

34.

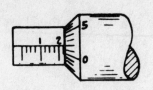

The micrometer above reads

(A) .2270
(B) .2120
(C) .2252
(D) .2020

35. Wheels should be balanced

(A) only when they are new
(B) whenever air pressure is low
(C) every 10,000 miles
(D) whenever tires are put onto rims

36.

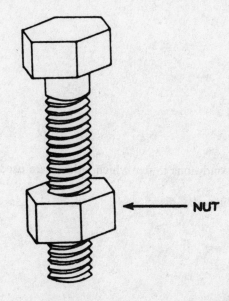

← **NUT**

Which of the following statements is true?

(A) If the nut is held stationary and the head turned clockwise, the bolt will move up.
(B) If the head of the bolt is held stationary and the nut is turned clockwise, the nut will move down.
(C) If the head of the bolt is held stationary and the nut is turned clockwise, the nut will move up.
(D) If the nut is held stationary and the bolt is turned counterclockwise, the nut will move up.

37. A good lubricant for locks is

(A) graphite
(B) grease
(C) mineral oil
(D) motor oil

38. The *least* likely result of a severe electric shock is

(A) unconsciousness
(B) a burn
(C) clenched muscles
(D) heavy breathing

39.

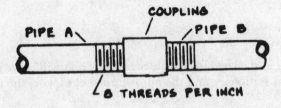

The tool shown above is a

(A) punch
(B) drill holder
(C) Phillips-type screwdriver
(D) socket wrench

40. Wood ladders should not be painted because

(A) paint will wear off rapidly due to the conditions under which ladders are used
(B) ladders are slippery when painted
(C) it is more effective to store the ladder in a dry place
(D) paint will hide defects in the ladder

41.

If pipe A is held in a vise and pipe B is turned ten revolutions with a wrench, the overall length of the pipes and coupling will decrease

(A) ⁵/₈ inch
(B) 2¹/₂ inches
(C) 1¹/₄ inches
(D) 3³/₄ inches

42. Which of the saws is used to make curved cuts?

(A)

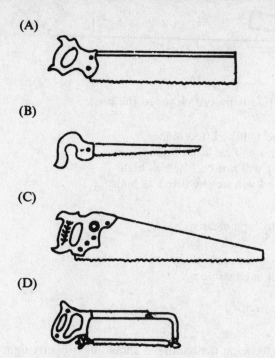

(B)

(C)

(D)

43. A method that can be used to prevent the forming of "skin" on a partially used can of oil paint is to

(A) turn the can upside down every few months
(B) pour a thin layer of solvent over the top of the paint
(C) store the paint in a well-ventilated room
(D) avoid shaking the can after it has been sealed

44. The reason that a lubricant prevents rubbing surfaces from becoming hot is that the oil

(A) is cold and cools off the rubbing metal surfaces
(B) is sticky, preventing the surfaces from moving over each other too rapidly
(C) forms a smooth layer between the two surfaces, preventing their coming into contact
(D) makes the surfaces smooth so that they move easily over each other

45.

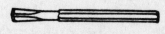

The tool shown above is used to

(A) set nails
(B) drill holes in concrete
(C) cut a brick accurately
(D) centerpunch for holes

46.

If the block on which the lever is resting is moved closer to the brick

(A) the brick will be easier to lift and will be lifted higher
(B) the brick will be harder to lift and will be lifted higher
(C) the brick will be easier to lift but will not be lifted as high
(D) the brick will be harder to lift and will not be lifted as high

47. The term "whipping" when applied to rope means

(A) binding the ends with cord to prevent unraveling
(B) coiling the rope in as tight a ball as possible
(C) lubricating the strands with tallow
(D) wetting the rope with water to cure it

48. If the head of a hammer has become loose on the handle, it should be properly tightened by

(A) driving the handle further into the head
(B) driving a nail alongside the present wedge
(C) using a slightly larger wedge
(D) soaking the handle in water

49.

The figure above represents a water tank containing water. The number 1 indicates an intake pipe and 2 indicates a discharge pipe. Of the following, the statement which is *least* accurate is that the

(A) tank will eventually overflow if water flows through the intake pipe at a faster rate than it flows out through the discharge pipe
(B) tank will empty completely if the intake pipe is closed and the discharge pipe is allowed to remain open
(C) water in the tank will remain at a constant level if the rate of intake is equal to the rate of discharge
(D) water in the tank will rise if the intake pipe is operating when the discharge pipe is closed

50. The tool that is best suited for use with a wood chisel is

(A)
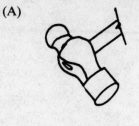

(B)

(C)
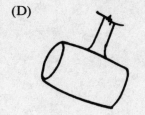

(D)

51.

The tool shown above is

(A) an Allen-head wrench
(B) a double scraper
(C) an offset screwdriver
(D) a nail puller

52.

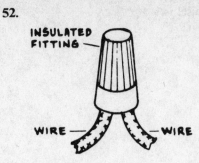

Wires are often spliced by the use of a fitting like the one shown above. The use of this fitting does away with the need for

(A) skinning
(B) cleaning
(C) twisting
(D) soldering

53. With which of these screw heads do you use an "Allen" wrench?

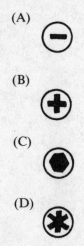

(A)

(B)

(C)

(D)

54.

RIVETED SPLICE

In the structural steel splice the different types of rivets are shown by different symbols. The number of different types of rivets is

(A) 6
(B) 4
(C) 5
(D) 3

55. The most important rule for a driver to remember in the care of an automobile battery is to

    (A) make certain that the points are properly adjusted in the spark plugs
    (B) burn the headlights or play the radio, occasionally, while the ignition is turned on
    (C) have the battery discharged at regular intervals, weekly in the winter, biweekly in the summer
    (D) keep the level of liquid above the plates

56. A brad is similar in shape to a

    (A) box nail
    (B) finishing nail
    (C) common nail
    (D) tack

57.

The tool shown above is used to measure

    (A) clearance
    (B) wire thickness
    (C) inside slots
    (D) screw pitch

58. When sanding wood by hand, best results are usually obtained in finishing the surface when the sanding block is worked

    (A) across the grain
    (B) in a diagonal to the grain
    (C) in a circular motion
    (D) with the grain

59.

The convenience outlet that is known as a *polarized* outlet is number

    (A) 1
    (B) 2
    (C) 3
    (D) 4

60. The tool used to measure the depth of a hole is

    (A)

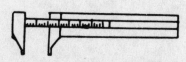

(B)

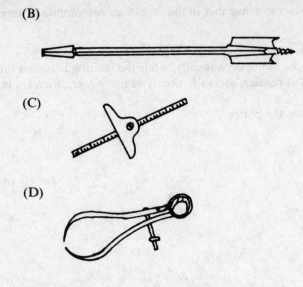

(C)

(D)

61. Neutral wire can be quickly recognized by the

(A) greenish color
(B) bluish color
(C) natural or whitish color
(D) black color

62.

The device shown above is a

(A) C-clamp
(B) test clip
(C) battery connector
(D) ground clamp

63. If the flush tank of a water-closet fixture overflows, the fault is likely to be

(A) failure of the ball to seat properly
(B) excessive water pressure
(C) defective trap in the toilet bowl
(D) waterlogged float

64. A governor is used on an automobile primarily to limit its

(A) rate of acceleration
(B) maximum speed
(C) fuel consumption
(D) stopping distance

65.

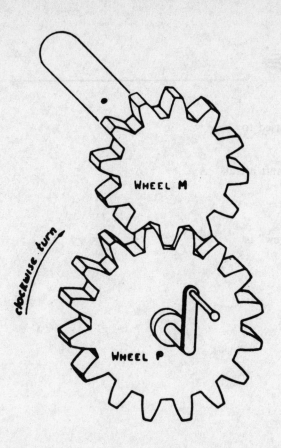

Study the gear wheels in the figure above, then determine which of the following statements is true.

(A) If you turn wheel M clockwise by means of the handle, wheel P will also turn clockwise.
(B) It will take the same time for a tooth of wheel P to make a full turn as it will for a tooth of wheel M.
(C) It will take less time for a tooth of wheel P to make a full turn than it will take a tooth of wheel M.
(D) It will take more time for a tooth of wheel P to make a full turn than it will for a tooth of wheel M.

66. Locknuts are frequently used in making electrical connections on terminal boards. The purpose of the locknuts is to

(A) eliminate the use of flat washers
(B) prevent unauthorized personnel from tampering with the connections
(C) keep the connections from loosening through vibration
(D) increase the contact area at the connection point

67. Glazier's points are used to

(A) hold glass in wooden window sash
(B) scratch glass so that it can be broken to size
(C) force putty into narrow spaces between glass and sash
(D) remove broken glass from a pane

68.

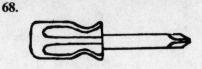

The tool shown above is used to

(A) ream holes in wood
(B) countersink holes in soft metals
(C) turn Phillips-head screws
(D) drill holes in concrete

69. The carpenter's "hand screw" is

(A)

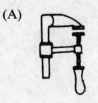

(B)

(C)

(D)

70. A squeegee is a tool that is used in

(A) drying windows after washing
(B) cleaning inside boiler surfaces
(C) the central vacuum cleaning system
(D) clearing stoppage in waste lines

71. Boxes and fittings intended for outdoor use should be of

(A) weatherproof type
(B) stamped steel of not less than No. 16
(C) standard gauge
(D) stamped steel plated with cadmium

72. The device used to change AC to DC is a

    (A) frequency changer
    (B) regulator
    (C) transformer
    (D) rectifier

73.

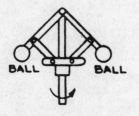

BALL   BALL

The figure above shows a governor on a rotating shaft. As the shaft speeds up, the governor balls will move

    (A) down
    (B) upward and inward
    (C) upward
    (D) inward

74. The purpose of an air valve in a heating system is to

    (A) prevent pressure from building up in a room due to the heated air
    (B) relieve the air from steam radiators
    (C) allow excessive steam pressure in the boiler to escape to the atmosphere
    (D) control the temperature in the room

75. If a fuse of higher than the required current rating is used in an electrical circuit

    (A) better protection will be afforded
    (B) the fuse will blow more often since it carries more current
    (C) serious damage may result to the circuit from overload
    (D) maintenance of the large fuse will be higher

76. A wood screw that can be tightened by a wrench is known as a

    (A) lag screw
    (B) carriage screw
    (C) Phillips screw
    (D) monkey screw

77. Paint is "thinned" with

    (A) linseed oil
    (B) varnish
    (C) turpentine
    (D) gasoline

78.

The tool shown above is

(A) an offset wrench
(B) a box wrench
(C) a spanner wrench
(D) an open end wrench

79.

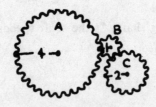

If gear A makes one clockwise revolution per minute, which of the following is true?

(A) Gear B makes one counterclockwise revolution every 4 minutes.
(B) Gear C makes two clockwise revolutions every minute.
(C) Gear B makes four clockwise revolutions every minute.
(D) Gear C makes one counterclockwise revolution every 8 minutes.

80.

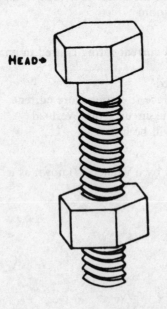

HEAD→

Referring to the figure above, which one of the following statements is true?

(A) If the nut is held stationary and the head turned clockwise, the bolt will move down.

(B) If the head of the bolt is held stationary and the nut is turned clockwise, the nut will move down.

(C) If the head of the bolt is held stationary and the nut is turned clockwise, the nut will move up.

(D) If the nut is held stationary and the head turned counter-clockwise, the bolt will move up.

81. If the float of a flush tank leaks and fills with water, the most probable result will be

(A) no water in tank
(B) ball cock will remain open
(C) water will flow over tank rim onto floor
(D) flush ball will not seat properly

82. If the temperature gauge indicates the engine is getting overheated

(A) allow it to cool down
(B) pour cold water in immediately
(C) pour hot water in immediately
(D) pour in a cooling antifreeze at once

83.

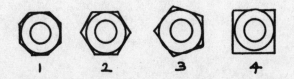

The shape of nut most commonly used on electrical terminals is

(A) 1
(B) 2
(C) 3
(D) 4

84.

The sketch shows a head-on view of a three-pronged plug used with portable electrical power tools. Considering the danger of shock when using such tools, it is evident that the function of the U-shaped prong is to

(A) insure that the other two prongs enter the outlet with the proper polarity
(B) provide a half-voltage connection when doing light work
(C) prevent accidental pulling of the plug from the outlet
(D) connect the metallic shell of the tool motor to ground

85. After brakes have been severely overheated, what should be checked for?

(A) water condensation in brake fluid
(B) glazed brake shoes
(C) wheels out of alignment
(D) crystallized wheel bearings

Answer Key

1.	B	18.	B	35.	D	52.	D	69.	B
2.	B	19.	C	36.	B	53.	C	70.	A
3.	B	20.	C	37.	A	54.	B	71.	A
4.	C	21.	C	38.	D	55.	D	72.	D
5.	D	22.	C	39.	D	56.	B	73.	C
6.	D	23.	D	40.	D	57.	D	74.	B
7.	C	24.	B	41.	C	58.	D	75.	C
8.	A	25.	A	42.	B	59.	A	76.	A
9.	D	26.	C	43.	B	60.	C	77.	C
10.	A	27.	D	44.	C	61.	C	78.	D
11.	C	28.	C	45.	B	62.	D	79.	B
12.	C	29.	C	46.	C	63.	D	80.	C
13.	D	30.	B	47.	A	64.	B	81.	B
14.	B	31.	C	48.	C	65.	D	82.	A
15.	C	32.	C	49.	B	66.	C	83.	B
16.	D	33.	C	50.	D	67.	A	84.	D
17.	C	34.	A	51.	C	68.	C	85.	B

Explanatory Answers

1. **(B)** The tool is a plunger. It is used by a plumber to dislodge material that is clogging drains.

2. **(B)** Since power from the engine is transmitted to both the front wheels and the back wheels, a four-wheel drive vehicle has much better traction. Four-wheel drive is especially useful in mud, sand, and very uneven terrain. While four-wheel drive tends to be a feature of durable vehicles (D), like jeeps, it is not the four-wheel drive that makes them durable.

3. **(B)** One-half the circumference of the sprocket wheel is 10 inches, therefore the entire circumference is twenty inches. In one complete revolution of the wheel, the chain will move 20 inches. As weight 2 moves up 20 inches, weight 1 will move down 20 inches. The difference between the heights of the two weights will be 40 inches.

4. **(C)** Once a victim of electric shock is no longer in contact with the electricity whether from a turn-off of the current or from falling or being thrown from the source of the shock, it is no longer dangerous to come into contact with that person. Breathing is absolutely essential to life. The oxygen-starved person will die within minutes. Therefore, if a person is not breathing, you must start artificial respiration *immediately*, no matter what the weather or the extent of his other injuries.

5. **(D)** The air from R must follow a route down through valve 7, up through G, then through valve 3, down through S, to the right through 5, then up and over to M. The air could not pass through valves 1 and 2 to G because it would escape through the opening between valves 1 and 2. If either valve 4 or valve 6 were to be opened, the air would be diverted from the appointed route.

6. **(D)** *Flux* is used during soldering. It removes oxide as the solder contacts metal, making the solder stick better. Oakum (A) is a tar-treated rope used for sealing joints. Lye (B) is a caustic agent used for heavy cleaning. Oil (C) would be the wrong substance to make solder stick.

7. **(C)** The generator or alternator is operated by the car's engine. It produces electricity which flows to the battery and keeps the battery charged.

8. **(A)** Water is not likely to spread an electrical fire (B), though it may be ineffective at putting it out. It is true that metal parts may rust if soaked with water (C) and fuses may blow (D), but these are not serious considerations. There is a real danger from fatal shock if water is used on an electrical fire. The danger is much greater still if that water comes from a hose, because there would then be a great deal of water on the floor and the firefighters would be standing in it. Electricity travels rapidly through water, and the people standing in the water would be its targets.

9. **(D)** A turning gear always turns the gear with which it interlocks in the opposite direction. If gear 1 turns clockwise, then gear 2 must turn counterclockwise. In turn, gears 3 and 4, since they are both turned by gear 2, must both turn clockwise.

10. **(A)** The answer to this question is pure common sense. Electrical wires should serve only one purpose—to supply electricity. If electrical wires are required to support weight there is danger of breakage in the wires and of damage to the fixture itself caused by tension at the connections.

11. **(C)** Asbestos is an excellent insulator against heat. If electric wires are to be used under conditions where they are exposed to very high, possibly damaging, heat, the wires are covered with asbestos for their own protection.

12. **(C)** In Case 2, each spring bears one-half of the weight of the 5-pound weight. Each spring in Case 2 is therefore stretched by 2½ pounds. In Case 1, each spring bears a full 5-pound load. Each spring in Case 1 must be stretched twice as much as each spring in Case 2.

13. **(D)** The *compass* shown would be very useful in marking out equal short distances on a board. The compass would not, of course, be of use in the actual drilling.

14. **(B)** Antifreeze is green. Power steering fluid (A) and transmission fluid (C) are pink. Crankcase oil (D) is brown.

15. **(C)** Since the circumference of the drum is one foot, one complete revolution of the drum will take up one foot of each rope. As each of the separate ropes supporting the weight is shortened by one foot, the weight will move up one foot.

16. **(D)** Lubrication of machinery is scheduled according to time elapsed and amount of use. All the other reasons offered are irrelevant.

17. **(C)** The tool is a drill bit. The smooth end is inserted into the drill chuck and is tightened so that it is very secure. When the drill motor goes, the bit turns very fast and drills holes wherever it is applied.

18. **(B)** If the coupling is turned, but the pipes are held firm so that they cannot turn, then the coupling will move along the length of one or the other pipes, but the overall length of the three pieces will remain the same.

19. **(C)** Yanking at an electric cord can cause hidden damage inside the plug. The force can cause terminals to loosen or bits of wire to break off and fray. Frayed wires can, in turn, come into contact with the opposite pole, causing fire or short circuit in the plug at some later date.

20. **(C)** The muffler is, in effect, a silencer. It cuts down exhaust noise. While the muffler is part of the exhaust system, it has no effect on the fumes (A).

21. **(C)** If water pressure in the pipe is increased, more water will flow into the water chamber. Since the chamber is enclosed, the air will be unable to escape. As more water enters the chamber, the existing air must be compressed into a smaller space.

22. **(C)** If the carburetor is sending to the engine a mixture that contains too much gasoline with too little air, the mixture is said to be too rich and black smoke will come from the muffler. If there is too much lubricating oil (A), bluish-gray smoke will come from the tailpipe.

23. **(D)** A nick in a wire can be dangerous because it weakens the wire at that point and can lead to breakage. If the wire is nicked during stripping, you should cut off the weakened portion and begin again. Later breakage from an unnoticed weakness can lead to a short circuit.

24. **(B)** Wood screws are usually *more* difficult to install than are nails, but they are often preferable because they generally *hold better* and are less likely to split the wood.

25. **(A)** The "clunk" of the universals is a very distinctive noise. Steam rising from the hood (B) most likely is caused by a broken water hose. The universals have nothing to do with the engine (C) and sticking doors (D) may need to have their hinges oiled.

26. **(C)** Common sense should give you the answer to this question. In order for the cutter to cut the block, the two must be in contact. Obviously, the block must move in direction 3 in order to make contact wth the cutter. For a cutter to cut, it must move in a direction so that the sharp edge of its teeth bites into the object being cut. In direction 1, the teeth will bite into the block. In direction 2, the back edge of the teeth would just slide off the block.

27. **(D)** Each division marks 2 units: 20 units/10 divisions = 2 units/division. The pointer is $3\frac{1}{2}$ divisions above 10 or 2 units/division $\times$ $3\frac{1}{2}$ divisions = 7 + 10 = 17.

28. **(C)** If an electric lightbulb is operated at more than its rated voltage, the extra surge of electricity will cause the bulb to burn more brightly. However, the same excess electrical force will weaken the filament and cause the bulb to burn out more quickly.

29. **(C)** is a pipe wrench. (A) is a crescent or expandable wrench; (B) is a ratchet wrench; and (D) is an open-end wrench.

30. **(B)** Kerosene and fine steel wool would effectively remove a thin layer of rust. Vaseline (A) can prevent rust, but not remove it. A sharp knife (C) might chip off thick crusty rust, but not a thin coat. A dry cloth (D) alone would not remove rust.

31. **(C)** Shock absorbers are the heavy duty springs in the suspension. The rear shock absorbers are very near the exhaust system (D), but they are not part of that system. While springs in the upholstery (B) may indeed absorb shocks, they are not called *shock absorbers.*

32. **(C)** Ten quarts of water have brought the water level to one foot. An additional fifteen quarts would raise the water level by $1^1/_2$ feet to a total height of $2^1/_2$ feet.

33. **(C)** This shape plug will only go into a similarly shaped socket. The other outlets are of different shapes.

34. **(A)** The measurements that can be made on the micrometer are: 1) 2 major divisions and 1 minor division on the ruler-type scale, or: $.2 + .025 = .225$; b) 2 minor divisions above 0 on the rotating scale, or .002. Summing, we find the final measurement is: $.225 + .002 = .227$.

35. **(D)** Each time a tire is put onto a rim the wheel must be balanced. If you take your regular tires off the rims and mount snow tires on those same rims, the wheels must be balanced, even though the tires may not be new (A). If you answered (C), you were probably thinking of tire rotation, which is done according to schedule.

36. **(B)** Clockwise is left to right, so if the nut moves, it follows the threads of the bolt downward.

37. **(A)** Graphite, which is powdered carbon, is very slippery and will not bind the small springs and metal parts of a lock.

38. **(D)** Severe electrical shock may cause serious bodily injury including unconsciousness, burns, and clenched muscles. Electrical shock is more likely to cause the victim to stop breathing than to cause him to breathe heavily.

39. **(D)** Although this tool looks like a screwdriver, the head will fit into a hex nut and works like a socket wrench.

40. **(D)** One would not be able to see a defect in a painted ladder, such as a knot or a split in the wood. A ladder should *never* be painted.

41. **(C)** The overall length of pipes and coupling could decrease or increase, depending upon the direction in which pipe B is turned. However, as stated in this question, pipe B is turned so as to disappear into the coupling. Since there are 8 threads to the inch, eight complete revolutions of the pipe would shorten the pipes and coupling by one inch. An additional two turns, for a total of ten, would shorten the pipes and coupling by an additional two-eighths or one-fourth of an inch.

42. **(B)** is a keyhole saw used to make curved cuts. (A) is a backsaw; (C) is a rip or crosscut saw; and (D) is a hacksaw.

43. **(B)** "Skin" forms when air combines with paint. To stop this from happening, pour a thin layer of solvent over the paint. The air will then be prevented from reaching the paint and forming a layer of skin.

44. **(C)** When two pieces of metal rub together, the friction causes a great deal of heat. Oil reduces the friction between the two pieces of metal.

45. **(B)** The tool shown is a "star drill." It is hit with a hammer to make a hole in concrete.

46. **(C)** If the block is moved toward the brick, the moment for a given force exerted will increase (being further from the force) making it easier to lift; the height will be made smaller, hardly raising the brick when moved to the limit (directly underneath it).

47. **(A)** A rope is made from many separate strands of hemp or synthetic fiber, such as nylon. When a rope is cut, the strands can unravel if the ends are not whipped or wrapped with cord.

48. **(C)** If you look at the top of a hammer where it is joined to the handle, you will see either the top of a wooden wedge or the top of a metal wedge. Driving another wedge into the handle will tighten the hammer.

49. **(B)** If pipe 2 is open while pipe 1 is closed, then the level will drop to the lowest level of 2, leaving the volume below 2 still filled, having no way to discharge. All other statements are true.

50. **(D)** A wooden mallet is used in woodworking. The other hammers are made of steel. They are too hard and might crack a wood chisel. Choice (A) is a ball peen hammer, choice (B) is a straight peen hammer, and choice (C) is a brick hammer.

51. **(C)** The tool is an offset screwdriver. It is used for tightening screws in hard-to-reach places where a regular screwdriver cannot turn in a complete revolution.

52. **(D)** This is a mechanical or solderless connector. It does away with the need to solder wires and is found in house wiring.

53. **(C)** An "Allen" wrench is hexagon-shaped and will fit into screw C.

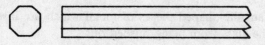

54. **(B)** There are only four different types of symbols shown in the pictures:

55. **(D)** The liquid in a battery is called an electrolyte. It acts as a chemical that allows electrons to pass. If the liquid runs out, electricity will not be made by the battery.

56. **(B)** A brad is similar to a finishing nail in that both nails do not have flat heads and can be countersunk into the wood.

57. **(D)** When a blade in the gauge matches the threads in the screw, the measure is the screw pitch.

58. **(D)** The smoothest finish can be obtained by sanding the wood with the grain.

59. **(A)** The plug can go into the outlet in only one way in a polarized outlet. In the other outlets, the plug can be reversed.

60. **(C)** The flattened part of the (C) rests at the top of the hole and the ruler is then pushed down into the hole until it reaches the bottom. The depth of the hole is then read from the ruler.

61. **(C)** The neutral wire is whitish in color; the hot lead is black; and the ground wire is green.

62. **(D)** This object is a ground clamp. It will be tightened around a cold water pipe. A grounding wire will be attached to the screw and thus stray electricity will be grounded.

63. **(D)** The water shut-off valve on a flush tank is closed by the force of a lightweight ball rising inside the tank. If this float becomes waterlogged, it will not rise and shut off the water.

64. **(B)** The governor is a device which is used to limit the maximum speed of an auto. It is used as a safety device.

65. **(D)** Wheel P has 16 teeth; wheel M has 12 teeth. When wheel M makes a full turn, wheel P will still have 4 more teeth to turn. So wheel P is slower, and will take more time to turn.

66. **(C)** Locknuts are bent so that their metal edges will bite into the terminal board and will require the use of a wrench to loosen them.

67. **(A)** Glazier's points are triangular-shaped pieces of metal which are inserted into a window frame to prevent the glass from being pushed out.

68. **(C)** This is a Phillips-head screwdriver. It will turn Phillips-head screws with this shape:

69. **(B)** The carpenter's "hand screw" is shown in figure (B).

70. **(A)** A squeegee is a rubber wiper that removes water from a wet window.

71. **(A)** Outdoor boxes and fittings must be weatherproof to withstand any problems caused by moisture.

72. **(D)** A rectifier or diode is a device that changes AC to DC.

73. **(C)** The centrifugal force acts to pull the balls outward. Since the two balls are connected to a yolk around the center bar, this outward motion pulls the balls upward.

74. **(B)** An air valve on a radiator removes air from the steam pipes. If air is trapped in the pipes, it prevents the steam from going to the radiator. This would prevent the radiator from producing heat.

75. **(C)** Never use a fuse having a higher rating than that specifically called for in the circuit. A fuse is a safety device used to protect a circuit from serious damage caused by too high a current.

76. **(A)** The diagram below shows how a lag screw can fit into the head of a wrench.

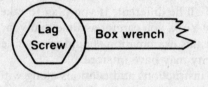

77. **(C)** Paint is made thinner or easier to apply by diluting it with turpentine. Linseed oil and varnish are not used as paint thinners.

78. **(D)** The opened face on this tool shows that it is an open end wrench.

79. **(B)** Gear A turns in the opposite direction from gear B. A clockwise turn of A results in a counterclockwise revolution of B. Since the distance traversed by A (perimeter = $\pi \times$ diameter = $\pi \times 4$) is twice that of C (perimeter = $\pi \times 2$), the speed of C is doubled.

80. **(C)** To tighten the bolt, turn it counterclockwise. To tighten the nut on the bolt the reverse is true—turn it clockwise.

81. **(B)** The ball cock will remain open, the float will not rise with the incoming water, and the tank will continue to fill because the shut-off valve will not be shut off.

82. **(A)** If an engine starts to overheat, you must stop the car immediately. Otherwise, the metal will expand and damage the engine. If you pour anything on an overheated engine, the rapid cooling may cause the block to crack.

83. **(B)** The shape of the nut most commonly used on electrical terminals is hexagonal, nut 2.

84. **(D)** The third prong in the plug is the grounding wire.

85. **(B)** Overheating the brake shoe will cause the brake material to glaze and become slippery. Slippery brakes are dangerous because they take longer to stop a car.

PRACTICE EXAMINATION 1

The examination that follows is an actual, recent firefighter examination given in the City of New York and reprinted here by permission of the city. If you will be taking a New York City exam for firefighter, it will NOT be the same as this one. The format may be similar, but the questions will all be different. If you plan to take a firefighter exam anywhere outside the City of New York, you cannot assume that your exam will look like this at all. However, practice with this examination should give you excellent preparation for whatever your own town or city may have in store.

We are including all of the instructions and cautions along with the exam itself so that you can see how seriously this exam must be taken. Keep in mind how important this exam is as you study and prepare yourself.

122

122

City of New York
Department of Personnel

EXAMINATION NO. 7022
FIREFIGHTER

Social Security No. _____
Seat No. _____
Room No. _____
School _____

Written Test: Weight 50, pass mark to be determined.

December 12, 1987

GENERAL INSTRUCTIONS FOR CANDIDATES—READ NOW

<u>GENERAL INFORMATION:</u> Read the General Information and the Test Instructions now, and fill out the information requested at the top of this booklet. Following are explanations of the information requested.

> The Soc. Sec. No. is your correct Social Security Number. If your Social Security Number is not correct on your admission card, fill out the pink correction form (DP-148), so that the correction can be made after the test. Your Seat No. is the number shown on the sheet of scrap paper on your desk. Your Room No. and School are written on the blackboard.

<u>THE FORMS:</u> You must <u>now</u> fill out your fingerprint card and answer sheet as explained below. Use pencil only.

> Fingerprint Card: Print in pencil all information requested including your Social Security Number. Your fingerprints will be taken approximately 15 minutes after the Seventh Signal.

> Answer Sheet: Print in pencil all the information requested on the Answer Sheet including your Social Security No. and the Exam. No. in the boxes at the top of the columns, ONE NUMBER IN A BOX. In each column, darken (with pencil) the oval containing the number in the box at the top of the column; only ONE OVAL in each COLUMN should be darkened. Then, using your pencil, print the Exam, Title, School or Building, Room No., Seat No., and Today's Date (December 12, 1987) in the appropriate spaces.

> After filling in this information at the top of the Answer Sheet, please read the Confidential Questionnaire at the bottom of the Answer Sheet, and answer the questions accurately.

After answering the Confidential Questionnaire, please complete the Survey Questionnaire.

You are not permitted to use a calculator on this test.

TEST INSTRUCTIONS

EIGHT SIGNALS WILL BE USED DURING THIS TEST. Following is an explanation of these signals.

<u>FIRST SIGNAL:</u>	First Memory Booklet is distributed. Write your Room, Seat, and Social Security Numbers on the cover. DO <u>NOT</u> OPEN THE BOOKLET UNTIL THE SECOND SIGNAL.
<u>SECOND SIGNAL:</u>	OPEN FIRST MEMORY BOOKLET. You will be given <u>five</u> minutes to remember as many details of the scene as you can. You may <u>not</u> write or make any notes while studying this material. The first eight questions will be based on this material.
<u>THIRD SIGNAL:</u>	CLOSE FIRST MEMORY BOOKLET. The monitor will collect this booklet and distribute the SECOND MEMORY BOOKLET. Write Room, Seat Number, and Social Security Numbers on top of Second Memory Booklet. DO <u>NOT</u> OPEN THE SECOND MEMORY BOOKLET OR WRITE ANYTHING ELSE UNTIL THE NEXT SIGNAL.
<u>FOURTH SIGNAL:</u>	Open the SECOND MEMORY BOOKLET and answer questions 1 to 8. You will be given <u>five</u> minutes to answer the questions. DO NOT TURN TO ANY OTHER PART OF THE BOOKLET UNTIL THE NEXT SIGNAL. After you have answered questions 1 to 8 on your Answer Sheet, you may, for future reference, circle your answers in the SECOND MEMORY BOOKLET.
<u>FIFTH SIGNAL:</u>	Turn to the DIAGRAM in the SECOND MEMORY BOOK. You will be given <u>five</u> minutes to try to remember as many details about the diagram as you can. You may <u>not</u> write or make any notes during this time. <u>DO NOT TURN TO ANY OTHER PAGE DURING THIS TIME.</u>
<u>SIXTH SIGNAL:</u>	Close Second Memory Booklet. Monitors will collect this booklet and distribute the Question Booklet. Do <u>not</u> open the Question Booklet until the next signal. <u>DO NOT WRITE ANYTHING DURING THIS PERIOD.</u>
<u>SEVENTH SIGNAL:</u>	Open Question Booklet and answer questions 9 to 16 dealing with the second memory section and then continue answering the rest of the questions in the Question Booklet. This test consists of 100 questions. After you finish the memory questions, check to make sure that the Question Booklet has all the questions from 9 to 100 and is <u>not</u> defective. If you believe that your booklet is defective, inform the monitor. You will have <u>4</u> hours from this signal to complete all the questions. YOU WILL BE FINGERPRINTED DURING THIS PERIOD.

<u>EIGHTH SIGNAL:</u>	END OF TEST. STOP WORKING. If you finish before this signal and wish to leave, raise your hand.

<u>ANYONE DISOBEYING ANY OF THESE INSTRUCTIONS MAY BE DIS-QUALIFIED–THAT IS, YOU MAY RECEIVE A SCORE OF ZERO FOR THE ENTIRE TEST!</u>

<u>DIRECTIONS FOR ANSWERING QUESTIONS:</u>	Answer all the questions on the Answer Sheet before the Eighth Signal is given. ONLY YOUR ANSWER SHEET WILL BE MARKED although you should also circle your answers in the Question Booklet before the end of the test for future reference. Use a soft pencil (No. 2) to mark your answers. If you want to change an answer, erase it and then mark your new answer. For each question, pick the best answer. Then, on your Answer Sheet, in the row with the same number as the question, blacken the oval with the same letter as your answer. Do not make any stray pencil dots, dashes, or marks any place on the Answer Sheet. Do <u>NOT</u> fold, roll, or tear the Answer Sheet.

Here is a sample of how to mark your answers:

SAMPLE 0: The sum of 5 and 3 is

(A) 11   (B) 9   (C) 8   (D) 2

Since the answer is 8, your Answer Sheet is marked like this:

SAMPLE 0: (A) ◯   (B) ◯   (C) ⬤   (D) ◯

<u>WARNING:</u>	You are not allowed to copy answers from anyone, or to use books or notes. It is against the law to take the test for somebody else or to let somebody else take the test for you. There is to be <u>NO SMOKING</u> anywhere in the building.
<u>LEAVING:</u>	After the test starts, candidates may not leave the building until after they are fingerprinted. No one may leave the building until after 10:45 A.M. No one may come in after 10:45 A.M. During the test, you may not leave the room unless accompanied by a monitor. If you want to drop out of the test and not have your answers marked, write "I withdraw" on your Answer Sheet and sign your name.
	You may take your Question Booklet and two Memory Booklets with you when you finish. The room monitor will return your Memory Booklets to you when you hand in your Answer Sheet.

Answer Sheet
New York City Exam —
December 12, 1987

TEAR HERE

1. Ⓐ Ⓑ Ⓒ Ⓓ	21. Ⓐ Ⓑ Ⓒ Ⓓ	41. Ⓐ Ⓑ Ⓒ Ⓓ	61. Ⓐ Ⓑ Ⓒ Ⓓ	81. Ⓐ Ⓑ Ⓒ Ⓓ
2. Ⓐ Ⓑ Ⓒ Ⓓ	22. Ⓐ Ⓑ Ⓒ Ⓓ	42. Ⓐ Ⓑ Ⓒ Ⓓ	62. Ⓐ Ⓑ Ⓒ Ⓓ	82. Ⓐ Ⓑ Ⓒ Ⓓ
3. Ⓐ Ⓑ Ⓒ Ⓓ	23. Ⓐ Ⓑ Ⓒ Ⓓ	43. Ⓐ Ⓑ Ⓒ Ⓓ	63. Ⓐ Ⓑ Ⓒ Ⓓ	83. Ⓐ Ⓑ Ⓒ Ⓓ
4. Ⓐ Ⓑ Ⓒ Ⓓ	24. Ⓐ Ⓑ Ⓒ Ⓓ	44. Ⓐ Ⓑ Ⓒ Ⓓ	64. Ⓐ Ⓑ Ⓒ Ⓓ	84. Ⓐ Ⓑ Ⓒ Ⓓ
5. Ⓐ Ⓑ Ⓒ Ⓓ	25. Ⓐ Ⓑ Ⓒ Ⓓ	45. Ⓐ Ⓑ Ⓒ Ⓓ	65. Ⓐ Ⓑ Ⓒ Ⓓ	85. Ⓐ Ⓑ Ⓒ Ⓓ
6. Ⓐ Ⓑ Ⓒ Ⓓ	26. Ⓐ Ⓑ Ⓒ Ⓓ	46. Ⓐ Ⓑ Ⓒ Ⓓ	66. Ⓐ Ⓑ Ⓒ Ⓓ	86. Ⓐ Ⓑ Ⓒ Ⓓ
7. Ⓐ Ⓑ Ⓒ Ⓓ	27. Ⓐ Ⓑ Ⓒ Ⓓ	47. Ⓐ Ⓑ Ⓒ Ⓓ	67. Ⓐ Ⓑ Ⓒ Ⓓ	87. Ⓐ Ⓑ Ⓒ Ⓓ
8. Ⓐ Ⓑ Ⓒ Ⓓ	28. Ⓐ Ⓑ Ⓒ Ⓓ	48. Ⓐ Ⓑ Ⓒ Ⓓ	68. Ⓐ Ⓑ Ⓒ Ⓓ	88. Ⓐ Ⓑ Ⓒ Ⓓ
9. Ⓐ Ⓑ Ⓒ Ⓓ	29. Ⓐ Ⓑ Ⓒ Ⓓ	49. Ⓐ Ⓑ Ⓒ Ⓓ	69. Ⓐ Ⓑ Ⓒ Ⓓ	89. Ⓐ Ⓑ Ⓒ Ⓓ
10. Ⓐ Ⓑ Ⓒ Ⓓ	30. Ⓐ Ⓑ Ⓒ Ⓓ	50. Ⓐ Ⓑ Ⓒ Ⓓ	70. Ⓐ Ⓑ Ⓒ Ⓓ	90. Ⓐ Ⓑ Ⓒ Ⓓ
11. Ⓐ Ⓑ Ⓒ Ⓓ	31. Ⓐ Ⓑ Ⓒ Ⓓ	51. Ⓐ Ⓑ Ⓒ Ⓓ	71. Ⓐ Ⓑ Ⓒ Ⓓ	91. Ⓐ Ⓑ Ⓒ Ⓓ
12. Ⓐ Ⓑ Ⓒ Ⓓ	32. Ⓐ Ⓑ Ⓒ Ⓓ	52. Ⓐ Ⓑ Ⓒ Ⓓ	72. Ⓐ Ⓑ Ⓒ Ⓓ	92. Ⓐ Ⓑ Ⓒ Ⓓ
13. Ⓐ Ⓑ Ⓒ Ⓓ	33. Ⓐ Ⓑ Ⓒ Ⓓ	53. Ⓐ Ⓑ Ⓒ Ⓓ	73. Ⓐ Ⓑ Ⓒ Ⓓ	93. Ⓐ Ⓑ Ⓒ Ⓓ
14. Ⓐ Ⓑ Ⓒ Ⓓ	34. Ⓐ Ⓑ Ⓒ Ⓓ	54. Ⓐ Ⓑ Ⓒ Ⓓ	74. Ⓐ Ⓑ Ⓒ Ⓓ	94. Ⓐ Ⓑ Ⓒ Ⓓ
15. Ⓐ Ⓑ Ⓒ Ⓓ	35. Ⓐ Ⓑ Ⓒ Ⓓ	55. Ⓐ Ⓑ Ⓒ Ⓓ	75. Ⓐ Ⓑ Ⓒ Ⓓ	95. Ⓐ Ⓑ Ⓒ Ⓓ
16. Ⓐ Ⓑ Ⓒ Ⓓ	36. Ⓐ Ⓑ Ⓒ Ⓓ	56. Ⓐ Ⓑ Ⓒ Ⓓ	76. Ⓐ Ⓑ Ⓒ Ⓓ	96. Ⓐ Ⓑ Ⓒ Ⓓ
17. Ⓐ Ⓑ Ⓒ Ⓓ	37. Ⓐ Ⓑ Ⓒ Ⓓ	57. Ⓐ Ⓑ Ⓒ Ⓓ	77. Ⓐ Ⓑ Ⓒ Ⓓ	97. Ⓐ Ⓑ Ⓒ Ⓓ
18. Ⓐ Ⓑ Ⓒ Ⓓ	38. Ⓐ Ⓑ Ⓒ Ⓓ	58. Ⓐ Ⓑ Ⓒ Ⓓ	78. Ⓐ Ⓑ Ⓒ Ⓓ	98. Ⓐ Ⓑ Ⓒ Ⓓ
19. Ⓐ Ⓑ Ⓒ Ⓓ	39. Ⓐ Ⓑ Ⓒ Ⓓ	59. Ⓐ Ⓑ Ⓒ Ⓓ	79. Ⓐ Ⓑ Ⓒ Ⓓ	99. Ⓐ Ⓑ Ⓒ Ⓓ
20. Ⓐ Ⓑ Ⓒ Ⓓ	40. Ⓐ Ⓑ Ⓒ Ⓓ	60. Ⓐ Ⓑ Ⓒ Ⓓ	80. Ⓐ Ⓑ Ⓒ Ⓓ	100. Ⓐ Ⓑ Ⓒ Ⓓ

DEPARTMENT OF PERSONNEL

Social Security No. _____
Room No. _____
Seat No. _____
School _____

First Memory Booklet   **FIREFIGHTER**

December 12, 1987   **EXAMINATION NO. 7022**

<u>DO NOT OPEN THIS BOOKLET UNTIL THE SECOND SIGNAL IS GIVEN!</u>

Write your Social Security Number, Room Number, Seat Number, and School in the appropriate spaces at the top of this page.

You <u>must</u> follow the instructions found in the <u>CANDIDATES' INSTRUCTION BOOKLET</u>.

<u>ANYONE DISOBEYING ANY OF THE INSTRUCTIONS FOUND IN THE CANDIDATES' INSTRUCTION BOOKLET MAY BE DISQUALIFIED– RECEIVE A ZERO ON THE ENTIRE TEST.</u>

This booklet contains one scene. Try to remember as many details in the scene as you can.

<u>DO **NOT** WRITE OR MAKE **ANY** NOTES WHILE STUDYING THE SCENE.</u>

<u>DO NOT OPEN THIS BOOKLET UNTIL THE SECOND SIGNAL IS GIVEN!</u>

<u>Note:</u> You will have five (5) minutes to study the scene shown.

The New York City Department of Personnel makes no commitment, and no inference is to be drawn, regarding the content, style, or format of any future examination for the position of firefighter.

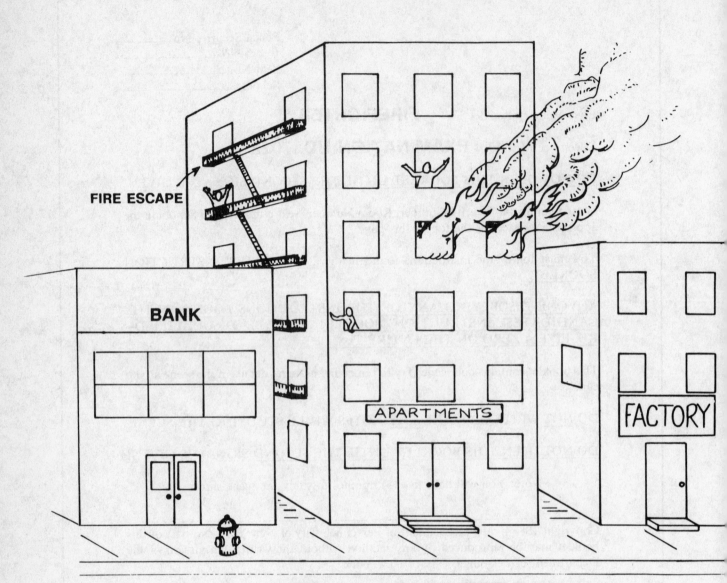

FIRE ESCAPE

BANK

APARTMENTS

FACTORY

NOTE: The ground floor is the first floor.

CITY OF NEW YORK
DEPARTMENT OF PERSONNEL

Social Security No. _____
Room No. _____
Seat No. _____
School _____

Second Memory Booklet

December 12, 1987

FIREFIGHTER
EXAMINATION No. 7022

DO NOT OPEN THIS BOOKLET UNTIL THE FOURTH SIGNAL IS GIVEN!

Write your Social Security Number, Room Number, Seat Number, and School in the appropriate spaces at the top of this page.

You must follow the instructions found in the CANDIDATES' INSTRUCTION BOOKLET.

ANYONE DISOBEYING ANY OF THE INSTRUCTIONS FOUND IN THE CANDIDATES' INSTRUCTION BOOKLET MAY BE DISQUALIFIED—RECEIVE A ZERO ON THE ENTIRE TEST.

This booklet contains Questions 1 through 8, which are based on the Scene shown in the First Memory Booklet. Answer Questions 1 through 8 after the fourth signal.

At the FIFTH SIGNAL, you should turn to the Second Memory Diagram. You will have 5 minutes to memorize the floor plan presented in the diagram.

DO NOT OPEN THIS BOOKLET UNTIL THE FOURTH SIGNAL IS GIVEN!

The New York City Department of Personnel makes no commitment, and no inference is to be drawn, regarding the content, style, or format of any future examination for the position of firefighter.

DIRECTIONS: Answer questions 1 through 8 based on the scene in the first memory booklet.

1. The fire is located on the

   (A) first floor
   (B) fourth floor

   (C) fifth floor
   (D) top floor

2. The smoke and flames are blowing

   (A) up and to the left
   (B) up and to the right

   (C) down and to the left
   (D) down and to the right

3. There is a person on a fire escape on the

   (A) second floor
   (B) third floor
   (C) fourth floor
   (D) fifth floor

4. Persons are visible in windows at the front of the building on fire on the

   (A) second and third floors
   (B) third and fifth floors
   (C) fourth and sixth floors
   (D) fifth and sixth floors

5. The person who is closest to the flames is in a

   (A) front window on the third floor
   (B) front window on the fifth floor
   (C) side window on the fifth floor
   (D) side window on the third floor

6. A firefighter is told to go to the roof of the building on fire. It would be correct to state that the firefighter can cross directly to the roof from

   (A) the roof of the bank
   (B) the roof of the factory
   (C) either the bank or the factory
   (D) neither the bank nor the factory

7. On which side of the building on fire are fire escapes visible?

   (A) left   (B) front   (C) right   (D) rear

8. The hydrant on the sidewalk is

   (A) in front of the bank
   (B) between the bank and the apartments
   (C) in front of the apartments
   (D) between the apartments and the factory

DO <u>NOT</u> TURN PAGE UNTIL NEXT SIGNAL

DIRECTIONS: LOOK AT THIS FLOOR PLAN OF AN APARTMENT. IT
IS ON THE 3RD FLOOR OF THE BUILDING. THE FLOOR PLAN ALSO
INDICATES THE PUBLIC HALLWAY.

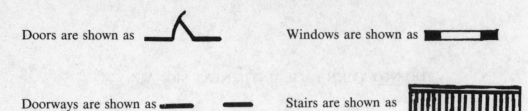

Doors are shown as ⌐   Windows are shown as ▭

Doorways are shown as — —   Stairs are shown as ▦

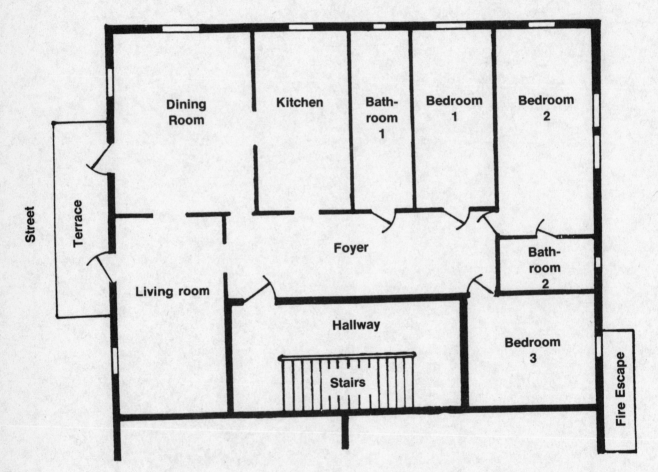

CITY OF NEW YORK
DEPARTMENT OF PERSONNEL

Social Security No. _____
Room No. _____
Seat No. _____
School _____

Question Booklet

FIREFIGHTER

December 12, 1987

EXAMINATION NO. 7022

Written Test: Weight 50

Time allowed: 4 Hours

DO NOT OPEN THIS BOOKLET UNTIL THE SEVENTH SIGNAL IS GIVEN!

Record your answers on the Official Answer Sheet before the last signal. If you wish, you may also record your answers in the Question Booklet before the last signal is given. This will be your only record of answers.

This Question Booklet contains procedures and definitions which are to be used to answer certain questions. THESE PROCEDURES AND DEFINITIONS ARE NOT NECESSARILY THOSE OF THE NEW YORK CITY FIRE DEPARTMENT. HOWEVER, YOU ARE TO ANSWER THESE QUESTIONS SOLELY ON THE BASIS OF THE MATERIAL GIVEN.

After the seventh signal is given, open this Question Booklet and begin work. You will have 4 hours to complete this test.

YOU MAY WRITE IN THIS BOOKLET (INCLUDING THE DIAGRAMS) AND USE THE SCRAP PAPER ON YOUR DESK. If you need additional scrap paper, ask the monitor.

Remember, only your Official Answer Sheet will be rated, so be sure to mark all your answers on the Official Answer Sheet before the eighth signal. No additional time will be given for marking your answers after the test has ended.

DO NOT OPEN THIS BOOKLET UNTIL THE SEVENTH SIGNAL IS GIVEN!

The New York City Department of Personnel makes no commitment, and no inference is to be drawn, regarding the content, style, or format of any future examination for the position of firefighter.

DIRECTIONS: Answer questions 9 through 16 on the basis of the floor plan you have just studied.

9. Which room is farthest from the fire escape?

   (A) bedroom 2   (C) kitchen
   (B) bedroom 3   (D) dining room

10. Which one of the following rooms has only one door or doorway?

   (A) living room   (C) kitchen
   (B) bedroom 1   (D) dining room

11. Which room can firefighters reach directly from the fire escape?

   (A) dining room   (C) bedroom 1
   (B) living room   (D) bedroom 3

12. Which room does not have a door or doorway leading directly to the foyer?

   (A) bathroom 1   (C) bedroom 1
   (B) bathroom 2   (D) dining room

13. A firefighter leaving bathroom 2 would be in

   (A) bedroom 1   (C) bedroom 3
   (B) bedroom 2   (D) the foyer

14. Firefighters on the terrace would be able to enter directly into which rooms?

   (A) bedroom 1 and bathroom 1   (C) dining room and kitchen
   (B) bedroom 2 and bathroom 2   (D) dining room and living room

15. Which rooms have at least one window on two sides of the building?

   (A) bedroom 2 and dining room
   (B) bedroom 2 and bedroom 3
   (C) dining room and living room
   (D) dining room, bedroom 2, and bedroom 3

16. Firefighters can enter the kitchen directly from the foyer and

   (A) bedroom 1   (C) bathroom 1
   (B) living room   (D) dining room

Please check your Answer Sheet to make sure that you have written in and blackened your Social Security number correctly. If you have not yet written in and blackened your Social Security number, please do so immediately. If you have incorrectly written in or blackened your Social Security number, please correct the mistake immediately. Failure to correctly blacken your Social Security number may result in your Answer Sheet NOT being rated.

17. Firefighters are often required to rescue individuals from a fire. The greatest possibility of a firefighter having to rescue someone in a private home occurs between the hours of

   (A) 7 A.M. and 11 A.M.   (C) 2 P.M. and 6 P.M.
   (B) 10 A.M. and 2 P.M.   (D) 2 A.M. and 6 A.M.

18. At a fire in an apartment building, a firefighter is told to inform the lieutenant if she finds any dangerous conditions in the basement. Which one of the following is the most dangerous condition?

(A) Gas is leaking from a broken pipe.
(B) The sewer pipe is broken.
(C) Water is seeping into the basement.
(D) The electricity has been turned off.

19. Firefighters are required to use portable ladders to rescue people. When firefighters are positioning a portable ladder for a rescue, which one of the following would present the greatest threat to the firefighters' safety?

(A) A person to be rescued who is standing near an open window.
(B) Tree branches which are close to the ladder.
(C) A person to be rescued who is dressed in a long robe.
(D) Overhead electrical wires which are close to the ladder.

20. Firefighters are instructed to notify an officer whenever they attempt to rescue someone who is seriously endangered by fire or smoke. Firefighters respond to a fire in a 6-story apartment building. The fire is in a fourth-floor apartment in the front of the building. Firefighters should notify an officer that they are attempting to rescue

(A) a person who disappears from a smoke-filled window on the fourth floor
(B) a person who is on the roof
(C) three persons on the third floor rear fire escape who appear to be very frightened
(D) two children who are locked in their apartment on the first floor

21. Firefighters who forcibly enter an apartment on fire may find conditions that indicate that they should immediately search for victims. Of the following conditions in an apartment on fire, which one would most clearly indicate to firefighters that they should immediately search for victims?

(A) There is a pot on the stove.
(B) The apartment door was chain-locked from the inside.
(C) Water is dripping into a pail.
(D) All the windows in the apartment are closed.

DIRECTIONS: Answer questions 22 through 24 based solely on the following passage.

When there is a fire in a subway train, it may be necessary for firefighters to evacuate people from the trains by way of the tunnels. In every tunnel, there are emergency exit areas that have stairways that can be used to evacuate people to the street from the track area. All emergency exits can be recognized by an exit sign near a group of five white lights.

There is a Blue Light Area that is located every 600 feet in the tunnel. These areas contain a power removal box, a telephone, and a fire extinguisher. Removal of power from the third rail is the first step firefighters must take when evacuating people through tunnels. When a firefighter uses the power removal box to turn off electrical power during evacuation procedures, the firefighter must immediately telephone the trainmaster and explain the reason for the power removal. Communication between the firefighter and the trainmaster is essential. If the trainmaster does not receive a phone call within four minutes after power removal, the power will be restored to the third rail.

22. When evacuating passengers through the subway tunnel, firefighters must *first*

    (A) telephone the trainmaster for assistance
    (B) remove electrical power from the third rail
    (C) locate the emergency exit in the tunnel
    (D) go to the group of five white lights

23. Immediately after using the power removal box to turn off the electrical power, a firefighter should

    (A) wait four minutes before calling the trainmaster
    (B) begin evacuating passengers through the tunnel
    (C) call the trainmaster and explain why the power was turned off
    (D) touch the third rail to see if the electrical power has been turned off

24. A group of five white lights in a subway tunnel indicates that

    (A) a telephone is available
    (B) the electrical power is off in the third rail
    (C) a fire extinguisher is available
    (D) an emergency exit is located there

DIRECTIONS: Answer questions 25 and 26 based on the following passage.

A new firefighter learns the following facts about his Company's response area: all the factories are located between 9th Avenue and 12th Avenue, from 42nd Street to 51st Street; all the apartment buildings are located between 7th Avenue and 9th Avenue, from 47th Street to 51st Street; all the private houses are located between 5th Avenue and 9th Avenue, from 42nd Street to 47th Street; and all the stores are located between 5th Avenue and 7th Avenue, from 47th Street to 51st Street.

The firefighter also learns that the apartment buildings are all between 4 and 6 stories; the private houses are all between 1 and 3 stories; the factories are all between 3 to 5 stories; and the stores are all either 1 or 2 stories.

25. An alarm is received for a fire located on 8th Avenue between 46th Street and 47th Street. A firefighter should assume that the fire is in a

(A) private house between 1 and 3 stories
(B) private house between 4 and 6 stories
(C) factory between 3 to 5 stories
(D) factory between 4 and 6 stories

26. The company responds to a fire on 47th Street between 6th Avenue and 7th Avenue. The firefighter should assume that he would be responding to a fire in

(A) a store of either 1 or 2 stories
(B) a factory between 3 and 5 stories
(C) an apartment building between 4 and 6 stories
(D) a private house between 4 and 6 stories

DIRECTIONS: Answer question 27 based on the following passage.

During a recent day tour with an engine company, firefighter Sims was assigned to the control position on the hose. The company responded to the following alarms during this tour:

Alarm 1: At 9:30 A.M., the company responded to a fire on the first floor of an apartment building. At the fire scene, firefighter Sims pulled the hose from the fire engine and assisted the driver in attaching the hose to the hydrant.

Alarm 2: At 11:00 A.M., the company responded to a fire on the third floor of a vacant building. Firefighter Sims pulled the hose from the fire engine and went to the building on fire.

Alarm 3: At 1:00 P.M., the company responded to a fire in a first-floor laundromat. Firefighter Sims pulled the hose from the fire engine and assisted the driver in attaching the hose to the hydrant.

Alarm 4: At 3:00 P.M., the company responded to a fire on the fourth floor of an apartment building. Firefighter Sims pulled the hose from the fire engine and went to the building on fire.

Alarm 5: At 5:45 P.M., the company responded to a fire on the second floor of a private house. Firefighter Sims pulled the hose from the fire engine and assisted the driver in attaching the hose to the hydrant.

27. The firefighter assigned to the control position assists the driver in attaching the hose to a hydrant when the fire is

(A) in an apartment building
(B) above the second floor
(C) in a vacant building
(D) below the third floor

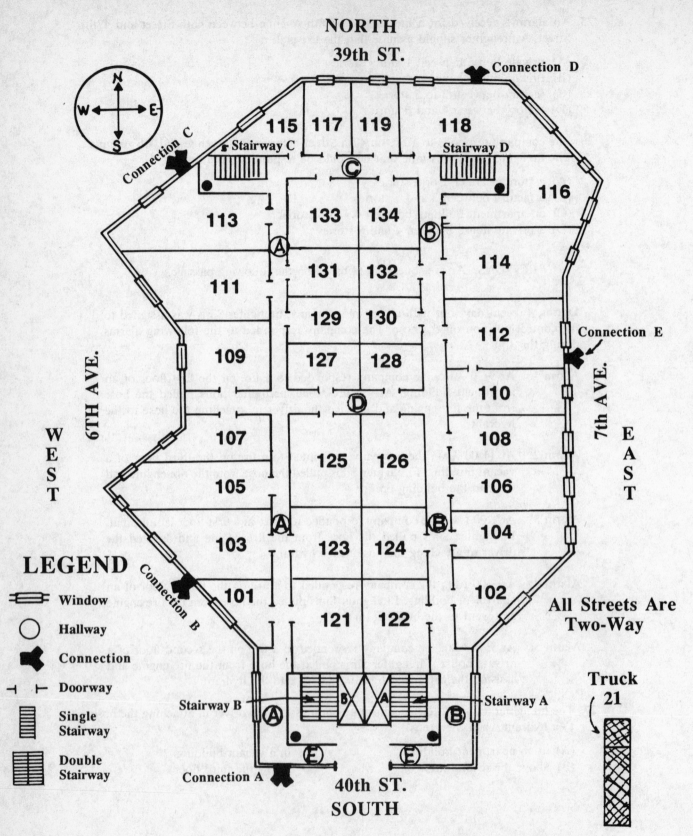

NORTH
39th ST.

Connection D

115 117 119 118

Stairway C Stairway D

116

C

113 133 134 B

A

131 132 114

111

129 130 112 Connection E

109 127 128 110

D

107 125 126 108 7th AVE.

105 106

A B 104

103 123 124

LEGEND 101 102 All Streets Are Two-Way

121 122

Window Truck 21

Hallway

Connection Stairway B B A Stairway A

Doorway

Single Stairway A B

Double Stairway E E

Connection A

40th ST.
SOUTH

Connection C

6TH AVE.

WEST

EAST

Connection B

DIRECTIONS: Answer questions 28 through 30 based on the floor plan shown on page 140 and on the following information.

The floor plan represents a typical high-rise office building in midtown Manhattan. Numbers shown indicate room numbers. The pipe connections for the water supply system are outside the building at street level. Firefighters attach hoses to those connections to send water into the pipes in the building.

Questions 28 through 30 refer to a fire on the first floor in Room 111.

28. After fighting the fire in Room 111, firefighters are instructed to go immediately to the east–west hallway in the center of the building and search for victims in that hallway. Which one of the following lists all of the rooms that the firefighters should search?

   (A) 115, 117, 118, 119, 133, and 134
   (B) 125, 126, 127, 128, and 129
   (C) 107, 109, 125, 126, 127, and 128
   (D) 121, 122, 123, 124, 125, and 126

29. Firefighters are told to search Room 134. They enter the building from 40th Street. What is the shortest route for the firefighters to take to reach this room?

   (A) west in hallway E, north in hallway A, then east in hallway C
   (B) west in hallway E, north in hallway A, east in hallway D, north in hallway B, then west in hallway C
   (C) east in hallway E, north in hallway B, then west in hallway C
   (D) east in hallway E, north in hallway B, west in hallway D, north in hallway A, then east in hallway C

30. Firefighters in Truck 21 have been ordered to attach a hose to a connection outside the building. The firefighters cannot use connection A because 40th Street is blocked by traffic. What is the first connection the firefighters can drive to?

   (A) Connection B   (B) Connection C   (C) Connection D   (D) Connection E

DIRECTIONS: Answer questions 31 through 33 based on the following passage.

Firefighters often know the appearance and construction features of apartments by recognizing the general features on the outside of the building. The following are some general features of different types of buildings in New York City.

   1. Old Law Tenements
      Height — 5 to 7 stories
      Width — 25 feet
      Fire Escapes — There will be a rear fire escape if there are two apartments per floor. There will be front and rear fire escapes if there are four apartments per floor.

   2. Row Frames
      Height — 2 to 5 stories
      Width — 20 feet to 30 feet
      Fire Escapes — There will be a rear fire escape if the building is higher than 2 stories.

3. Brownstones
   Height — 3 to 5 stories
   Width — 20 feet to 25 feet
   Fire Escapes — If the Brownstone has been changed from a private home to a multiple dwelling, there will be a rear fire escape. Unchanged Brownstones have no fire escapes.

31. Upon arrival at a fire, a firefighter observes that the building is 3 stories high and 25 feet wide. There are fire escapes only in the rear of the building. The firefighter should conclude that the building is either a

(A) row frame or an unchanged brownstone
(B) row frame or an old law tenement with two apartments per floor
(C) changed brownstone or an old law tenement with four apartments per floor
(D) row frame or a changed brownstone

32. At another fire, the building is 5 stories high and 25 feet wide. There is a front fire escape. The firefighters should conclude that this building has

(A) a rear fire escape because the building is a row frame higher than two stories
(B) a rear fire escape because the building is an old law tenement with four apartments per floor
(C) no rear fire escape because the building is a brownstone that has been changed into a multiple dwelling
(D) no rear fire escape because the building has a front fire escape

33. At another fire, the building is 4 stories high and 30 feet wide. The building has no front fire escape. The firefighter should conclude that the building is

(A) a row frame which has no rear fire escape
(B) an old law tenement which has 4 apartments per floor
(C) a row frame which has a rear fire escape
(D) a brownstone which has been changed from a private home to a multiple dwelling

DIRECTIONS: Answer questions 34 through 36 based on the following passage.

Firefighters use two-way radios to alert other firefighters of dangerous conditions and of the need for help. Messages should begin with "MAY DAY" or "URGENT." "MAY DAY" messages have priority over "URGENT" messages. Following is a list of specific emergencies and the messages which should be sent.

"MAY DAY" Messages:

1. When a collapse is probable in the area where the firefighters are working: "MAY DAY–MAY DAY, collapse probable, GET OUT".

2. When a collapse has occurred in the area where the firefighters are working: "MAY DAY–MAY DAY, collapse occurred." The firefighter should also give the location of the collapse. If there are trapped victims, the number and condition of the trapped victims is also given.

3. When a firefighter appears to be a heart attack victim: "MAY DAY–MAY DAY, CARDIAC." The location of the victim is also given.

4. When anyone has a serious, life-threatening injury: "MAY DAY–MAY DAY." The firefighter also describes the injury and gives the condition and the location of the victim.

"URGENT" Messages:

1. When anyone has a less serious injury which requires medical attention (for example, a broken arm): "URGENT–URGENT." The firefighter also gives the type of injury and the location of the victim.

2. When the firefighters should leave the building and fight the fire from the outside: "URGENT–URGENT, back out." The firefighter also indicates the area to be evacuated.

3. "URGENT" messages should also be sent when firefighters' lives are endangered due to a drastic loss of water pressure in the hose.

34. Firefighters are ordered to extinguish a fire on the third floor of an apartment building. As the firefighters are operating the hose on the third floor, the stairway collapses and cuts the hose. What message should the firefighters send?

(A) "URGENT–URGENT, back out"
(B) "URGENT–URGENT, we have a loss of water on the third floor"
(C) "MAY DAY–MAY DAY, collapse occurred on third floor stairway"
(D) "MAY DAY–MAY DAY, collapse probable, GET OUT"

35. Two firefighters on the second floor of a vacant building are discussing the possibility of the floor's collapse. One of the firefighters clutches his chest and falls down. What message should the other firefighter send?

(A) "MAY DAY–MAY DAY, firefighter collapse on the second floor"
(B) "MAY DAY–MAY DAY, CARDIAC on the second floor"
(C) "URGENT–URGENT, firefighter unconscious on the second floor"
(D) "URGENT–URGENT, collapse probable on the second floor"

36. A firefighter has just decided that a collapse of the third floor is probable when he falls and breaks his wrist. What is the first message he should send?

(A) "URGENT–URGENT, broken wrist on the third floor"
(B) "MAY DAY–MAY DAY, broken wrist on the third floor"
(C) "MAY DAY–MAY DAY, collapse probable, GET OUT"
(D) "URGENT–URGENT, back out, third floor"

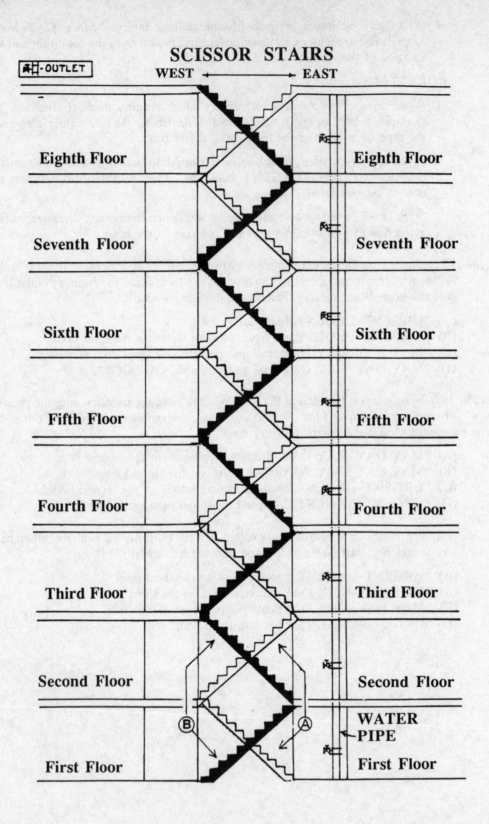

SCISSOR STAIRS

DIRECTIONS: Answer questions 37 and 38 based on the diagram shown on page 144 and on the following information.

An eight-story apartment building has scissor stairs beginning on the first floor and going to the roof. Scissor stairs are two separate stairways (Stairway A and Stairway B) that criss-cross each other and lead to opposite sides of the building on each floor. Once a person has entered either stairway, the only way to cross over to the other stairway on any floor is by leaving the stairway and using the hallway on that floor. A person entering Stairway A, which starts on the east side of the building on the first floor, would end up on the west side of the building on the second floor, and back on the east side on the third floor. Similarly, a person entering Stairway B, which starts on the west side of the building on the first floor, would end up on the east side of the building on the second floor, and back on the west side on the third floor.

The apartment building has one water pipe for fighting fires. This pipe runs in a straight line near the stairway on the east side of the building from the first floor to the roof. There are water outlets for this pipe on each floor.

Both of the following questions involve a fire in an apartment on the west side of the sixth floor.

37. Firefighters are ordered to connect a hose to the nearest outlet *below* the fire. Upon reaching this outlet they find that it is not usable. Where is the next available outlet?

    (A) fifth floor near stairway B   (C) fourth floor near stairway B
    (B) third floor near stairway A   (D) fourth floor near stairway A

38. A firefighter working on the west side of the seventh floor is ordered to search for victims on the west side of the eighth floor. The door leading to the stairway on the west side of the seventh floor is jammed shut. To reach the victims, the firefighter should take

    (A) Stairway A to the eighth floor, and then go across the hallway to the west side of the floor
    (B) Stairway B to the eighth floor, and then go across the hallway to the west side of the floor
    (C) the hallway to the east side of the seventh floor and go up Stairway A
    (D) the hallway to the east side of the seventh floor and go up Stairway B

39. Firefighters refer to the four sides of a building on fire as "exposures." The front of the fire building is referred to as Exposure 1. Exposures 2, 3, and 4 follow in clockwise order. Firefighters are working at a building whose front entrance faces south. A firefighter who is in the center of the roof is ordered to go to Exposure 3. To reach Exposure 3, the direction in which he must walk is

    (A) east   (B) west   (C) south   (D) north

DIRECTIONS: Answer questions 40 through 44 based solely on the following passage.

The most important activities that firefighters perform at fires are search, rescue, ventilation, and extinguishment. Ventilation is a vital part of firefighting because it prevents fire from spreading to other areas, and because it enables firefighters to search for victims and to bring hoses closer to the fire area. Two types of ventilation used by firefighters are natural venting and mechanical venting. Both types permit the vertical and horizontal movement of smoke and gas from a fire building.

Natural *vertical* ventilation is generally performed on the roof of the building on fire by making an opening. This allows the heat and smoke to travel up and out of the fire building. Opening windows in the fire area is an example of natural *horizontal* ventilation. This allows the heat and smoke to travel out of the windows.

Mechanical ventilation takes place when mechanical devices, such as smoke ejectors or hoses with nozzles, are used to remove heated gases from an area. A smoke ejector might be used in a cellar fire when smoke has traveled to the far end of the cellar, creating a heavy smoke condition that cannot be removed naturally. The smoke ejector would be brought into the area to draw the smoke out of the cellar. A nozzle is used with a hose to create a fine spray of water. When directed towards an open window, the water spray pushes smoke and heated gases out of the window.

Extinguishment means bringing a hose to the fire and operating the nozzle to put water on the fire. The proper positioning of hoses is essential to firefighting tactics. Most lives are saved at fires by the proper positioning of hoses.

At each fire, firefighters must use the quickest and best method of extinguishment. There are times when an immediate and direct attack on the fire is required. This means that the hose is brought directly to the fire itself. A fire in a vacant lot, or a fire in the entrance of a building, calls for an immediate and direct attack on the fire.

It is generally the ladder company that is assigned the tasks of venting, search, and rescue while the engine company performs the task of extinguishment.

40. Ventilation performed at the roof is generally

    (A) mechanical vertical ventilation
    (B) natural vertical ventilation
    (C) natural horizontal ventilation
    (D) mechanical horizontal ventilation

41. When an immediate and direct attack on the fire is required, the hose is

    (A) positioned between the building on fire and the building which the fire might spread to
    (B) brought to a window in order to push smoke and gases out
    (C) brought to the roof to push the smoke and gases out
    (D) brought directly to the fire itself

42. Ladder companies are generally assigned the tasks of

    (A) extinguishment, rescue, and search
    (B) extinguishment, venting, and search
    (C) venting, search, and rescue
    (D) venting, rescue, and extinguishment

43. Most lives are saved at fires by

    (A) a systematic search
    (B) the proper positioning of hoses
    (C) the proper performance of ventilation
    (D) the use of nozzles for ventilation and extinguishment

44. Ventilation enables firefighters to

    (A) bring hoses to the fire and search for victims
    (B) create a fine spray of water
    (C) use a nozzle to remove smoke and gases
    (D) use an ejector to draw smoke out of an area

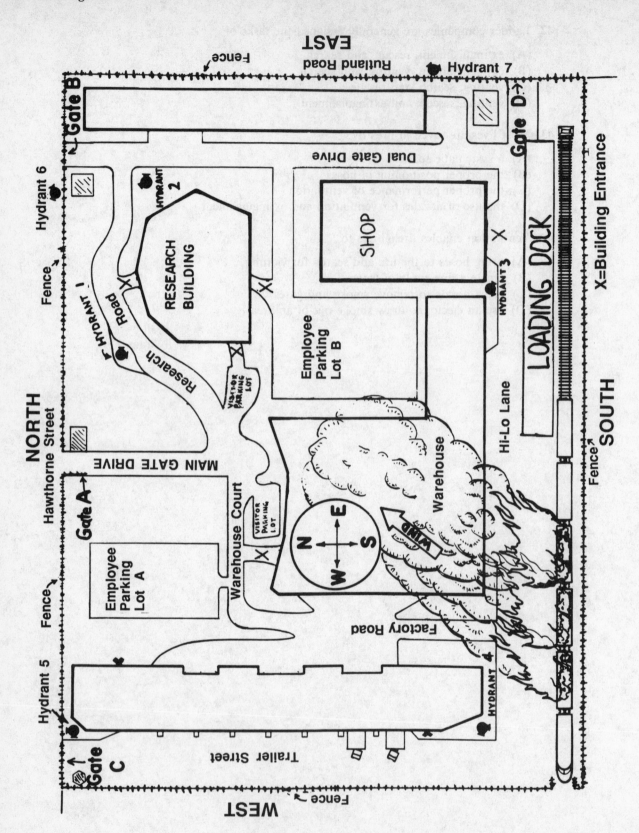

DIRECTIONS: Answer questions 45 through 50 based upon the diagram on page 148 and the following information.

At three o'clock in the morning, a fire alarm is received for the area shown in the diagram. A train loaded with highly flammable material is on fire. The entire area is surrounded by a 10-foot high fence. At the time of the fire, Gate A is open and Gates B, C, and D are locked.

45. The first engine company arrives at the fire scene. The security guard at Gate A informs the firefighters of the location of the fire. Firefighter Jensen knows the area. He should inform the lieutenant that the way to drive to a hydrant that is as close to the fire as possible without passing through the smoke and flames is by going

   (A) south on Main Gate Drive, east on Research Road, south on Dual Gate Drive, and west on Hi-Lo Lane to hydrant 3
   (B) south on Main Gate Drive, west on Warehouse Court, south on Factory Road, and west on Hi-Lo Lane to hydrant 4
   (C) south on Main Gate Drive and east on Research Road to hydrant 1
   (D) east on Hawthorne Street and south on Rutland Road to hydrant 7

46. Firefighters at Employee Parking Lot A are ordered to drive their truck to the fence outside Gate D. Which of the following is the shortest route the firefighters could take from Warehouse Court?

   (A) south on Factory Road, then west on Hi-Lo Lane and north on Trailer Street
   (B) east on Research Road and south on Dual Gate Drive
   (C) north on Main Gate Drive, east on Hawthorne Street, and south on Rutland Road
   (D) north on Main Gate Drive, west on Hawthorne Street, south on Trailer Street, ánd west on Hi-Lo Lane

47. The first ladder company arrives at the fire scene. As they are driving north on Rutland Road, firefighters see the fire through Gate D. They cut the locks and enter Gate D. The Lieutenant orders a firefighter to go on foot from Gate D to the Research Building and to search it for occupants. The entrance to the Research Building which is closest to the firefighter is

   (A) connected to the Visitor Parking Lot
   (B) located on Research Road
   (C) connected to Parking Lot B
   (D) located on Dual Gate Drive

48. The second engine company to arrive is ordered to attach a hose to a hydrant located outside of the fenced area and then to await further orders. The hydrant outside of the fenced area that is closest to the flames is

   (A) hydrant 6   (B) hydrant 3   (C) hydrant 4   (D) hydrant 7

49. The second ladder company to arrive at the fire scene is met at Gate C by a security guard who gives them the keys to open all the gates. They drive south on Trailer Street to the corner of Hi-Lo Lane and Trailer Street. The company is then ordered to drive

to the corner of Research Road and Dual Gate Drive. Which is the shortest route for the company to take without being exposed to the smoke and flames?

(A) east on Hi-Lo Lane, north on Factory Road, and east on Warehouse Court to Research Road
(B) east on Hi-Lo Lane and north on Dual Gate Drive
(C) north on Trailer Street, east on Hawthorne Street, and south on Dual Gate Drive
(D) north on Trailer Street, east on Hawthorne Street, south on Main Gate Drive, and east on Research Road

50. The heat from the fire in the railroad cars ignites the warehouse on the other side of Hi-Lo Lane. The officer of the first ladder company orders two firefighters who are on the west end of the loading dock to break the windows on the north side of the warehouse. Of the following, the shortest way for the firefighters to reach the northwest corner of the warehouse without passing through the smoke and flames is to go

(A) east on Hi-Lo Lane, north on Dual Gate Drive, and then west on Research Road to the entrance on Warehouse Court
(B) west on Hi-Lo Lane, north on Factory Road, and then east on Warehouse Court to the Visitor Parking Lot on Warehouse Court
(C) east on Hi-Lo Lane, north on Rutland Road, west on Hawthorne Street, and then south on Main Gate Drive to the Visitor Parking Lot on Warehouse Court
(D) east on Hi-Lo Lane, north on Dual Gate Drive, west on Hawthorne Street, and then south on Main Gate Drive to the entrance on Warehouse Court

DIRECTIONS: Answer question 51 based on the following passage.

Firefighters may have to use tools to force open an entrance door. Before the firefighters use the tools, they should turn the doorknob to see if the door is unlocked. If the door is locked, one firefighter should use an axe and another firefighter should use the halligan tool. This tool is used to pry open doors and windows. Firefighters must take the following steps in the order given to force open the door:

1. Place the prying end of the halligan tool approximately six inches above or below the lock. If there are two locks, the halligan tool should be placed between them.

2. Tilt the halligan tool slightly downward so that a single point on the prying end is at the door's edge.

3. Strike the halligan tool with the axe until the first point is driven in between the door and the door frame.

4. Continue striking with the axe, until the door and frame are spread apart and the lock is broken.

5. Apply pressure toward the door and the door will spring open.

51. Firefighters respond to a fire on the second floor of a three-story apartment building. Two firefighters, one equipped with an axe and the other with a halligan tool, climb the stairs to the apartment on fire. They see that there are two locks on the apartment door. The firefighters should now

(A) place the prying end of the halligan tool about six inches above or below the locks
(B) turn the doorknob to determine whether the door is locked

(C) tilt the halligan tool towards the floor before striking

(D) place the prying end of the halligan tool between the locks

DIRECTIONS: Answer questions 52 and 53 based on the following passage.

The firefighter who is assigned to the roof position at a fire in a brownstone building should perform the following steps in the order given:

1. Go to the roof using one of the following ways:

   a. First choice—The aerial ladder.

   b. Second choice—An attached building of the same height as the fire building.

   c. Third choice—A rear fire escape.

   d. Fourth choice—A thirty-five foot portable ladder.

2. Upon arrival at the roof, look around to determine if any people are trapped who cannot be seen from the street.

   a. If a trapped person is observed, notify the officer and the driver that a life-saving rope rescue is required. While waiting for assistance to conduct this rescue, assure the victim that help is on the way and proceed to Step 3.

   b. If no trapped persons are visible, proceed directly to Step 3.

3. Remove the cover from the opening in the roof.

   a. If there is no smoke or very little smoke coming from the opening, report to the officer for further orders.

   b. If heavy smoke comes from the opening, proceed to Step 4.

4. Remove the glass from the skylight.

52. Firefighters arriving at a fire in a brownstone are using the aerial ladder to make an immediate rescue. The firefighter assigned to the roof position should go to the roof of the building on fire by

(A) a 35-foot portable ladder
(B) a rear fire escape
(C) an attached building of the same height
(D) the inside stairway of the fire building

53. The firefighter assigned to the roof position at a fire in a brownstone arrives at the roof and finds that no persons are trapped. He then removes the roof cover from the opening in the roof. Which one of the following steps should be performed next?

(A) He should remove the glass from the skylight if heavy smoke is coming from the opening.
(B) He should remove the glass from the skylight if no smoke is coming from the roof opening.
(C) He should go to the top floor to assist in the search for trapped persons if heavy smoke is coming from the roof opening.
(D) He should report to the officer if heavy smoke is coming from the roof opening.

DIRECTIONS: Answer questions 54 and 55 based on the following passage.

Firefighters are often required to remove people who are trapped in elevators. At this type of emergency, firefighters perform the following steps in the order given:

1. Upon entering the building, determine the location of the elevator involved.

2. Reassure the trapped occupants that the Fire Department is on the scene and that firefighters are attempting to free them.

3. Determine if there are any injured people in the elevator.

4. Determine if all the doors from the hallways into the elevator shaft are closed.

5. If all the doors are closed, call for an elevator mechanic.

6. Wait until a trained elevator mechanic arrives before attempting to remove any trapped persons from the elevator, unless they can be removed through the door to the hallway. However, firefighters must remove the trapped persons by any safe method if any one of the following conditions exists:

    a. There is a fire in the building.

    b. Someone in the elevator is injured.

    c. The people trapped in the elevator are in a state of panic.

54. Firefighters arrive at an elevator emergency in an office building. When they arrive, a maintenance man directs them to an elevator that is stuck between the fourth and fifth floors. He informs the firefighters that there is a young man in the elevator who apparently is calm and unhurt. Which one of the following steps should the firefighters perform next?

(A) Determine if the young man is injured.
(B) Reassure the young man that the Fire Department is on the scene and that firefighters are attempting to free him.
(C) Check to make sure that all the doors to the elevator and hallways are closed.
(D) Call for an elevator mechanic and await his arrival.

55. Firefighters are called to an elevator emergency at a factory building. The freight elevator has stopped suddenly between floors. The sudden stop caused heavy boxes to fall on the elevator operator, breaking his arm. Upon arrival, the firefighters determine the location of the elevator. They tell the trapped operator that they are on the scene, are aware of his injury, and are attempting to free him. They determine that all the hallway doors leading into the elevator shaft are closed. The firefighters' next step should be to

(A) call for an ambulance and wait until it arrives
(B) remove the trapped person through the door to the hallway
(C) call for an elevator mechanic
(D) remove the trapped person by any safe method

DIRECTIONS: Answer question 56 based on the following passage.

The preferred order of actions for firefighters to take when removing a victim from an apartment on fire is as follows:

1. First choice—Remove the victim to the street level through the public hallway.

2. Second choice—Remove the victim to the street level by the fire escape.

3. Third choice—Remove the victim to the street level using either a portable ladder or an aerial ladder.

4. Fourth choice—Lower the victim to the street level with a lifesaving rope.

56. Firefighters answering an alarm are not able to use the entrance to a building on fire to reach a victim on the third floor because there is a fire in the public hallway. The victim is standing at the front window of an apartment on fire which has a fire escape. A firefighter places a portable ladder against the building, climbs the ladder, and enters the window where the victim is standing. The firefighter is carrying the lifesaving rope and a two-way radio. The radio allows her to communicate with the firefighter who operates the aerial ladder. The firefighter should then remove this victim from the apartment by using the

(A) fire escape

(B) aerial ladder

(C) portable ladder

(D) lifesaving rope

PIERCE HIGH SCHOOL

DIRECTIONS: Answer questions 57 through 61 based on the floor plan on page 154 and on the following information.

Since many children may need to be rescued in the event of a school fire, New York City firefighters must become familiar with the floor layouts of public schools. Firefighters can develop this familiarity by conducting training drills at the schools.

A ladder company and an engine company recently conducted a drill at Pierce High School. The firefighters determined that the room layout is the same on all floors.

Several days after the drill, the ladder and the engine companies report to a fire at Pierce High School in classroom 304, which is on the third floor. The fire has spread into the hallway in front of Room 304, blocking the hallway.

57. A firefighter is instructed to search for victims in the southwest area of the third floor. He wants to search as many rooms as possible and start his search as close to the fire as possible without passing through the fire. From the street the firefighter should use his ladder to enter

(A) Room 302   (B) Room 306   (C) Room 312   (D) Room 352

58. A firefighter goes to the third floor by way of the southwest building stairway. In Room 317, he finds a child who has been overcome by smoke. Upon returning to the hallway he finds that the stairway he came up is now blocked by fire hoses. Which is the closest stairway that the firefighter can use to bring the child to the street level?

(A) Stairway A   (B) Stairway B   (C) Stairway C   (D) Stairway D

59. Fire has spread from Room 304 to the room directly across the hall. As a result of heavy smoke, firefighters are ordered to break the windows of this room from the closest room on the floor above. Which room should the firefighters go to?

(A) Room 403   (B) Room 305   (C) Room 313   (D) Room 413

60. The fire is in Rooms 303 and 305. Firefighters are told to go to rooms in the north corridor facing the courtyard that are directly opposite 303 and 305. Which rooms should the firefighters go to?

(A) Rooms 323 and 325   (C) Rooms 333 and 335
(B) Rooms 326 and 328   (D) Rooms 355 and 357

61. Another fire breaks out in Room 336, blocking the entire hallway. Firefighters have brought a hose up the northeast stairway to fight this fire. Another hose must be brought up another stairway so that firefighters can approach the fire from the same direction. What is the closest stairway that the firefighters could use?

(A) Stairway A   (B) Stairway B   (C) Stairway C   (D) Stairway D

62. When attempting to rescue a person trapped at a window, firefighters frequently use an aerial ladder with a rotating base. The rotating base allows the ladder to move up and down and from side to side. In a rescue situation, the ladder should be placed so that both sides of the ladder rest fully on the window sill after it has been raised into position. Which one of the following diagrams shows the proper placement of the rotating base in relation to the window where the person is trapped?

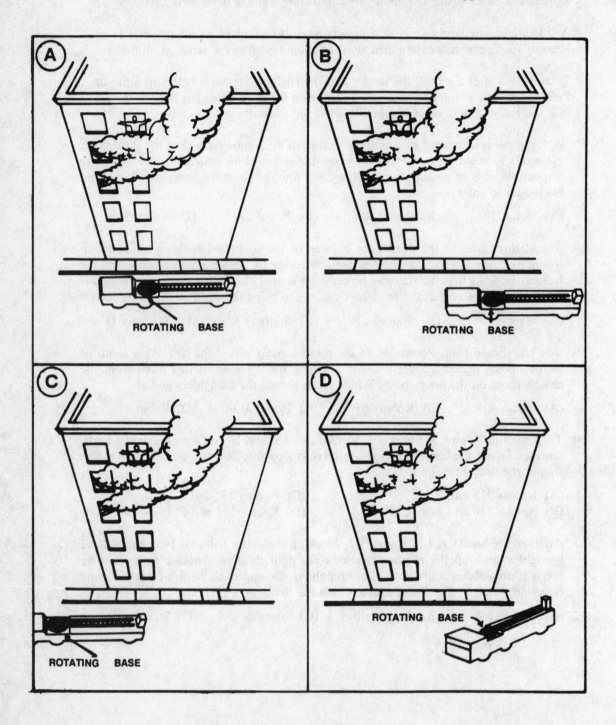

DIRECTIONS: Answer question 63 based solely on the following passage.

Firefighters must regularly inspect office buildings to determine whether fire prevention laws have been obeyed. Some of these fire prevention laws are as follows:

DOORS: Doors should be locked as follows:

1. Doors on the ground floor may be locked on the street side to prevent entry into the stairway.

2. Doors in office buildings that are less than 100 feet in height may be locked on the stairway side on each floor above the ground floor.

3. Doors in office buildings that are 100 feet or more in height may be locked on the stairway side except for every fourth floor.

63. The doors in an office building which is less than 100 feet in height may be locked on the stairway side

(A) on all floors including the ground floor
(B) on all floors above the ground floor
(C) except for every fourth floor
(D) on all floors above the fourth floor

DIRECTIONS: Answer question 64 based solely on the following passage.

SIGNS: Signs concerning stairways should be posted in the following manner:

1. A sign shall be posted near the elevator on each floor, stating "IN CASE OF FIRE, USE STAIRS UNLESS OTHERWISE INSTRUCTED." The sign shall contain a diagram showing the location of the stairs and the letter identification of the stairs.

2. Each stairway shall be identified by an alphabetical letter on a sign posted on the hallway side of the stair door.

3. Signs indicating the floor number shall be attached to the stairway side of each door.

4. Signs indicating whether re-entry can be made into the building, and the floors where re-entry can be made, shall be posted on the stairway side of each door.

64. Which one of the following correctly lists the information which should be posted on the stairway side of a door?

(A) A sign will indicate the floor number, whether re-entry can be made into the building, and the floors where re-entry can be made.
(B) A sign will indicate the alphabetical letter of the stairway, whether re-entry can be made into the building, and the floors where re-entry can be made.
(C) A sign will indicate the alphabetical letter of the stairway and the floor number.
(D) A sign will indicate the alphabetical letter of the stairway, the floor number, whether re-entry can be made into the building, and the floors where re-entry can be made.

DIRECTIONS: Answer question 65 based solely on the following passage.

Every firefighter must know the proper first-aid procedures for treating injured people when there is a subway fire. In general, anyone suffering from smoke inhalation or heat exhaustion should be removed to fresh air and given oxygen immediately. Heart attack victims should be kept calm and should receive oxygen and medical attention immediately. Persons suffering from broken bones should not be moved until a splint is applied to the injury. However, in a situation where there is a smoky fire in the subway and the passengers needing immediate first aid are in danger from the fire, firefighters must first evacuate the passengers and perform first aid later, regardless of the injury.

65. The proper first-aid procedure for a man who has apparently suffered a heart attack on the station platform is to

(A) have the man take the next train to the nearest hospital
(B) remove the man to the street and administer oxygen
(C) turn off the electrical power and evacuate the man through the tunnel
(D) keep the man calm and administer oxygen

DIRECTIONS: Answer questions 66 and 67 based solely on the following passage.

A firefighter is responsible for a variety of duties other than fighting fires. One such duty is housewatch.

A firefighter's primary responsibility during housewatch is to properly receive alarm information. This enables firefighters to respond to alarms for fires and emergencies. The alarms are received at the firehouse by one of the following methods: computer teleprinter messages, Fire Department telephone, or verbal alarm. The computer teleprinter and the telephone are used to alert the fire companies. These two types of alarms are transmitted by a dispatcher from a central communication office to the firehouse closest to the fire. The verbal alarm occurs when someone comes to the firehouse or stops the fire truck on the street to report a fire. Once an alarm has been received, the firefighter on housewatch duty alerts the rest of the firefighters to respond to the alarm.

Other housewatch responsibilities include keeping the appearance of the housewatch area neat and orderly, keeping the front of the firehouse clear of all vehicles and obstructions, and receiving telephone calls and visitors with complaints about fire hazards. The firefighter on housewatch duty also keeps an accurate and complete record of all administrative matters in a journal.

66. The methods a dispatcher uses to transmit alarms to the firehouse are the

(A) computer teleprinter, Fire Department telephone, and verbal alarm
(B) verbal alarm and computer teleprinter
(C) Fire Department telephone and verbal alarm
(D) computer teleprinter and Fire Department telephone

67. The primary responsibility of a firefighter on housewatch duty is to

(A) properly assign firefighters to specific duties

(B) properly receive alarm information
(C) keep the housewatch area neat and orderly
(D) write all important information in the company journal

DIRECTIONS: Answer questions 68 and 69 based solely on the following passage.

One duty of a firefighter on housewatch is to ensure that the computer teleprinter is working properly. A company officer should be notified immediately of any equipment problems. The firefighter on housewatch should check on the amount of paper in the teleprinter and should refill it when necessary. The firefighter should also check the selector panel on the computer. This selector panel has a series of buttons which are used by the firefighter to let the dispatcher know that an alarm has been received and that the fire company is responding. These buttons have lights. To check that the computer is functioning properly, the firefighter should press the button marked "test" and then release the button. If the computer lights go on, and then go off after the "test" button has been released, the computer is working properly. In addition, the light next to the "test" button should always be blinking.

68. In order to check that the selector panel of the computer is working properly, the firefighter on housewatch duty presses the button marked "test," and then releases the button. The firefighter should conclude that the computer is working properly if the

(A) computer lights stay on
(B) computer lights keep blinking
(C) computer lights go on and then off
(D) "test" light stays on

69. A firefighter on housewatch duty notices that the teleprinter is almost out of paper. In this situation, the firefighter should

(A) test the computer panel by pushing the "test" button
(B) notify the officer to replace the paper
(C) place a new supply of paper in the teleprinter
(D) notify the dispatcher that the paper is being changed

DIRECTIONS: Answer questions 70 and 71 based on the following passage.

Following is a list of rules for fire extinguishers which are required in different types of public buildings in New York City:

Rule 1: Hospitals, nursing homes, hotels, and motels must have one 2½-gallon water extinguisher for every 2500 square feet, or part thereof, of floor area on each floor.

Rule 2: Stores with floor areas of 1500 square feet or less must have one 2½-gallon water extinguisher. Stores with floor areas of over 1500 square feet must have one 2½-gallon water extinguisher for every 2500 square feet, or part thereof, of floor area on each floor.

Rule 3: Kitchens must have one 2½-gallon foam extinguisher or one 5-pound dry chemical extinguisher for every 1250 square feet, or part thereof, of floor area on each floor. For kitchen areas, this rule is in addition to Rules 1 and 2.

70. A firefighter is inspecting a one-story nursing home which has a total of 3000 square feet of floor area. This includes a kitchen, which is 1500 square feet in area, in the rear of the floor. Of the following, the firefighter should conclude that the nursing home should be equipped with

(A) one water extinguisher and one foam extinguisher
(B) one water extinguisher and one dry chemical extinguisher
(C) two water extinguishers and two foam extinguishers
(D) two foam extinguishers and one dry chemical extinguisher.

71. A firefighter is inspecting a store which has two floors. The first floor has 2600 square feet. The second floor has 1450 square feet. The store should be equipped with at least

(A) two $2\frac{1}{2}$-gallon water extinguishers, one for each floor
(B) three $2\frac{1}{2}$-gallon water extinguishers, two for the first floor and one for the second floor
(C) two $2\frac{1}{2}$-gallon foam extinguishers, one for each floor
(D) two $2\frac{1}{2}$-gallon extinguishers, either foam or water, one for each floor.

DIRECTIONS: Answer question 72 based on the following passage.

Firefighters from the first arriving ladder company work in teams while fighting fires in private homes. The inside team enters the building through the first-floor entrance and then searches the first floor for victims. The outside team uses ladders to enter upper level windows for a quick search of the bedrooms on the second floor and above. The assignments for the members of the outside team are as follows:

Roof person—This member places a ladder at the front porch and enters the second floor windows from the roof of the porch.

Outside vent person and driver—These members work together and place a portable ladder at a window on the opposite side of the house from which the roof person is working. However, if the aerial ladder can be used, the outside vent person and driver climb the aerial ladder in the front of the house and the roof person places a portable ladder on the left side of the house.

In order to search all four sides of a private home on the upper levels, firefighters from the second arriving ladder company place portable ladders at the sides of the house not covered by the first ladder company, and enter the home through the upper level windows.

72. The second ladder company to arrive at a fire in a two-story private home sees the aerial ladder being raised to the front porch roof. In this situation, the firefighters should place their portable ladders

(A) to the left and right sides of the house since there is a front porch
(B) to the rear and right sides of the house since the aerial ladder is being used
(C) to the rear and left sides of the house since there is a front porch
(D) to the left and right sides of the house since the aerial ladder is being used

DIRECTIONS: Answer question 73 based on the following passage.

The priority for the removal of a particular victim by aerial ladder depends on the following conditions:

If two victims are at the same window and are not seriously endangered by spreading fire, the victim who is easier to remove is taken down the ladder first and helped safely to the street. In general, the term "easier to remove" refers to the victim who is more capable of being moved and more able to cooperate. After the easier removal is completed, time can be spent on the more difficult removal.

If there are victims at two different windows, the aerial ladder is first placed to remove the victims who are the most seriously endangered by the fire. The ladder is then placed to remove the victims who are less seriously exposed to the fire.

73. Assume that you are working at a fire and that there are a total of three victims at two windows. Victims 1 and 2 are at the same window, which is three floors above the fire and shows no evidence of heat or smoke. Victim 1 is a disabled, 23-year-old male and Victim 2 is a 40-year-old woman. Victim 3, a 16-year-old male, is at a window of the apartment on fire. From your position in the street, you can see heavy smoke coming from this window and flames coming out of the window next to it. Which one of the following is the proper order for victim removal?

   (A) Victim 3, Victim 2, Victim 1.
   (B) Victim 1, Victim 2, Victim 3.
   (C) Victim 1, Victim 3, Victim 2.
   (D) Victim 3, Victim 1, Victim 2.

DIRECTIONS: Answer questions 74 and 75 based on the following passage.

The four different types of building collapses are as follows:

1. *Building Wall Collapse*—An outside wall of the building collapses but the floors maintain their positions.

2. *Lean-to Collapse*—One end of the floor collapses onto the floor below it. This leaves a sheltered area on the floor below.

3. *Floor Collapse*—An entire floor falls to the floor below it but large pieces of machinery in the floor below provide spaces that can provide shelter.

4. *Pancake Collapse*—A floor collapses completely onto the floor below it, leaving no spaces. In some cases, the force of this collapse causes successive lower floors to collapse.

74. The most serious injuries are likely to occur at

   (A) pancake collapses
   (B) lean-to collapses
   (C) floor collapses
   (D) building wall collapses

75. Of the following, a floor collapse is most likely to occur in a

   (A) factory building
   (B) private home
   (C) apartment building
   (D) hotel

DIRECTIONS: Answer question 76 based solely on the following passage.

Many subway tunnels contain a set of three rails used for train movement. The subway trains run on two rails. The third rail carries electricity and is the source of power for all trains. Electricity travels from the third rail through metal plates called contact shoes which are located near the wheels on every train car. Electricity then travels through the contact shoes into the train's motor. Firefighters must be very careful when operating near the third rail because contact with the third rail can result in electrocution.

76. The source of power for subway trains is the

(A) third rail
(B) contact shoes

(C) motor
(D) metal plates

DIRECTIONS: Answer questions 77 through 81 based solely on the following passage.

Firefighters receive an alarm for an apartment fire on the fourth floor of a 14-story housing project at 1191 Park Place. One firefighter shouts the address as the other firefighters are getting on the fire truck. Knowledge of the address helps the firefighters decide which equipment to pull off the fire truck when they reach the fire scene.

The firefighters know where the water outlets are located in the building on fire. There is an outlet in every hallway. Firefighters always attach the hose at the closest outlet on the floor below the fire.

As they arrive at 1191 Park Place, three firefighters immediately take one length of hose each and go into the building. Since an officer has been told by the dispatcher that two children are trapped in the rear bedroom, the officer and two firefighters begin searching for victims and opening windows immediately upon entering the apartment on fire.

As in all housing project fires, the roof person goes to the apartment above the apartment of fire. From this position, he attaches a tool to a rope in order to break open the windows of the apartment on fire. From this position, the roof person could also make a rope rescue of a victim in the apartment on fire.

77. A firefighter shouted the address of the fire when the alarm was received so that the firefighters would

(A) know which equipment to take from the truck at the fire scene
(B) be more alert when they arrived at the fire
(C) be prepared to make a rope rescue
(D) know that two children were reported to be trapped

78. The hose should be attached to an outlet on the

(A) floor above the fire
(B) ground floor

(C) fire floor
(D) floor below the fire

79. Because of the information given to the officer by the dispatcher, the officer and two firefighters

    (A) entered the apartment above the fire for a rope rescue
    (B) began immediately to search for victims and to open windows
    (C) opened all the windows before the hose was moved in
    (D) attached a hose and moved to the origin of the fire

80. The roof person broke the windows of the apartment on fire with

    (A) an axe while leaning out of the windows
    (B) an axe attached to the end of a rope
    (C) a tool while standing on the roof
    (D) a tool attached to a rope

81. The proper location for a rescue by the roof person at a fire in a housing project is the

    (A) hallway
    (B) apartment below the apartment on fire
    (C) apartment above the apartment on fire
    (D) fire escape

DIRECTIONS: Answer question 82 based on the following passage.

You are a firefighter who is inspecting a building for violations. You must perform the following steps in the order given:

1. Find the manager of the building and introduce yourself.

2. Have the manager accompany you during the inspection.

3. Start the inspection by checking the Fire Department permits which have been issued to the building. The permits are located in the office of the building.

4. Inspect the building for violations of the Fire Prevention Laws. Begin at the roof and work down to the basement or cellar.

5. As you inspect, write on a piece of paper any violations you find and explain them to the building manager.

82. You are inspecting a supermarket. After entering the building, you identify yourself to the store manager and ask him to come along during the inspection. Which one of the following actions should you take next?

    (A) Start inspecting the supermarket, beginning at the basement.
    (B) Start inspecting the supermarket, beginning at the roof.
    (C) Ask to see the Fire Department permits which have been issued to the supermarket.
    (D) Write down any violations that you notice while introducing yourself to the manager.

DIRECTIONS: Answer questions 83 and 84 based on the following passage.

Engine company firefighters are responsible for putting water on a fire and extinguishing it. To do this properly, they should perform their tasks in the following order:

1. Once an apartment door has been forced open, the engine company officer orders the driver to start the flow of water through the hose.

2. As the water starts to flow into the hose, it pushes trapped air ahead of it. To clear this air from the hose, the nozzle is pointed away from the fire area, opened, and then closed before water starts to flow from it. This is done to prevent a rush of fresh air from the hose which will intensify the fire.

3. When the fire is found, the nozzle is directed at the ceiling to allow water to rain down on the fire. As the fire becomes smaller, the nozzle is aimed directly at the burning object.

83. When engine company firefighters enter an apartment where there is a fire, the occupant takes the firefighters to a fire in the bedroom. Once air has been cleared from the hose, the firefighter operating the nozzle should

(A) wait for the ladder company to open the door
(B) aim water directly onto the bed
(C) spray the water across the floor
(D) direct the water at the ceiling and allow it to run down on the fire

84. After the locked door at an apartment fire has been forced open and water starts to flow into the hose, the firefighter operating the nozzle sees an intense fire just inside the apartment doorway. She should then point the nozzle

(A) away from the fire
(B) at any burning object
(C) directly at the fire
(D) at the apartment floor

DIRECTIONS: Answer question 85 based on the following passage.

When a leak occurs in an oil pipeline, the Fire Department dispatcher sends a message for specific companies to respond to the leak. There is a letter code on the message indicating the possible location of the leak. After a firefighter reads the letter code on the message, the following steps are performed in the order given to properly close the valves on the pipeline.

1. The firefighter looks up the alarm assignment card which gives the location of the valve.

2. The firefighter makes sure that a short wrench and a long wrench are placed on the fire truck.

3. Upon arrival at the location of the valve, the firefighter removes the steel valve cover with the short wrench.

4. The long wrench is then used to completely close the valve.

85. A firefighter receives a message from the dispatcher about a leak in an oil pipeline. After noting the letter code on the message, the firefighter should

(A) give the alarm card immediately to the officer in charge
(B) make sure the proper wrenches for closing the valve have been placed on the fire apparatus
(C) look up the alarm assignment card to determine the location of the valve
(D) remove the steel valve cover

DIRECTIONS: Answer question 86 based on the following passage.

When administering first aid to a person who is severely bleeding, firefighters should perform the following steps in the order given:

1. Direct Pressure

   (a) Press a dressing directly on the wound.

2. Elevation

   a. Lift the injured body part above the heart level, while continuing direct pressure. This will slow down the movement of blood to the wound.

   b. Never elevate a body part if it might be fractured.

   c. If the bleeding continues while the injured body part is elevated and is receiving direct pressure, pressure should be applied to an artery or other pressure point.

86. A firefighter arrives at an automobile accident and finds a woman who is bleeding severely from a cut just above the ankle. She also has a fracture of the upper arm. After placing a dressing on the ankle wound, the firefighter should next

(A) apply pressure to a pressure point to stop the bleeding
(B) elevate the upper arm and apply direct pressure to it
(C) apply direct pressure to an artery in the arm
(D) apply direct pressure to the leg while elevating it

DIRECTIONS: Answer question 87 based solely on the following passage.

One firefighter of the first company to arrive at a fire in a private house is assigned to the roof position. During the beginning stages of a fire, the roof person is part of a team which enters and searches the building for victims.

The roof person's duties are performed in the following order:

1. Climb a portable ladder to the front porch roof, break a window, and enter the building through this opening if there is no victim in immediate danger visible at another window. If there is such a victim, place the ladder at that window.

2. If there is no front porch, use the portable extension ladder at a side of the house.

3. Enter the house after breaking a window.

87. The roof person at a fire in a private house sees a boy at a top floor window on the right side of the house. There is a porch in the front of the house. The roof person should now place the portable ladder at the

(A) front of the porch
(B) top floor front window
(C) top floor window, right side of the house
(D) top floor window, left side of the house

DIRECTIONS: Answer question 88 based on the following passage.

The following is a description of the actions taken by the outside vent person (O.V.) at different types of fires:

Fire 1: At a fire at 3:00 A.M. on the third floor of a five-story apartment building, the O.V. climbed the fire escape to break the windows in the fire area.

Fire 2: At a fire at 1:00 P.M. on the fourth floor of a four-story apartment building, the O.V. went to the roof to assist the firefighter assigned to the roof position.

Fire 3: At a fire at 2:00 P.M. on the third floor of a three-story factory building, the O.V. went to the roof to assist the firefighter assigned to the roof position.

Fire 4: At a fire at 6:00 A.M. on the first floor of a two-story clothing store, the O.V. broke the first-floor windows from the outside of the building.

Fire 5: At a fire at 1:00 A.M. on the fifth floor of a five-story apartment building, the O.V. went to the roof to assist the firefighter assigned to the roof position.

88. A firefighter should conclude that the O.V. assists the firefighter assigned to the roof position when the fire

(A) is in an apartment building
(B) occurs at night
(C) is located on the top floor of the building
(D) is in a commercial building

DIRECTIONS: Answer question 89 based on the following passage.

Firefighters perform both vertical and horizontal ventilation. In vertical ventilation, an opening is made on the upper levels of the fire building so that the natural air currents can assist in the removal of smoke and heated gases. In horizontal ventilation, windows are opened on the same level as the fire to allow fresh air into the fire area.

A ladder company recently responded to the following fires at different six-story apartment buildings and performed the actions indicated:

Fire 1: At a fire in a mattress in a second-floor apartment, firefighters immediately opened the roof door. After water began flowing through the hose, firefighters opened the windows in the apartment. The fire was quickly extinguished.

Fire 2: At a fire in the top-floor bedroom, firefighters cut a hole in the roof as soon as possible and then used the water from the hose to shatter the windows. The fire was quickly extinguished.

Fire 3: At an apartment fire on the second floor, firefighters broke the windows in the apartment as the hose was being brought from the street. The fire spread before water could be applied to the fire.

89. Based upon the types of ventilation performed at the three fires, which one of the following statements is correct?

(A) Vertical ventilation must be performed by opening the roof door.
(B) Horizontal ventilation should be delayed until firefighters are ready to apply water to the fire.
(C) Horizontal ventilation must be performed before vertical ventilation.
(D) Vertical ventilation must be delayed until water is applied to the fire.

DIRECTIONS: Answer questions 90 and 91 based on the following passage.

A newly appointed firefighter is studying the proper use of foam, water, or dry chemicals to extinguish a fire. The firefighter looks over past fire reports to see whether any patterns exist.

Fire 1: A gasoline fire near a car was extinguished by foam.

Fire 2: A fire in a television set with a disconnected power cord was extinguished by water.

Fire 3: A fire in a fuse box of a private home was extinguished by dry chemicals.

Fire 4: An oil fire near an oil burner in a private home was extinguished by foam.

Fire 5: A fire near the electrical rail at a subway was extinguished by dry chemicals.

Fire 6: A fire involving an electric range was extinguished by dry chemicals.

Fire 7: A fire in the front seat of an automobile was extinguished by water.

90. The firefighter should conclude that dry chemicals are used to extinguish fires which involve

(A) automobiles   (C) oil and gasoline
(B) private homes   (D) live electrical equipment

91. The firefighter should conclude that foam is used to extinguish

(A) car fires   (C) stove fires
(B) oil and gasoline fires   (D) electrical fires

92. In situations where there is no fire, firefighters must make immediate rescue attempts when they come upon persons in danger. Which one of the following persons to be rescued is in the greatest danger?

    (A) A sleeping baby is inside a car that has a leaking gasoline tank.
    (B) A woman is inside an elevator stuck on the thirty-third floor.
    (C) A teenage boy is drifting on a lake in a boat.
    (D) A woman confined to a wheel chair is locked in her apartment.

93. Firefighters are inspecting a furniture factory. During the inspection, they find employees smoking cigarettes in various areas. In which area does smoking pose the greatest danger of causing a fire?

    (A) employee lounge
    (B) woodworking shop
    (C) a private office
    (D) a rest room

94. Firefighters should observe the sidewalk and street area in front of the firehouse and inform the officer of any conditions that may cause a significant delay when fire trucks must go out in response to an alarm. Which one of the following conditions should the firefighters report to the officer?

    (A) A car is stopped in front of the firehouse for a red light.
    (B) People are standing and talking in front of the firehouse.
    (C) A truck is parked in front of the firehouse.
    (D) There is a small crack in the sidewalk in front of the firehouse.

95. Which of the following conditions observed by a firefighter inspecting an automobile repair garage is the most dangerous fire hazard?

    (A) A mechanic is repairing a hole in a half-filled gas tank while smoking a cigarette.
    (B) A heater is being used to heat the garage.
    (C) A mechanic is changing a tire on a van while smoking a pipe.
    (D) There is no fire extinguishing equipment in the garage.

96. During a fire prevention inspection, a firefighter may find a condition which could be the immediate cause of death in the event of a fire. Which one of the following conditions in a restaurant is the most dangerous?

    (A) blocked exit doors
    (B) a crack in the front door
    (C) a window that does not open
    (D) a broken air conditioning system

97. It is very important to get fresh air into a closed area that is filled with gas or smoke. In which one of the following situations should such an action be taken *first*?

    (A) a smell of gas coming through an open apartment window
    (B) an unoccupied car with the motor running
    (C) a strong odor coming from a closed refrigerator
    (D) a gas leak in the closed basement of an occupied building

98. A firefighter who is assigned to the roof position at a fire must notify the officer of dangerous conditions that can be seen from the roof. Which one of the following conditions is the most dangerous?

    (A) The roof is sagging and may collapse because of the fire on the top floor.
    (B) Rubbish is visible on the roof of the building next door.

(C) The stairway to the roof of the building has poor lighting.
(D) An automobile accident in the street is causing a traffic jam.

99. Firefighters must often deal with people who need medical assistance. In life-threatening situations, firefighters must perform first aid until an ambulance arrives. In less serious situations, firefighters should make the person comfortable and wait for the ambulance personnel to give first aid. A firefighter should give first aid until an ambulance arrives when a person

   (A) appears to have a knee injury
   (B) is bleeding heavily from a stomach wound
   (C) has bruises on his head
   (D) has a broken ankle

100. At a fire on the west side of the third floor of a ten-story office building, a firefighter is responsible for rescuing trapped persons by means of the aerial ladder. The trapped person who is in the most dangerous location should be removed first. Of the following, the firefighter should *first* remove the trapped person who is

   (A) on the rear fire escape on the east side of the second floor
   (B) on the roof
   (C) at a window on the west side of the fourth floor
   (D) at a window on the tenth floor

Answer Key for Practice Examination 1

Proposed Final Key Answers for Written Test Held on December 12, 1987.

1. **B**	26. **A**	51. **B**	76. **A**
2. **B**	27. **D**	52. **C**	77. **A**
3. **D**	28. **C**	53. **A**	78. **D**
4. **B**	29. **C**	54. **B**	79. **B**
5. **B**	30. **D**	55. **C**	80. **D**
6. **D**	31. **D**	56. **A**	81. **C**
7. **A**	32. **B**	57. **B**	82. **C**
8. **A**	33. **C**	58. **B**	83. **D**
9. **D**	34. **C**	59. **A**	84. **A**
10. **B**	35. **B**	60. **A**	85. **C**
11. **D**	36. **C**	61. **D**	86. **D**
12. **B**	37. **C**	62. **A**	87. **Delete**
13. **B**	38. **C**	63. **B**	88. **C**
14. **D**	39. **D**	64. **A**	89. **B**
15. **A**	40. **B**	65. **D**	90. **D**
16. **D**	41. **D**	66. **D**	91. **B**
17. **D**	42. **C**	67. **B**	92. **A**
18. **A**	43. **B**	68. **C**	93. **B**
19. **D**	44. **A**	69. **C**	94. **C**
20. **A**	45. **A**	70. **C**	95. **A**
21. **B**	46. **C**	71. **B**	96. **A**
22. **B**	47. **C**	72. **B**	97. **D**
23. **C**	48. **D**	73. **A**	98. **A**
24. **D**	49. **C**	74. **A**	99. **B**
25. **A**	50. **A**	75. **A** and/or **B** and/or **C** and/or **D**	100. **C**

Explanatory Answers
for Practice Examination 1

1. **(B)** The note at the bottom of the picture states that the ground floor is the first floor. Count up from the sidewalk and find the fire coming out the fourth-floor windows.

2. **(B)** As you stand in front, facing the buildings, the bank is on the left side of the page. The smoke and flames are blowing from the left to the right and up.

3. **(D)** The fire escapes are on the left side of the apartment building and are indicated by the words "fire escape." Count up. The person is on the fire escape at the fifth floor.

4. **(B)** The ground floor is the first floor. Count up again.

5. **(B)** The front window, fifth floor is directly over the fire and closest to the fire.

6. **(D)** The building on fire is an "isolated" building. This means that there is no direct access to the roof from the roofs of the adjacent buildings.

7. **(A)** As you stand facing the front of the buildings, the fire escapes are on your left-hand side. This is considered the left side of the building for purposes of this exam.

8. **(A)** The fire hydrant is directly in front of the bank.

9. **(D)** The fire escape is on the lower right-hand corner of the diagram. The dining room is on the upper left corner. The dining room is the greatest distance from the fire escape.

10. **(B)** Bedroom 1 has only one door. Of the living room, dining room, and kitchen, each has at least two door openings.

11. **(D)** As indicated on the diagram, the fire escape is directly outside bedroom 3.

12. **(B)** The only door from bathroom 2 leads directly to bedroom 2.

13. **(B)** This is indicated by the door symbol between bathroom 2 and bedroom 2.

14. **(D)** The terrace on the left side of the building has two door symbols to indicate that access from the terrace can be gained to the living room and to the dining room.

15. **(A)** The window symbols are visible in the black outline of the floor-plan. This shows that there are two windows in the dining room and three windows in bedroom 2.

16. **(D)** Doorways are shown by the break in the solid lines within the floor plan. The kitchen can be entered from the doorway in the dining room as well as through the doorway from the foyer.

17. **(D)** During the hours of 2:00 A.M. to 6:00 A.M., most people are at home and are sleeping. At the other hours, they are more likely to be out of the house or at least awake and aware of the fire so that they can escape in time on their own.

18. **(A)** Leaking gas can be ignited, causing a fire. If a large amount of gas collects in the basement and is ignited, an explosion and fire are likely.

19. **(D)** Portable ladders used in the fire service are usually metal ladders. If a ladder is placed too close to electrical wires, a person on the ladder or the ladder itself may touch the live wires, causing serious burns or even electrocution to those involved.

20. **(A)** The person at the smoke-filled window and the fire are on the same floor. This person is in great danger. The other people are a greater distance from the fire and are not as directly exposed to it. Upon notification, the officer will be aware of the location of his firefighter in the event that conditions worsen and the rescuer needs assistance.

21. **(B)** A secured chain lock on the inside of the door generally indicates that the occupant is in the apartment; most are designed to be locked and unlocked from the inside only.

22. **(B)** As stated in the third sentence of the second paragraph, removal of power from the third rail is the FIRST step firefighters must take when evacuating people through

tunnels. By removing the power, the chances of people's being electrocuted are removed. This also stops the trains from moving so that people walking on the tracks are not hit by trains.

23. **(C)** This is done to inform the Train Master of the type of emergency and to insure that he keeps the power turned off.

24. **(D)** The five white lights in a vertical cluster can be easily seen in the darkened tunnel and can be recognized as an exit from a distance.

25. **(A)** The fire cannot be in a factory because all factories are located between 9th Avenue and 12th Avenue. Private houses are all between one and three stories tall.

26. **(A)** There are both stores and private houses on 47th Street between 6th Avenue and 7th Avenue, but the private houses are at most three stories high so the fire must be in a one- or two-story store.

27. **(D)** At alarms 1, 3, and 5, all fires below the third floor of the building, Firefighter Sims, in the control position, assisted the driver in attaching the hose to the hydrant. At the fires on the third floor or above, he helped in getting the hoseline into the building. The control man is also responsible for making certain that enough hose has been taken to reach the fire.

28. **(C)** The compass directional on the upper left corner of the diagram indicates that travel from 7th Avenue to 6th Avenue is travel in the east–west direction. Corridor D is the east–west hallway in the center of the building. Rooms 107, 109, 125, 126, 127, and 128 are in corridor D.

29. **(C)** The firefighters enter the building on 40th Street which is the south side of the building. Upon entering the lobby, they would have to turn east in hallway E; north in Hallway B, and west in hallway C.

30. **(D)** From the diagram, we are able to see that Truck 21 is facing north on 7th Avenue. The nearest hose connection is E, about midway between 39th and 40th Streets.

31. **(D)** The description of row frames and brownstones indicates that either can be three stories high and 25 feet wide. The key is in the fire escapes clauses. Fire escapes are required on row frame buildings higher than two stories. Rear fire escapes are required in changed brownstones. This cannot be an old law tenement because it is less than five stories high.

32. **(B)** As stated in the description of old law tenements, fire escapes are required on the rear of the building if there are two apartments on a floor and fire escapes are required on both front and rear if there are four apartments on a floor. This is done to provide two means of escape from every apartment. Row frames and brownstones do not have front fire escapes.

33. **(C)** If the building is 30 feet wide, it must be a row frame. A row frame of more than two stories must have a rear fire escape.

34. **(C)** This is stated in example 2 of the May Day messages. Depending on the extent of the collapse, it is extremely important that an immediate roll call be taken to make sure that all firefighters have escaped injury and that none are trapped.

35. **(B)** This is explained in example 3 of the May Day messages.

36. **(C)** The probable collapse at this point is more serious than the broken wrist. "May Day" should be used instead of "Urgent, Urgent."

37. **(C)** Stairway B is the dark-colored stairway. It enters the fourth floor on the east side. With the fifth floor outlet not usable, this is the nearest usable outlet below the fire.

38. **(C)** By using stairway A on the east side of the seventh floor, he would come out on the eighth floor west.

39. **(D)** If the front entrance of the building faces south and the front of the building is Exposure 1, then, in clockwise order, Exposure 2 would face west, Exposure 3 north, and Exposure 4 east.

40. **(B)** This is stated in the first sentence of the second paragraph. As the products of combustion become heated, they rise and accumulate at the top of a closed structure. When the roof is opened or ventilated, the heated smoke and gasses rush out.

41. **(D)** As explained in paragraph five, by attacking the fire directly in the entrance of the building, you can prevent its extension into the building. The fire in the vacant lot must be extinguished rapidly to prevent its spreading to houses on either side.

42. **(C)** The answer is in paragraph six. Firefighters in ladder companies carry tools which enable them to open doors, windows, roofs, etc. Firefighters in engine companies stretch hoselines to extinguish the fire.

43. **(B)** This is clearly stated in the last sentence of the fourth paragraph. Positioning of hoselines is critical in, for example:

    1. *An Apartment House*—Such a fire must be attacked from the public hall so as to contain the fire in the individual apartment. The fire must be kept from spreading into the public hall.

    2. *An Automobile Fire*—The multiple considerations here are to position the line between victims and gas tanks, to get hoseline and water between the fire and the people or property you are attempting to save, and to position the hose between the car and the exposed buildings nearby.

44. **(A)** See the second sentence. A fire area is generally filled with smoke and hot gasses, with practically no visibility at all. The firefighter must try to coordinate ventilation of the area with the arrival of water at the nozzle so that, as the area is vented, the smoke can rise and can give firefighters better opportunity to locate the victims and the fire itself.

45. **(A)** Hydrants 3 and 4 are both considerably closer to the fire than hydrants 1 and 7. Hydrant 4 cannot be reached without passing through smoke and fire. Hydrant 3 is in the best position for protecting the warehouse from the fire.

46. **(C)** Try each route and see for yourself. This route will also get them out of the complex and away from danger.

47. **(C)** This is indicated on the diagram for the question.

48. **(D)** Of the choices offered, only hydrants 6 and 7 are outside the fence. Hydrant 7 is closer to the fire.

49. **(C)** Hi-Lo Lane is engulfed in smoke and flame. The route in choice (C) is shorter than that in choice (D).

50. **(A)** Try each route. This is the shortest way.

51. **(B)** If the door is unlocked, it is the simplest and preferred way to enter an apartment. Sometimes, in the anxious moments of arrival at a fire, the firefighter presumes the door to be locked and immediately starts forcible entry, a mistake in procedure.

52. **(C)** Since the aerial ladder is in use, the firefighter must go by step 1b.

53. **(A)** This is proper procedure as stated in step 3b and step 4. Opening the skylight will immediately ventilate the interior stairs, causing pent-up smoke and gasses to escape, thereby making it easier for firefighting interior forces to operate.

54. **(B)** This is proper procedure as stated in step 2. By doing this, you generally will relieve any fears the people inside may have and may even secure their assistance from their position inside.

55. **(C)** In general, a broken arm is not a life-threatening injury. Rather than trying to remove the person by any means possible, firefighters should adhere to the instruction at step 5 and send for an elevator mechanic who is better equipped to extricate the victim. They should, however, maintain verbal contact with the operator to be sure he does not go into shock. If the injury situation appears to worsen, then they must move directly to step 6b and take measures before the elevator mechanic arrives.

56. **(A)** This is the second choice. Victims to be rescued are generally in a very nervous state. Use of the fire escape would be more familiar to them and would be less stressful than climbing over the railing and then down a portable ladder.

57. **(B)** Room 304 is on the south side of the building. Room 306 is the nearest room to the west of Room 304.

58. **(B)** Stairway B as shown on the diagram is the closest. The stairway used by the firefighter to gain access to the third floor was Stairway A.

59. **(A)** Room 303 is directly across the hallway from Room 304. Room 403 is directly above Room 303.

60. **(A)** By checking the compass points in the center of the diagram, we know that Rooms 303 and 305 are on the south side of the courtyard. Rooms 323 and 325 are on the opposite side of the courtyard, which is the north side of the building.

61. **(D)** Stairway D is the correct answer. The northeast stair is Stairway C. By coming up Stairway D, both hoselines can operate from the same direction to extinguish the fire.

62. **(A)** Diagram A is correct because the rotating base is directly in front of the victim at the window. Ladders are generally positioned with both beams on the sill to prevent twisting of the ladder under the weight of people.

63. **(B)** This is stated in Rule 2. Doors are locked on the stairway side to force people to exit at the ground level in case of fire. In buildings over 100 feet in height, people can seek refuge several floors below the fire floor. Doors on street level are locked from the street side for security purposes.

64. **(A)** The correct answer is stated in Rules 3 and 4. The re-entry signs are posted in order to safely evacuate occupants to floors below the fire without having occupants walk down many flights of stairs in very tall buildings.

65. **(D)** The answer is directly stated in the third sentence. First removing the victim to street level would only delay the administration of oxygen.

66. **(D)** The question asks about transmittal of alarms, not about receipt. The fourth and fifth sentences give the answer.

67. **(B)** See the first sentence of the first paragraph. The firefighter on housewatch has other duties, such as greeting visitors, answering the telephone, and maintaining security, but his or her primary responsibility is to properly receive alarm information.

68. **(C)** Read carefully. The answer is in the next to last sentence.

69. **(C)** See the third sentence.

70. **(C)** See Rules 1 and 3. Because the home is 3000 square feet in area, one 2½-gallon water extinguisher is required for the first 2500 square feet and another 2½-gallon water extinguisher for the additional 500 square feet. The kitchen needs one foam extinguisher for the first 1250 square feet and another foam extinguisher for the additional 250 square feet.

71. **(B)** The first floor (2600 square feet) requires one water extinguisher for the first 2500 square feet and another water extinguisher for the additional 100 square feet. The second floor is smaller than 1500 square feet so requires only one 2½-gallon water extinguisher.

72. **(B)** The aerial ladder is at the front of the house and the portable ladder at the left. The second company must now ladder right and rear. Ladder company personnel are assigned specific areas to search in order that no area of the building is left unsearched.

73. **(A)** Victim 3 is removed first because he is the most seriously exposed to the fire. Victim 2 is easier to remove because she is not disabled. Victim 1 is the last one removed because this removal will be more difficult and will take more time due to the disability.

74. **(A)** A pancake collapse leaves little or no space beneath that is capable of providing refuge for trapped victims.

75. Upon realizing that the wording of this question might lead to different interpretations of its meaning, the Department of Personnel agreed to accept any answer. By the reading probably originally intended, the answer should be (A). The definition of a floor collapse requires that there be large pieces of machinery on the floor below that hold up sections of floor and create spaces that provide shelter. Large pieces of machinery on any or every floor are most likely to be found in a factory building. In other

buildings, large pieces of machinery are more likely to be found only in the basement. This limits the number of floor collapses possible.

76. **(A)** The third rail carries the electricity that provides power. The contact shoes and metal plates are one and the same. The shoes or plates carry electricity to the motor.

77. **(A)** The answer is in the third sentence.

78. **(D)** See the second paragraph. The reason is that if the hoseline is taken from the outlet on the fire floor and the fire gains control of the floor and hall, there is the possibility that the line will have to be abandoned. Using the outlet below the fire enables firefighters to work in a smoke-free atmosphere. Should the fire get out into the hallway, they can still maintain control of the water supply.

79. **(B)** Search and rescue, especially with knowledge of trapped victims as in this case, must be the first order of business.

80. **(D)** See the last paragraph. The tool attached to the rope is likely to be an axe, but this is not specified in the passage. The tool attached to the rope gives the firefighter greater reach when breaking windows from above.

81. **(C)** Almost by definition, the roof person would effect a rescue from above the fire. Fourteen-story buildings do not have fire escapes; therefore, rescue from above, except on the highest floors where it might be made from the roof, must be made from an apartment window above the fire.

82. **(C)** The answer is stated in step 3. Checking the permits first makes you aware of:

1. Whether or not the permits are up to date.

2. The areas in the building that will require special attention, such as the fuel oil storage, refrigeration equipment, paint storage.
The permit will note such information as: the amount to be stored; the number and type of refrigerants being used; the grade and amount of fuel oil being used and stored.

83. **(D)** This is stated in item 3. When the stream of water from the nozzle strikes the ceiling, the stream is broken into smaller droplets creating a spraying effect which covers a larger area.

84. **(A)** The firefighter must follow directions given in item 2 and point the nozzle away from the fire as air is being pushed out of the hose.

85. **(C)** See step 1. The alarm assignment card generally will also have a map of the pipeline to show location of that valve and adjacent valves. The card may also tell what Fire Department units are responding to this location and specific duties and assignments of the units responding.

86. **(D)** See step 2a.

87. This question was deleted following candidates' appeal.

88. **(C)** At fires 2, 3, and 5, the outside vent (O.V.) person went to the roof. In each of these instances, the fire was on the top floor so the vent man and the roof man vented

windows on the fire floor by leaning over from the roof. A fire on the ground floor can be vented from the ground. A fire on a floor other than ground or top is vented from the fire escape.

89. **(B)** At fires 1 and 2, the firefighters first performed vertical ventilation, then delayed horizontal ventilation until water was available. At the third fire, the fire spread when horizontal ventilation was performed before water was available.

90. **(D)** Dry chemical is used to extinguish live electrical fires because, unlike water, it does not have a continuous chain of molecules that can conduct electricity back to the extinguisher operator.

91. **(B)** Firefighting foam is composed of gas- or air-filled bubbles that are capable of floating on oil and gasoline. Foam extinguishes an oil or gasoline fire by blanketing and cooling the fuel and by separating the fuel from its oxygen supply.

92. **(A)** A leaking gasoline tank is an extremely dangerous condition. A source of ignition, even at a great distance from the actual leak, can cause the flame to travel back to the tank setting it on fire or causing an explosion.

93. **(B)** Woodworking areas present the greatest danger of fire due to the large accumulation of sawdust, woodshavings, flammable lacquers, and lacquer thinners that may easily be ignited by a cigarette or discarded match.

94. **(C)** A truck or car parked in front of the firehouse could seriously delay the response of the firefighters to a fire.

95. **(A)** A partially filled gasoline tank with a hole is extremely dangerous because an explosive gas–air mixture is present in the vapor space of the tank. The fumes coming from the tank can be ignited by the cigarette, travel back into the tank and cause a disasterous explosion.

96. **(A)** Blocked exit doors in any establishment where people work, live, play, or congregate create an extremely serious condition. A fire or emergency requiring the immediate evacuation of the area could result in people being trapped, trampled, and killed while trying to escape.

97. **(D)** A gas leak in a closed basement could result in an explosion. The closed basement would allow the gas to accumulate until it reached a gas–air mixture that would explode if a source of ignition were present. That source of ignition might be the furnace being fired or an electrical spark.

98. **(A)** A firefighter assigned to the roof position who sees the roof sagging and in danger of collapse should immediately give a "May Day" signal. This would alert all firefighters operating on the top floor to evacuate the collapse area at once.

99. **(B)** A person bleeding heavily from a stomach wound is in serious condition. If the bleeding is not halted, the victim can go into shock and die from loss of blood.

100. **(C)** At this fire, the occupant of the fourth floor should be rescued first. The floor directly above the fire is the most seriously exposed floor. Dangerous products of combustion, mainly carbon monoxide, rise rapidly through convection to fill the area, excluding oxygen and causing death or serious injury to trapped people.

Answer Sheet for Practice Examination 2

1. Ⓐ Ⓑ Ⓒ Ⓓ
2. Ⓐ Ⓑ Ⓒ Ⓓ
3. Ⓐ Ⓑ Ⓒ Ⓓ
4. Ⓐ Ⓑ Ⓒ Ⓓ
5. Ⓐ Ⓑ Ⓒ Ⓓ
6. Ⓐ Ⓑ Ⓒ Ⓓ
7. Ⓐ Ⓑ Ⓒ Ⓓ
8. Ⓐ Ⓑ Ⓒ Ⓓ
9. Ⓐ Ⓑ Ⓒ Ⓓ
10. Ⓐ Ⓑ Ⓒ Ⓓ
11. Ⓐ Ⓑ Ⓒ Ⓓ
12. Ⓐ Ⓑ Ⓒ Ⓓ
13. Ⓐ Ⓑ Ⓒ Ⓓ
14. Ⓐ Ⓑ Ⓒ Ⓓ
15. Ⓐ Ⓑ Ⓒ Ⓓ
16. Ⓐ Ⓑ Ⓒ Ⓓ
17. Ⓐ Ⓑ Ⓒ Ⓓ
18. Ⓐ Ⓑ Ⓒ Ⓓ
19. Ⓐ Ⓑ Ⓒ Ⓓ
20. Ⓐ Ⓑ Ⓒ Ⓓ

21. Ⓐ Ⓑ Ⓒ Ⓓ
22. Ⓐ Ⓑ Ⓒ Ⓓ
23. Ⓐ Ⓑ Ⓒ Ⓓ
24. Ⓐ Ⓑ Ⓒ Ⓓ
25. Ⓐ Ⓑ Ⓒ Ⓓ
26. Ⓐ Ⓑ Ⓒ Ⓓ
27. Ⓐ Ⓑ Ⓒ Ⓓ
28. Ⓐ Ⓑ Ⓒ Ⓓ
29. Ⓐ Ⓑ Ⓒ Ⓓ
30. Ⓐ Ⓑ Ⓒ Ⓓ
31. Ⓐ Ⓑ Ⓒ Ⓓ
32. Ⓐ Ⓑ Ⓒ Ⓓ
33. Ⓐ Ⓑ Ⓒ Ⓓ
34. Ⓐ Ⓑ Ⓒ Ⓓ
35. Ⓐ Ⓑ Ⓒ Ⓓ
36. Ⓐ Ⓑ Ⓒ Ⓓ
37. Ⓐ Ⓑ Ⓒ Ⓓ
38. Ⓐ Ⓑ Ⓒ Ⓓ
39. Ⓐ Ⓑ Ⓒ Ⓓ
40. Ⓐ Ⓑ Ⓒ Ⓓ

41. Ⓐ Ⓑ Ⓒ Ⓓ
42. Ⓐ Ⓑ Ⓒ Ⓓ
43. Ⓐ Ⓑ Ⓒ Ⓓ
44. Ⓐ Ⓑ Ⓒ Ⓓ
45. Ⓐ Ⓑ Ⓒ Ⓓ
46. Ⓐ Ⓑ Ⓒ Ⓓ
47. Ⓐ Ⓑ Ⓒ Ⓓ
48. Ⓐ Ⓑ Ⓒ Ⓓ
49. Ⓐ Ⓑ Ⓒ Ⓓ
50. Ⓐ Ⓑ Ⓒ Ⓓ
51. Ⓐ Ⓑ Ⓒ Ⓓ
52. Ⓐ Ⓑ Ⓒ Ⓓ
53. Ⓐ Ⓑ Ⓒ Ⓓ
54. Ⓐ Ⓑ Ⓒ Ⓓ
55. Ⓐ Ⓑ Ⓒ Ⓓ
56. Ⓐ Ⓑ Ⓒ Ⓓ
57. Ⓐ Ⓑ Ⓒ Ⓓ
58. Ⓐ Ⓑ Ⓒ Ⓓ
59. Ⓐ Ⓑ Ⓒ Ⓓ
60. Ⓐ Ⓑ Ⓒ Ⓓ

61. Ⓐ Ⓑ Ⓒ Ⓓ
62. Ⓐ Ⓑ Ⓒ Ⓓ
63. Ⓐ Ⓑ Ⓒ Ⓓ
64. Ⓐ Ⓑ Ⓒ Ⓓ
65. Ⓐ Ⓑ Ⓒ Ⓓ
66. Ⓐ Ⓑ Ⓒ Ⓓ
67. Ⓐ Ⓑ Ⓒ Ⓓ
68. Ⓐ Ⓑ Ⓒ Ⓓ
69. Ⓐ Ⓑ Ⓒ Ⓓ
70. Ⓐ Ⓑ Ⓒ Ⓓ
71. Ⓐ Ⓑ Ⓒ Ⓓ
72. Ⓐ Ⓑ Ⓒ Ⓓ
73. Ⓐ Ⓑ Ⓒ Ⓓ
74. Ⓐ Ⓑ Ⓒ Ⓓ
75. Ⓐ Ⓑ Ⓒ Ⓓ
76. Ⓐ Ⓑ Ⓒ Ⓓ
77. Ⓐ Ⓑ Ⓒ Ⓓ
78. Ⓐ Ⓑ Ⓒ Ⓓ
79. Ⓐ Ⓑ Ⓒ Ⓓ
80. Ⓐ Ⓑ Ⓒ Ⓓ

81. Ⓐ Ⓑ Ⓒ Ⓓ
82. Ⓐ Ⓑ Ⓒ Ⓓ
83. Ⓐ Ⓑ Ⓒ Ⓓ
84. Ⓐ Ⓑ Ⓒ Ⓓ
85. Ⓐ Ⓑ Ⓒ Ⓓ
86. Ⓐ Ⓑ Ⓒ Ⓓ
87. Ⓐ Ⓑ Ⓒ Ⓓ
88. Ⓐ Ⓑ Ⓒ Ⓓ
89. Ⓐ Ⓑ Ⓒ Ⓓ
90. Ⓐ Ⓑ Ⓒ Ⓓ
91. Ⓐ Ⓑ Ⓒ Ⓓ
92. Ⓐ Ⓑ Ⓒ Ⓓ
93. Ⓐ Ⓑ Ⓒ Ⓓ
94. Ⓐ Ⓑ Ⓒ Ⓓ
95. Ⓐ Ⓑ Ⓒ Ⓓ
96. Ⓐ Ⓑ Ⓒ Ⓓ
97. Ⓐ Ⓑ Ⓒ Ⓓ
98. Ⓐ Ⓑ Ⓒ Ⓓ
99. Ⓐ Ⓑ Ⓒ Ⓓ
100. Ⓐ Ⓑ Ⓒ Ⓓ

TEAR HERE

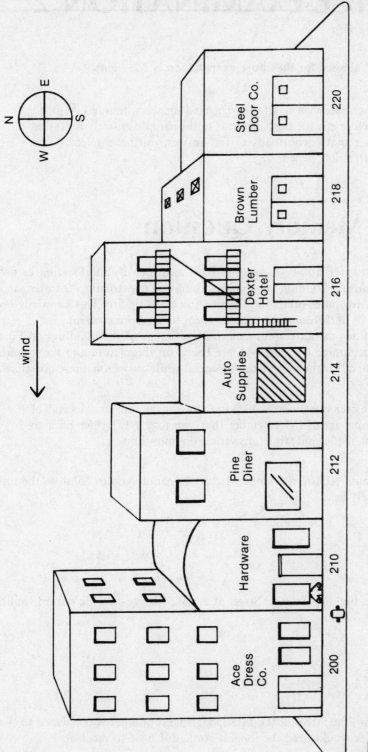

N E W S

wind

Steel Door Co.

220

Brown Lumber

218

Dexter Hotel

216

Auto Supplies

214

PINE STREET

Pine Diner

212

Hardware

210

Ace Dress Co.

200

⊠ Skylight

🛞 Sprinkler siamese

▨ Roll-down metal shutters

⟷ Fire hydrant

PRACTICE EXAMINATION 2

The time allowed for the entire examination is 3¹/₂ hours.

DIRECTIONS: Each question has four suggested answers, lettered A, B, C, and D. Decide which one is the best answer. On the sample answer sheet, find the question number that corresponds to the answer you have selected and darken the area with a soft pencil.

Memory Question

Firefighters must possess the ability to readily recall and identify the location of fire protection appliances, unusual building occupancies, building construction features, and other important physical attributes present at the site of a building fire. This knowledge is important for the safety of firefighters and enables them to be more efficient.

You will have five minutes to study and memorize the details of the buildings on Pine Street. Questions 1–13, beginning on page 182, are based on the picture and the legend. You may not refer back to either the picture or the legend while answering these questions.

DIRECTIONS: Answer questions 1–13 based on the picture and legend of Pine Street found on page 181. Remember that you may NOT refer back to either the picture or the legend while answering the questions.

1. The building that would require the longest ladder to enable you to climb to the top floor would most likely be

   (A) 212 Pine Street
   (B) 216 Pine Street
   (C) 200 Pine Street
   (D) 220 Pine Street

2. If a fire occurred in a building on Pine Street at 3 A.M., the greatest life hazard would most likely be in the building at

   (A) 212 Pine Street
   (B) 218 Pine Street
   (C) 220 Pine Street
   (D) 216 Pine Street

3. For a building fire on Pine Street, if the Fire Department pumper is connected to the hydrant, the firefighter would need the longest stretch of hose to reach the

   (A) Ace Dress Co.
   (B) Pine Diner
   (C) auto supply store
   (D) Steel Door Co.

4. A firefighter is opening the skylights to remove smoke and heat from the building. You would be correct to assume that he is on the roof of the

    (A) lumber company
    (B) Ace Dress Co.
    (C) hardware store
    (D) auto supply store

5. As a firefighter, you are ordered by Lieutenant Brown to connect a hoseline to the automatic sprinkler siamese connection. You would find this connection in front of

    (A) 220 Pine Street
    (B) 216 Pine Street
    (C) 200 Pine Street
    (D) 210 Pine Street

6. Roll-down metal shutters are installed on store fronts to deter burglars. However, they may cause a fire to burn undetected until it has gained considerable headway. The tenant on this block with roll-down metal shutters is the

    (A) hardware store
    (B) Pine Diner
    (C) lumber company
    (D) auto supply store

7. A firefighter is cutting a hole in the roof close to the front wall of 218 Pine Street. Severe conditions on the roof require that the firefighters make a fast exit from the roof. The safest and quickest exit may be made by

    (A) walking across the roof to 216 Pine Street
    (B) stepping over the parapet to the roof of 220 Pine Street
    (C) jumping across to the building behind 218 Pine Street
    (D) waiting until a ladder is placed against the building

8. The best protection against fire in its beginning stage is the automatic wet sprinkler system. The tenant on Pine Street that has the automatic wet sprinkler system is the

    (A) Ace Dress Co.
    (B) hardware store
    (C) lumber company
    (D) Dexter Hotel

9. When extinguishing a smoky fire, it is often advantageous to operate from the windward side (wind at your back) of the fire. From the diagram of Pine Street, you have observed that the wind is blowing from the

    (A) north
    (B) south
    (C) east
    (D) west

10. Hose stream penetration is important when extinguishing fires. The building on Pine Street that would require the deepest penetration of a hose stream is occupied by

    (A) the Steel Door Co.
    (B) Brown Lumber
    (C) Pine Diner
    (D) the auto supply store

11. Generally speaking, a curved roof is a trussed roof. Trussed roofs are notorious for rapid failure when subjected to high heat and fire for a short period of time. The building on Pine Street with a trussed roof is the

    (A) hardware store
    (B) lumber company
    (C) Steel Door Co.
    (D) Ace Dress Co.

12. Baled material and fibers become extremely heavy when wet. This causes overloading of floors which can result in collapse of the building. Effective use of water would be necessary for a fire at

    (A) 200 Pine Street
    (B) 212 Pine Street
    (C) 214 Pine Street
    (D) 218 Pine Street

13. Protecting exposures is an important part of firefighting strategy. If a heavy fire condition exists in Brown Lumber and your strategy is to protect the adjoining building, you would most likely have your firefighters stretch a hoseline into the

    (A) Dexter Hotel
    (B) Ace Dress Co.
    (C) Pine Diner
    (D) hardware store

14. A problem likely to be found in a hotel fire that is not usually found in fires involving other types of residences, such as apartment houses, is

    (A) obstructions in hallways and other passageways
    (B) large numbers of persons in relation to the number of rooms
    (C) delay in the transmission of the fire alarm
    (D) many occupants who don't know the location of exits

15. On the way to work one morning, a firefighter notices a high-tension wire which has blown down and is lying across the sidewalk and into the road. The most appropriate course of action for the firefighter to take would be to

    (A) move the wire to one side by means of a stock or branch of a tree and continue on his way
    (B) continue on his way to the firehouse and report the situation to the officer on duty
    (C) call the public utility company from the first public telephone he passes on his way to the firehouse
    (D) stand by the wire to warn away passersby and ask one of them to call the public utility company

16. Generally, firefighters on prevention inspection duty do not inspect the living quarters of private dwellings unless the occupants agree to the inspection. The best explanation why private dwellings are excluded from compulsory inspections is that

    (A) private dwellings seldom catch fire
    (B) fires in private dwellings are more easily extinguished than other types of fires
    (C) people may resent such inspections as an invasion of privacy
    (D) the monetary value of private dwellings is lower than that of other types of occupancies

17. Fire lines are usually established by the police in order to keep bystanders out of the immediate vicinity while firefighters are fighting the fire. Of the following, the best justification for the establishment of these fire lines is to

    (A) prevent theft of property from partially destroyed apartments or stores
    (B) prevent interference with firefighter operations
    (C) give privacy to the victims of the fire
    (D) help apprehend arsonists in the crowd

18. All probationary firefighters are trained in both engine company and ladder company operations. The most important reason for training in both types of operations is that

    (A) at any fire it may be necessary for engine company personnel to perform ladder company duties and vice versa
    (B) engine company personnel can perform their own functions better if they are also trained in ladder company functions and vice versa
    (C) after such training, the probationary firefighter can decide which type of company he or she prefers
    (D) after such training, the fire department can decide which people are best suited for each type of company

19. While on inspection duty, a firefighter discovers the superintendent of a tenement just starting to remove boxes and other materials which were blocking the hallways. Apparently, the superintendent had started removal when he saw the firefighter approach. In this situation, it is most important that the firefighter

    (A) warn the superintendent of the penalties for violation of the Fire Prevention Code
    (B) help the superintendent remove the material blocking the hallways
    (C) commend the superintendent for efforts to maintain a safe building
    (D) check again, after completing the inspection, to see whether the material has been removed completely

20. At 2 A.M., a passerby observes a fire in an office building. He runs to a nearby corner to look for a fire alarm box. After running a few steps, he notices a police telephone box on a lamp post and uses it to report the fire. This action is

    (A) good, chiefly because the police as well as the fire department respond to fire alarms
    (B) bad, chiefly because police telephone boxes are not to be used by civilians under any circumstances
    (C) good, chiefly because the fire department will receive prompt notification of the fire
    (D) bad, chiefly because police telephone boxes may be used by the public only for police emergencies

21. A fire insurance inspector suggested to the manager of a fireproof warehouse that bags of flour be stacked on skids (wooden platforms 6 inches high, 6 × 6 feet in area). Of the following, the best justification for this suggestion is that in the event of a fire, the bags on skids are less likely to

    (A) topple
    (B) be damaged by water used in extinguishment
    (C) catch fire
    (D) be ripped by fire equipment

22. Permitting piles of scrap paper cuttings to accumulate in a factory building is a bad practice chiefly because they may

    (A) ignite spontaneously
    (B) interfere with fire extinguishment operations
    (C) catch fire from a spark or smoldering match
    (D) interfere with escape of occupants if a fire occurs

23. High grass and weeds should not be permitted to grow near a building chiefly because, in the event of a grass fire, the weeds and grass may

    (A) give off toxic fumes
    (B) limit maneuverability of firefighters
    (C) interfere with the escape of occupants from the building
    (D) bring the fire to the building and set it on fire

24. Visitors near patients in oxygen tents are not permitted to smoke. The best of the following reasons for this prohibition is that

    (A) the flame of the cigarette or cigar may flare dangerously
    (B) smoking tobacco is irritating to persons with respiratory diseases
    (C) smoking in bed is one of the major causes of fire
    (D) diseases may be transmitted by means of tobacco smoke

25. At a hot and smoky fire, a lieutenant ordered firefighters to work in pairs. Of the following, the best justification for this order is that

    (A) better communications result since one person can bring messages back to the lieutenant
    (B) more efficient operation results since many vital activities require two people
    (C) better morale results since firefighters are more willing to face danger in pairs
    (D) safer operations result since one firefighter can help if the other becomes disabled

26. Firefighters frequently open windows, doors, and skylights of a building on fire in a planned or systematic way in order to ventilate the building. Least likely to be accomplished by ventilation is the

    (A) increase in visibility of firefighters
    (B) slowdown in the rate of burning of the building
    (C) reduction in the danger from toxic gases
    (D) control of the direction of travel of the fire

27. At the first sign of a fire, the manager of a movie theater had the lights turned on and made the following announcement: "Ladies and gentlemen, the management has found it necessary to dismiss the audience. Please remain seated until it is time for your aisle to file out. In leaving the theater, follow the directions of the ushers. There is no danger involved." The manager's action in this situation was

    (A) proper
    (B) improper, chiefly because the manager did not tell the audience the reason for the dismissal
    (C) improper, chiefly because the manager did not permit all members of the audience to leave at once
    (D) improper, chiefly because the manager misled the audience by saying that there was no danger

28. Generally, sprinkler heads must be replaced each time they are used. The best explanation for this is that the sprinkler heads

(A) are subject to rusting after discharging water
(B) may become clogged after discharging water
(C) have a distorted pattern of water discharge after use
(D) are set off by the effect of heat on metal and cannot be reset

29. After a fire in an apartment had been brought under control, firefighters were engaged in extinguishing the last traces of fire. A firefighter who noticed an expensive vase in the room in which activities were concentrated moved it to an empty closet in another room. The firefighter's action was

(A) proper, chiefly because the owners would realize that extreme care was taken to avoid damage to their possessions
(B) improper, chiefly because disturbance of personal possessions should be kept to a minimum.
(C) proper, chiefly because the chance of damage is reduced
(D) improper, chiefly because the firefighter should have devoted this time to putting out the fire

30. While standing in front of a firehouse a firefighter is approached by a woman with a baby carriage. The woman asks the firefighter if he will "keep an eye" on the baby while she visits a doctor in a nearby building. In this situation, the best course of action for the firefighter to take is to

(A) agree to the woman's request but warn her that it may be necessary to leave to answer an alarm
(B) refuse politely after explaining that he is on duty and may not become involved in other activities
(C) refer the woman to the officer on duty
(D) ask the officer on duty for permission to grant the favor

31. Suppose that, while cooking, a pan of grease catches fire. The method which would be most effective, if available, in putting out the fire is to

(A) dash a bucket of water on the fire
(B) direct a stream of water on the fire from a fire extinguisher
(C) pour a bottle of household ammonia water on the fire
(D) empty a box of baking soda on the fire

32. Persons engaged in certain hazardous activities are required to obtain a fire department permit or certificate for which a fee is charged. The main reason for requiring permits or certificates is to

(A) obtain revenue for the city government
(B) prevent unqualified persons from engaging in these activities
(C) obtain information about these activities in order to plan for fire emergencies
(D) warn the public of the hazardous nature of these activities

33. A firefighter traveling to work is stopped by a citizen who complains that the employees of a nearby store frequently pile empty crates and boxes in a doorway, blocking passage. The most appropriate action for the firefighter to take is to

(A) assure the citizen that the fire department's inspectional activities will eventually catch up with the store

(B) obtain the address of the store and investigate to determine whether the citizen's complaint is justified

(C) obtain the address of the store and report the complaint to a superior officer

(D) ask the citizen for specific dates on which this practice occurred to determine whether the complaint is justified

34. The crime of arson is defined as the willful burning of a house or other property. The best illustration of arson is a fire

(A) in an apartment started by a four-year-old boy playing with matches
(B) in a barn started by a drunken man who overturns a lantern
(C) in a store started by the bankrupt owner in order to collect insurance
(D) in a house started by a neighbor who carelessly burned leaves in the garden

35. The fire department now uses companies on fire duty with their apparatus for fire prevention inspection in commercial buildings. The change most important in making this inspection procedure practicable was the

(A) reduction of hours of work for firefighters
(B) use of two-way radio equipment
(C) use of enclosed cabs on fire apparatus
(D) increase in property value in the city

36. Many fires are caused by improper use of oxyacetylene torches. The main cause of such fires is the

(A) high pressure under which the gases are stored
(B) failure to control or extinguish sparks
(C) high temperatures generated by the equipment
(D) explosive nature of the gases used

37. The most important reason for having members of the fire department wear uniforms is to

(A) indicate the semimilitary nature of the fire department
(B) build morale and *esprit de corps* of members
(C) identify members on duty to the public and other members
(D) provide clothing suitable for the work performed

38. Of the following types of fires, the one which is likely to have the *least* amount of damage from water used in extinguishment is a fire in

(A) a rubber toy factory
(B) a retail hardware store
(C) an outdoor lumberyard
(D) a furniture warehouse

39. "In case of suspected arson, it is important that firefighters engaged in fighting the fire remember conditions that existed at the time of their arrival. Particular attention should be given to all doors and windows." The main justification for this statement is that knowledge of the condition of the doors and windows may indicate

(A) where the fire started
(B) who set the fire
(C) the best way to ventilate the building
(D) that someone wanted to prevent extinguishment

40. "After a fire has been extinguished, every effort should be made to determine how the fire started." Of the following, the chief reason for determining the origin of the fire is to

(A) reduce the amount of damage caused by the fire
(B) determine how the fire should have been fought
(C) eliminate causes of fire in the future
(D) explain delays in fighting the fire

41. A firefighter inspecting buildings in a commercial area came to one building whose outside surface appeared to be of natural stone. The owner told the firefighter that it was not necessary to inspect the building since it was "fireproof." The firefighter, however, completed inspection of the building. Of the following, the best reason for continuing the inspection is that

(A) stone buildings catch fire as readily as wooden buildings
(B) the fire department cannot make exceptions in its inspection procedures
(C) the building may have been built of imitation stone
(D) interiors and contents of stone buildings can catch fire

42. The *least* valid reason for the fire department to investigate the causes of fire is to

(A) determine whether the fire was the result of arson
(B) estimate the amount of loss for insurance purposes
(C) gather information useful in fire prevention
(D) discover violations of the Fire Prevention Code

43. While on duty at a fire, a probationary firefighter receives an order from the lieutenant which appears to conflict with the principles of firefighting taught at the fire school. Of the following, the best course of action for the firefighter to take is to follow the order and at a convenient time, after the fire, to

(A) discuss the apparent inconsistency with the lieutenant
(B) discuss the apparent inconsistency with another officer
(C) mention this apparent inconsistency in an informal discussion
(D) ask a more experienced firefighter about the apparent inconsistency

44. When fighting fires on piers, the fire department frequently drafts salt water from the harbor. The chief advantage of using harbor water instead of relying on water from street mains is that harbor water is

(A) less likely to cause water damage
(B) available in unlimited quantities
(C) more effective in extinguishing fires due to its salt content
(D) less likely to freeze in low temperatures due to its salt content

DIRECTIONS: Questions 45–48 are based on the following paragraph.

The canister type gas mask consists of a tight-fitting face piece connected to a canister containing chemicals which filter toxic gases and smoke from otherwise breathable air. These masks are of value when used with due regard to the fact that two or three percent of gas in air is about the highest concentration that the chemicals in the canister will absorb and that these masks do not provide the oxygen necessary for the support of life. In general, if flame is visible, there is sufficient oxygen for firefighters although toxic gases may be present. Where there is heavy smoke and no flame, an oxygen deficiency may exist. Fatalities have occurred where filter type canister masks have been used in attempting rescue from manholes, wells, basements, or other locations deficient in oxygen.

45. If the mask described in the paragraph is used in an atmosphere containing oxygen, nitrogen, and carbon monoxide, we would expect the mask to remove from the air breathed

   (A) the nitrogen only
   (B) the carbon monoxide only
   (C) the nitrogen and the carbon monoxide
   (D) none of these gases

46. According to this paragraph, when firefighters are wearing these masks at a fire where flame is visible, the firefighter can generally feel that, as far as breathing is concerned, they are

   (A) safe, since the mask will provide them with sufficient oxygen to live
   (B) unsafe, unless the gas concentration is below two or three percent
   (C) safe, provided the gas concentration is above two or three percent
   (D) unsafe, since the mask will not provide them with sufficient oxygen to live

47. According to this paragraph, fatalities have occurred to persons using this type of gas mask in manholes, wells, and basements because

   (A) the supply of oxygen provided by the mask ran out
   (B) the air in those places did not contain enough oxygen to support life
   (C) heavy smoke interfered with the operation of the mask
   (D) the chemicals in the canister did not function properly

48. The following formula may be used to show, in general, the operation of the gas mask described in the preceding paragraph:

(Chemicals in canister) → (air + gases) = breathable air

The arrow in the formula, when expressed in words, means most nearly

   (A) replace
   (B) are changed into
   (C) act upon
   (D) give off

DIRECTIONS: Questions 49–51 are based on the following paragraph.

The only openings permitted in fire partitions, except openings for ventilating ducts, shall be those required for doors. There shall be but one such door opening unless the provision of additional openings would not exceed in total width of all doorways 25 percent of the length of the wall. The minimum distance between openings shall be three feet. The maximum area for such a door opening shall be 80 square feet, except that such openings for the passage of trucks may be a maximum of 140 square feet.

49. According to the paragraph, openings in fire partitions are permitted only for

   (A) doors
   (B) doors and windows
   (C) doors and ventilation ducts
   (D) doors, windows, and ventilation ducts

50. In a fire partition 22 feet long and 10 feet high, the maximum number of doors 3 feet wide and 7 feet high is

(A) 1
(B) 2
(C) 3
(D) 4

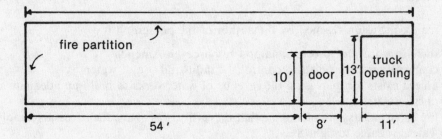

51. The most accurate statement about the layout shown above is that the

(A) total width of the openings is too large
(B) truck opening is too large
(C) truck and door openings are too close together
(D) layout is acceptable

52. Of the following, the basic purpose of Fire Department inspection of private property is to

(A) make sure that Fire Department equipment is properly maintained
(B) secure proper maintenance of all entrace and exit facilities
(C) obtain correction of conditions creating undue fire hazards
(D) make sure that walls are properly plastered and painted in order to prevent the spread of fire

53. Of the following items commonly found in a household, the one that uses the most electric current is

(A) a 150-watt light bulb
(B) a toaster
(C) a door buzzer
(D) an 8-inch electric fan

DIRECTIONS: Questions 54–56 are based on the following paragraph.

The average daily flow of water through public water systems in American cities ranges generally between 40 and 250 gallons per capita, depending upon the underground leakage in the system, the amount of waste in domestic premises, and the quantity used for industrial purposes. The problem of supplying this water has become serious in many cities. Supplies, once adequate, in many cases have become seriously deficient, due to greater demands with increased population and growing industrial use of water. Waterworks, operating on fixed schedules of water charges, in many cases have not been able to afford the heavy capital expenditures necessary to provide adequate supply, storage, and distribution facilities. Thus, the adequacy of a public water supply for fire protection in any given location cannot properly be taken for granted.

54. The four programs listed below are possible ways in which American communities might try to reduce the seriousness of the water shortage problem. The one program which does *not* directly follow from the paragraph is that of

(A) regular replacement of old street water mains by new ones
(B) inspection and repair of leaky plumbing fixtures
(C) fire prevention inspection and education to reduce the amount of water used to extinguish fires
(D) research into industrial processes to reduce the amount of water used in those processes

55. The main conclusion reached by the author of the paragraph is

(A) there is a waste of precious natural resources in America
(B) communities have failed to control the industrial use of water
(C) a need exists for increasing the revenue of waterworks to build up adequate supplies of water
(D) fire departments cannot assume that they will always have the necessary supply in water available to fight fires

56. Per capita consumption of water in a community is determined by the formula

(A) $\dfrac{\text{Population}}{\text{Total consumption in gallons}}$ = Per capita consumption in gallons

(B) $\dfrac{\text{Total consumption in gallons}}{\text{Population}}$ = Per capita consumption in gallons

(C) Total consumption in gallons $\times$ Population = Per capita consumption in gallons

(D) Total consumption in gallons $-$ Population = Per capita consumption in gallons

DIRECTIONS: Questions 57–60 are based on the following paragraph.

An annual leave allowance, which combines leaves previously given for vacation, personal business, family illness, and other reasons shall be granted members. Calculation of credits for such leave shall be on an annual basis beginning January 1 of each year. Annual leave credits shall be based on time served by members during the preceding calendar year. However, when credits have been accrued and a member retires during the current year, additional annual leave credits shall, in this instance, be granted at the accrual rate of three days for each completed month of service, *excluding terminal leave.* If accruals granted for completed months of service extend into the following month, the member shall be granted an additional three days accrual for the completed month. This shall be the only condition where accruals in a current year are granted for a vacation period in such year.

57. According to the paragraph, if a firefighter's spouse were to become seriously ill so that it was necessary to take time off from work to be with him or her, such time off from work would be deducted from the firefighter's

(A) annual leave allowance
(B) vacation leave allowance
(C) personal business leave allowance
(D) family illness leave allowance

58. It may be inferred that terminal leave means leave taken

(A) at the end of the calendar year

(B) at the end of the vacation year
(C) immediately before retirement
(D) before actually earned because of an emergency

59. A firefighter appointed on July 1, 1983 will be able to take his first full or normal leave during the period

(A) July 1, 1983 to June 30, 1984
(B) January 1, 1984 to December 31, 1984
(C) July 1, 1984 to June 30, 1985
(D) January 1, 1985 to December 31, 1985

60. According to this paragraph, a member who retires on July 15 of this year will be entitled to receive leave allowance, based on this year, of

(A) 15 days
(B) 18 days
(C) 21 days
(D) 24 days

DIRECTIONS: Questions 61–64 are based on the following paragraph.

During fire operations, all members shall be constantly alert to the possibility of the crime of arson. In the event that conditions indicate this possibility, the officer in command shall promptly notify the fire marshal. Unauthorized persons shall be prohibited from entering the premises and actions of those authorized carefully noted. Members shall refrain from discussion of the fire and prevent disturbance of essential evidence. If necessary, the officer in command shall detail one or more members at the location with information for the fire marshal upon his or her arrival.

61. From the preceding paragraph, it may be inferred that the reason for prohibiting unauthorized persons from entering the fire premises when arson is suspected is to prevent such persons from

(A) endangering themselves in the fire
(B) interfering with the fire
(C) disturbing any evidence of arson
(D) committing acts of arson

62. The title which best describes the subject matter of the paragraph is

(A) Techniques of Arson Detection
(B) The Role of the Fire Marshal in Arson Cases
(C) Fire Scene Procedures in Cases of Suspected Arson
(D) Evidence in Arson Investigations

63. The statement that is most correct and complete is that the responsibility for detecting signs of arson at a fire belongs to the

(A) fire marshal
(B) fire marshal and officer in command
(C) fire marshal, officer in command, and any members detailed at the location with information for the fire marshal
(D) members present at the scene of the fire regardless of their rank or position

64. From the preceding paragraph, it may be inferred that the fire marshal usually arrives at the scene of a fire

(A) before the fire companies
(B) simultaneously with the fire companies
(C) immediately after the fire companies
(D) some time after the fire companies

65. Sand and ashes are frequently placed on icy pavements to prevent skidding. The effect of the sand and ashes is to increase

(A) inertia
(B) gravity
(C) momentum
(D) friction

66. The air near the ceiling of a room is usually warmer than the air near the floor because

(A) there is better air circulation at floor level
(B) warm air is lighter than cold air
(C) windows are usually nearer the floor than the ceiling
(D) heating pipes usually run along the ceiling

67. It is safer to use the ladder positioned as shown in diagram 1 than as shown in diagram 2 because in diagram 1

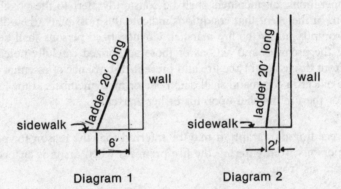

Diagram 1   Diagram 2

(A) less strain is placed on the center rungs of the ladder
(B) it is easier to grip and stand on the ladder
(C) the ladder reaches a lower height
(D) the ladder is less likely to tip over backwards

68. A substance which is subject to spontaneous heating is one that

(A) is explosive when heated
(B) is capable of catching fire without an external source of heat
(C) acts to speed up the burning of material
(D) liberates oxygen when heated

69. "Chief officers shall instruct their firefighters that when transmitting particulars of a fire, they shall include the fact that foods are involved if the fire involves premises where foodstuffs are sold or stored." Of the following, the best justification for this regulation is that, when foodstuffs are involved in a fire,

(A) the fire may reach serious proportions
(B) police protection may be desirable to prevent looting

(C) relatively little firefighting equipment may be needed

(D) inspection to detect contamination may be desirable

70. Fire can continue when fuel, oxygen from the air or another source, and a sufficiently high temperature to maintain combustion are present. The method of extinguishment of fire *most commonly* used is to

    (A) remove the fuel

    (B) exclude the oxygen from the burning material

    (C) reduce the temperature of the burning material

    (D) smother the flames of the burning material

71. The siphon arrangement shown below which would most quickly transfer a solution from the container on the left side to the one on the right side is numbered

    (A) 1

    (B) 2

    (C) 3

    (D) 4

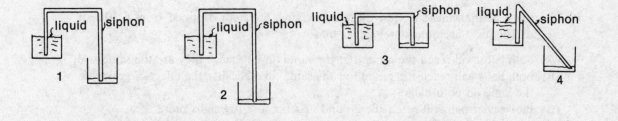

72. Static electricity is a hazard in industry chiefly because it may cause

    (A) dangerous or painful burns

    (B) chemical decomposition of toxic elements

    (C) sparks that can start an explosion

    (D) overheating of electrical equipment

73. A rowboat will float deeper in fresh water than in salt water because

    (A) in the salt water the salt will occupy part of the space

    (B) fresh water is heavier than salt water

    (C) salt water is heavier than fresh water

    (D) salt water offers less resistance than fresh water

74. It is easier to get the load onto the platform by using the ramp than it is to lift it directly onto the platform. This is true because the effect of the ramp is to

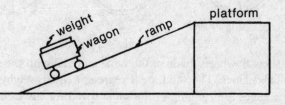

    (A) reduce the amount of friction so that less force is required

    (B) distribute the weight over a larger area

    (C) support part of the load so that less force is needed to move the wagon

    (D) increase the effect of the moving weight

75. More weight can be lifted by the method shown in diagram 2 than as shown in diagram 1 because

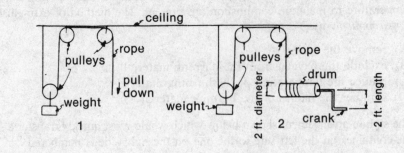

(A) it takes less force to turn a crank than it does to pull in a straight line
(B) the drum will prevent the weight from falling by itself
(C) the length of the crank is larger than the radius of the drum
(D) the drum has more rope on it, which eases the pull

76. Two balls of the same size, but different weights, are both dropped from a 10-foot height. The most accurate statement is that

(A) both balls will reach the ground at the same time because they are the same size
(B) both balls will reach the ground at the same time because the effect of gravity is the same on both balls
(C) the heavier ball will reach the ground first because it weighs more
(D) the lighter ball will reach the ground first because air resistance is greater on the heavier ball

77. "A partially filled gasoline drum is a more dangerous fire hazard than a full one." Of the following, the best justification for this statement is that

(A) a partially filled gasoline drum contains relatively little air
(B) gasoline is difficult to ignite
(C) when a gasoline drum is full, the gasoline is more explosive
(D) gasoline vapors are more explosive than gasoline itself

78.

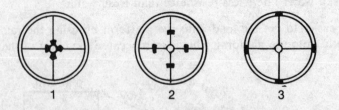

The diagrams above show flywheels made of the same material with the same dimensions and attached to similar engines. The solid areas represent equal weights attached to the flywheel. If all three engines are running at the same speed for the same length of time, and the power to the engines is shut off simultaneously,

(A) wheel 1 will continue turning longest
(B) wheel 2 will continue turning longest
(C) wheel 3 will continue turning longest
(D) all three wheels will continue turning for the same time

79. The substance which expands when freezing is

    (A) alcohol
    (B) ammonia
    (C) mercury
    (D) water

    DIRECTIONS: Questions 80 and 81 are to be answered in accordance with the information given in the following statement.

The electrical resistance of copper wires varies directly with their lengths and inversely with their cross section areas.

80. A piece of copper wire 30 feet long is cut into two pieces, 20 feet and 10 feet. The resistance of the longer piece, compared to the shorter, is

    (A) one-half as much
    (B) two-thirds as much
    (C) one and one-half as much
    (D) twice as much

81. Two pieces of copper wire are each 10 feet long but the cross sectional area of one is two-thirds that of the other. The resistance of the piece with the larger cross sectional area is

    (A) one-half of the resistance of the smaller
    (B) two-thirds the resistance of the smaller
    (C) one and one-half times the resistance of the smaller
    (D) twice the resistance of the smaller

82. When drilling a hole through a piece of wood with an auger bit, it is considered good practice to clamp a piece of scrap wood to the underside of the piece through which the hole is being drilled. The main reason for this is to

    (A) direct the auger bit
    (B) speed up the drilling operation
    (C) prevent the drill from wobbling
    (D) prevent splintering of the wood

83. "Firefighters holding a life net should keep their eyes on the person jumping from a burning building." Of the following, the best justification for this statement is that

    (A) a person attempting to jump into a life net may overestimate the distance of the net from the building
    (B) some persons will not jump into a life net unless given confidence
    (C) a person jumping into a life net may be seriously injured if the net is not allowed to give slightly at the moment of impact
    (D) firefighters holding a life net should be evenly spaced around the net at the moment of impact in order to distribute the shock

84. If boiling water is poured into a drinking glass, the glass is likely to crack. If, however, a metal spoon is first placed in the glass, it is much less likely to crack. The reason that the glass with the spoon is less likely to crack is that the spoon

    (A) distributes the water over a larger surface of the glass
    (B) quickly absorbs heat from the water
    (C) reinforced the glass
    (D) reduces the amount of water that can be poured into the glass

85. It takes more energy to force water through a long pipe than through a short pipe of the same diameter. The principal reason for this is

(A) gravity
(B) friction
(C) inertia
(D) cohesion

86. A pump discharging at 300 pounds per square inch pressure delivers water through 100 feet of pipe laid horizontally. If the valve at the end of the pipe is shut so that no water can flow, then the pressure at the valve is, for practical purposes,

(A) greater than the pressure at the pump
(B) equal to the pressure at the pump
(C) less than the pressure at the pump
(D) greater or less than the pressure at the pump, depending on the type of pump used

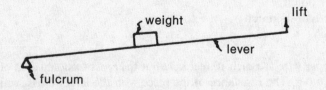

87. The leverage system in the sketch above is used to raise a weight. In order to reduce the amount of force required to raise the weight, it is necessary to

(A) decrease the length of the lever
(B) place the weight closer to the fulcrum
(C) move the weight closer to the person applying the force
(D) move the fulcrum farther from the weight

DIRECTIONS: Questions 88–91 relate to the information given in the following paragraph and are to be answered in accordance with this information.

Old buildings are liable to possess a special degree of fire risk merely because they are old. Outmoded electrical wiring systems and installation of new heating appliances for which the building was not designed may contribute to the increased hazard. Old buildings have often been altered many times; parts of the structure may antedate building codes; dangerous defects may have been covered up. On the average, old buildings contain more lumber than comparable new buildings which, in itself, makes old buildings more susceptible to fire. It is not true, though, that sound lumber in old buildings is drier than new lumber. Moisture content of lumber varies with that of the atmosphere to which it is exposed.

88. According to the paragraph, old buildings present a special fire hazard chiefly because of the

(A) poor planning of the buildings when first designed
(B) haphazard alteration of the buildings
(C) dryness of old lumber
(D) inadequate enforcement of the building codes

89. We may conclude from the paragraph that lumber

(A) should not be used in buildings unless absolutely necessary
(B) should not be used near electrical equipment
(C) is more inflammable than newer types of building materials
(D) tends to lose its moisture at a constant rate

90. According to the paragraph, the amount of moisture in the wooden parts of a building depends upon the

(A) age of the building
(B) moisture in the surrounding air
(C) type of heating equipment used in the building
(D) quality of lumber used

91. In regard to building codes, the paragraph implies that

(A) old buildings are exempt from the provisions of building codes
(B) some buildings now in use were built before building codes were adopted
(C) building codes usually don't cover electrical wiring systems
(D) building codes are generally inadequate

92. A firefighter must become familiar with first-aid methods. Of the following, the best reason for this rule is that

(A) it saves expense to have someone take care of simple injuries not requiring a doctor
(B) it may keep the Fire Department out of a lawsuit if a simple injury is taken care of promptly
(C) it is of great advantage that a firefighter be prepared to act quickly in an emergency
(D) in doing first aid, it is sometimes better to do nothing than to do the wrong thing

93. The method which is *not* a recommended procedure for treating shock is

(A) administering oxygen
(B) placing the victim in a horizontal position
(C) trying to prevent the loss of body heat by covering the victim with a blanket
(D) having the victim drink warm water

94. To control severe bleeding, a tourniquet should be applied only as a last resort. Once applied, it is left in place. Of the following, the statement that is *not* correct is:

(A) The tourniquet is placed two inches from the injury between the heart and the location of the injury.
(B) Tighten the touniquet only enough to stop the bleeding.
(C) Mark the person with the letters T–K and the time of day the tourniquet was applied.
(D) Loosen and retighten the tourniquet from time to time.

95. An open fracture generally means a fracture in which

(A) the broken bone has penetrated the skin
(B) the bone is broken at more than one point
(C) the bone has not only been broken but has been dislocated as well
(D) more than one bone has been broken

96. The newly appointed firefighter should realize that the principal function of an immediate superior officer is to

(A) inflict discipline for rule infractions
(B) direct the activities of subordinates so that they function in a proper and acceptable manner

(C) make recommendations to promote the most able subordinates under his or her command
(D) keep records that properly reflect the activities of subordinates

97. Firefighters who are eager for promotion and who thoroughly prepare themselves for promotion examinations are likely to become better firefighters. The most probable reason for this is which of the following statements?

(A) Only the better firefighters want to be promoted.
(B) Firefighters who do not seek promotions are convinced of their limitations.
(C) The study material necessary for promotion preparation is concerned mainly with the duties of the position.
(D) Firefighters who spend their free time preparing for promotion are likely to stay out of trouble.

98. A newly appointed firefighter should be aware that the main purpose of constructive criticism made by a superior officer would be to

(A) improve his or her performance
(B) enhance his or her opportunities for promotion
(C) display authority
(D) encourage the firefighter to participate in company social activities

99. During the course of fighting a fire, a superior officer notices a subordinate performing an important function in an improper manner. It would be most appropriate for the superior officer to correct the firefighter

(A) after the fire has been extinguished, in the privacy of the firehouse
(B) immediately
(C) immediately, only if the public is not in attendance and the firefighter would not be embarrassed
(D) at the next regularly scheduled training session

100. A firefighter should be aware that the main purpose of disciplinary action taken by superior officers in a fire department against their subordinates is to

(A) inflict punishment for rule infractions
(B) train subordinates to perform their duties properly
(C) show who is boss
(D) establish rules of acceptable behavior

Answer Key for Practice Examination 2

1. C	21. B	41. D	61. C	81. B
2. D	22. C	42. B	62. C	82. D
3. D	23. D	43. A	63. D	83. A
4. A	24. A	44. B	64. D	84. B
5. D	25. D	45. B	65. D	85. B
6. D	26. B	46. B	66. B	86. B
7. B	27. A	47. B	67. D	87. B
8. B	28. D	48. C	68. B	88. B
9. C	29. C	49. C	69. D	89. C
10. B	30. B	50. A	70. C	90. B
11. A	31. D	51. B	71. B	91. B
12. A	32. B	52. C	72. C	92. C
13. A	33. C	53. B	73. C	93. D
14. D	34. C	54. C	74. C	94. D
15. D	35. B	55. D	75. C	95. A
16. C	36. B	56. B	76. B	96. B
17. B	37. C	57. A	77. D	97. C
18. A	38. C	58. C	78. C	98. A
19. D	39. D	59. D	79. D	99. B
20. C	40. C	60. C	80. D	100. B

Explanatory Answers for Practice Examination 2

1. **(C)** 200 Pine Street is the tallest building on the block.

2. **(D)** The Dexter Hotel at 216 Pine Street has occupants that are likely to be asleep in the early morning hours. The area is commercial/industrial and most businesses are not open at 3 A.M.

3. **(D)** This is the building farthest from the hydrant.

4. **(A)** Note the skylights on the diagram.

5. **(D)** Refer to the legend and the diagram.

6. **(D)** Refer to the legend and the diagram.

7. **(B)** 218 and 220 Pine Street are both one-story buildings.

8. **(B)** The siamese connection on the front wall of the building indicates that the building has an automatic wet sprinkler system.

9. **(C)** The direction of wind is indicated by the arrow and compass points at the top of the diagram.

10. **(B)** The building housing Brown Lumber would require the deepest penetration of a hose stream because of the type of material stored there.

11. **(A)** Refer to the diagram and the definition of a trussed roof in the first sentence of the question.

12. **(A)** Baled material, fibers, and other materials are used in the manufacture of dresses and other clothing.

13. **(A)** The Dexter Hotel is immediately adjacent to the building which is afire. Fire burning through the roof of Brown Lumber would expose the hotel to flames, especially with the wind blowing from the east.

14. **(D)** Generally speaking, guests staying at a hotel are so preoccupied with their activities that they rarely take the time to familiarize themselves with the location of the fire exits and fire stairs.

15. **(D)** The firefighter knows the dangerous situation that the live wire presents and the tendency for it to recoil (wire has memory) without notice. The firefighter should protect the public until the fire department or utility company arrives.

16. **(C)** Entry into a private home or apartment is illegal unless permission to do so is given by the occupant. However, in the event of a fire, permission is not required.

17. **(B)** Bystanders in the immediate vicinity of the fire hinder effective firefighter operations. They impede movement of fire apparatus, the stretching of hoselines, and rescue efforts. There is also the danger of bystanders being injured by falling debris.

18. **(A)** In the event of a reduction of firefighting personnel in either a ladder or engine company, a firefighter can be detailed from one type of unit to another without loss of efficiency. It therefore provides for flexibility in assignment of firefighters.

19. **(D)** The accumulation of material may be a continuing condition. Therefore, a follow-up inspection should be made to assure that the superintendent is abiding by the fire code.

20. **(C)** Since the distance to the nearest fire alarm box is unknown to the passerby, calling in notice of the fire on the police box is a quick, positive action that should bring help.

21. **(B)** Raising sacks of flour from the floor lessens the risk to the insurer of total loss due to water damage.

22. **(C)** Scrap paper cuttings should be kept in closed covered metal containers and removed from the building daily, thus eliminating the possible fire hazard.

23. **(D)** Embers from grass and brush fires can be blown a considerable distance by the wind, causing the fire to spread to the buildings.

24. **(A)** Oxygen is a noncombustible gas; however, it will support combustion and will accelerate the fire in whatever is burning.

25. **(D)** Fires are always unpredictable. In hot, smoky conditions, the "buddy system" makes for greater firefighter safety.

26. **(B)** Unless the ventilation of a fire building is synchronized with advancing hoselines, the in-rush of fresh air can cause a fire to accelerate.

27. **(A)** Patrons in a theater are generally surprised when the lights are turned on suddenly. The manager calmed the patrons at once and informed them of how to make a safe and speedy exit. This was very well done.

28. **(D)** A sprinkler head has a fusible metal disc, which is used to hold back the water. The heat from the fire causes the disc to melt at a predetermined temperature and, in doing so, releases the water. It cannot be used again: Therefore, it must be replaced.

29. **(C)** The firefighter's job is to extinguish fires, save lives, and, at every opportunity, make an effort to hold property damage from fire or water to a minimum. With the fire out and no danger to life, the firefighter correctly saved property.

30. **(B)** The firefighter is on duty to respond to fires and emergencies. If he "kept an eye" on the child, he would not be available to respond with his unit.

31. **(D)** The baking soda would smother the fire. The other extinguishing agents would cause the grease to splatter and spread the fire from the pan to surrounding areas.

32. **(B)** Before obtaining a certificate or permit from the fire department, persons applying must show fire department examiners that they are trained and qualified to engage in the particular hazardous work.

33. **(C)** By reporting the complaint, the officer will then cause an official inspection to be made. If the doorway is blocked, a violation order will be issued to the person in charge.

34. **(C)** The bankrupt owner of the store deliberately set the fire in the store. The other three fires were started accidentally.

35. **(B)** The two-way radio enables the fire company to be away from the fire station and still be in radio contact with the fire alarm dispatcher and available to respond to alarms.

36. **(B)** Oxyacetylene torches generate a high intensity flame that is used to cut metal. As the metal is cut, the hot pieces and sparks bounce from the metal. Numerous fires are started by sparks and hot metal falling on combustible material.

37. **(C)** The firefighter in uniform serves as a beacon to a person seeking aid. The uniform also identifies the firefighter as a person with authority. This may be a deterrent to anyone intent on committing a criminal act.

38. **(C)** The majority of wood and wood products at an outdoor lumberyard are stored outdoors and are exposed to all weather conditions. Wetting by hose streams should have little adverse effect on the lumber.

39. **(D)** The arsonist will generally open windows to allow air in to accelerate the fire. The open or unlocked door usually indicates that someone was in the building prior to the arrival of the firefighters.

40. **(C)** Fire prevention is a very important aspect of the firefighter's role. Study of the causes of fires can lead to steps to eliminate those causes at other locations and thus to prevent future fires.

41. **(D)** The term "fireproof building" means that the walls, floors, stairways, partitions, and other structural components of the building are made of noncombustible material and will resist fire. However, the contents and furnishings are combustible and are likely to burn.

42. **(B)** The fire department would investigate a fire for the reasons stated in (A), (C), and (D). The amount of loss for insurance purposes would be investigated by an adjustor hired by the insurance company or the victim of the fire.

43. **(A)** The lieutenant may have been aware of other circumstances or conditions at the fire that the probationary firefighter, because of inexperience, was not aware of. The officer who gave the order is in the best position to explain it.

44. **(B)** Pier fires are usually fires of long duration which require enormous quantities of water to extinguish. By using the salt water, we have an unlimited supply and we can conserve our fresh-water resources.

45. **(B)** Carbon monoxide is a toxic gas. The atmospheric air we breathe is composed of 84 percent nitrogen and 16 percent oxygen; obviously nitrogen is not a toxic gas.

46. **(B)** The flame indicates that there is sufficient oxygen. Gas concentration is low enough to be handled by the mask.

47. **(B)** As stated in the last sentence.

48. **(C)** The canister type gas mask contains chemicals which filter toxic gases and smoke, resulting in breathable air.

49. **(C)** As stated in the first sentence.

50. **(A)** As stated in the second sentence. Doorway openings permitted in the wall are a maximum of 25 percent of 22 feet, or 5.5 feet. One opening would be 3 feet and two openings would be total of 6 feet. Therefore, (A) must be the answer.

51. **(B)** As stated in the third sentence. The maximum opening for a motor vehicle may be 140 square feet. The opening in the diagram is 13 feet × 11 feet, or 143 square feet. Therefore, it is too large.

52. **(C)** The purpose of fire department inspection of any and all property is to eliminate conditions creating undue fire hazards.

53. **(B)** The average household toaster uses 1400 watts of energy, much more than the other items mentioned in (A), (C), and (D).

54. **(C)** Choices (A), (B), and (D) are programs that investigate the causes of water usage as stated in the first sentence.

55. **(D)** As stated in the last sentence.

56. **(B)** To obtain the amount used by each person, take the total number of gallons consumed and divide this number by the total number of people (population).

57. **(A)** As stated in the first sentence.

58. **(C)** Self-explanatory, when active employment is terminated by retirement.

59. **(D)** Annual leave credits are accrued on the basis of the calendar year. The firefighter appointed July 1, 1983, begins accruing credits on January 1, 1984, and can begin to use those credits on January 1, 1985.

60. **(C)** As stated in the fourth sentence. 7 months × 3 days = 21 days.

61. **(C)** Unauthorized persons must be excluded so as to prevent disturbance of material or the removal of evidence.

62. **(C)** This title encompasses points mentioned throughout the paragraph.

63. **(D)** As stated in the first sentence.

64. **(D)** In the event that arson is suspected, the officer in command shall promptly notify the fire marshal. The officer in command shall also detail one or more members to the fire location until the fire marshal arrives. The fire marshal does not attend all fires, so will not come to the scene until summoned.

65. **(D)** Friction opposes motion. By placing a coarse substance between two smooth surfaces, friction is generated and skidding is reduced.

66. **(B)** Heated air expands; it becomes lighter and rises. This is called convection.

67. **(D)** The ladder with the gentler grade is more stable and is less likely to tip over. The horizontal distance between the wall and the base of the ladder is determined by dividing the length of the ladder by 5 and adding 2. Example: For a 20-foot ladder, dividing by 5 equals 4. 4 plus 2 equals 6. 6 feet is the proper distance from the wall for the 20-foot ladder.

68. **(B)** Spontaneous heating is the increase in temperature of a material or substance without drawing heat from its surroundings.

69. **(D)** Food which has been subjected to unusual heat and, perhaps, also to water may well have undergone changes which render it unfit for human consumption. For protection of the public, it is important that this possibility be recognized and that the food be inspected.

70. **(C)** The most common method of extinguishing a fire is to pour water on it. Water is a very effective cooling agent. The action of water is to reduce the temperature of the burning material.

71. **(B)** The principle that applies to this problem is: Pressure in a liquid is equal to the height multiplied by the density or $P = H \times D$. Since siphon 2 has a greater depth, the liquid at the bottom of the siphon 2 tube has a greater pressure than that of siphon 1. The greater pressure also produces greater velocity of the moving liquid. Therefore, the flow in siphon 2 would be faster, making it more efficient.

72. **(C)** Static electricity is the electrification of material through physical contact and separation. Sparks may result constituting a fire or explosion hazard.

73. **(C)** Salt water has greater density than fresh water. For this reason, it is easier to float in salt water than in fresh water.

74. **(C)** The inclined ramp (or plane) is simply a sloping platform that enables one to raise an object without having to lift it vertically. It is a simple machine that helps one perform work.

75. **(C)** A distinct mechanical advantage is gained by using a large lever (the crank) to move a small object (the drum).

76. **(B)** Both balls will reach the ground at the same time because the gravitational force on objects with the same surface size varies with the distance of the objects from the center of the earth. The weight of the objects does not matter.

77. **(D)** If you do not know the answer, you can reason out this question. Choices (B) and (C) are irrelevant. Choice (A) makes no sense. It is reasonable to assume that the space in a gasoline drum is filled with gasoline vapors.

78. **(C)** The explanation is based upon a well-known principle of centrifugal force.

79. **(D)** When water freezes, it increases its volume by one-eleventh. Example: When 11 cubic inches of water freeze, 12 cubic inches of ice form. This is the reason why water pipes burst and auto radiators without antifreeze are severely damaged.

80. **(D)** Electrical resistance varies directly with the length of the wire. The longer the wire, the greater the resistance, and since one piece is twice the length of the other, it stands to reason that the resistance would be twice as great.

81. **(B)** Resistance varies inversely with the cross-sectional area of the wire. The wire with the greater cross-sectional area has less resistance. Since the larger diameter wire is one-third greater than the other, the resistance is two-thirds that of the smaller wire.

82. **(D)** When resistance is placed against the drill bit, the tool will function satisfactorily and the hole is made. As the bit approaches the underside of the wood, the resistance is lessened and the bit is forced through the wood, splintering the underside.

83. **(A)** Jumping into a net is not an activity with which many people have had experience. The person jumping may be unable to gauge distance accurately. Firefighters must be alert so as to position the net under the person.

84. **(B)** The metal spoon is a good conductor and will rapidly conduct heat away from the water. This will reduce the amount of heat that would be absorbed by the glass and prevent it from breaking.

85. **(B)** Friction in a pipe or hose varies directly with the length. The longer the pipe or hose, the greater the friction. The immediate result of friction in a hose is to cut down the available pressure at the nozzle.

86. **(B)** This is a principle of pressure in fluids. Pressure applied to a confined fluid from without is transmitted in all directions without differentiation. Therefore, all points in the 100-foot pipe would have the same pressure.

87. **(B)** This is an example of a second-class lever. (The weight is between the fulcrum and the lift.) By moving the weight closer to the fulcrum, an increase in distance between the load and the lift point is achieved. This increases the mechanical advantage of the lever and less effort is necessary to lift the load.

88. **(B)** As stated in the third sentence. Alterations in old buildings frequently create voids or channels that allow fire to extend throughout a building. Removal of partitions and/or structural members have resulted in the collapse of buildings.

89. **(C)** This is the meaning of the fourth sentence.

90. **(B)** As stated in the last sentence.

91. **(B)** See the third sentence.

92. **(C)** Many times the firefighter is the first person at the scene of an accident. By being knowledgeable in first aid, he or she can give prompt and efficient care to an injured person. This is essential when attending to life-threatening emergencies.

93. **(D)** Proper treatment of shock is to avoid rough handling, control bleeding, assure breathing, and give nothing by mouth.

94. **(D)** Loosening the tourniquet may result in death by causing additional bleeding, dislodging clots, and causing tourniquet shock.

95. **(A)** An open fracture is, as its name implies, a fracture in which there is an open wound. Prompt and efficient care should be given to a closed fracture to prevent it from becoming an open or compound fracture.

96. **(B)** The primary objective of the supervisor is to get the job accomplished. Directing is instructing, informing, and ordering subordinates about what to do and sometimes how to do it in order to achieve the objectives of the department.

97. **(C)** The firefighter who prepares for promotion examinations will also develop a broad view of the many functions of the fire department in fire fighting, fire prevention, and administration. The promotion examination candidate acquires job knowledge by studying and through personal experience, thereby becoming a better firefighter.

98. **(A)** Constructive criticism, by definition, is criticism meant to lead to change and improvement in performance.

99. **(B)** Mistakes cannot be ignored and must be corrected immediately. By correcting the improper action quickly, the function would be performed properly, and the firefighter would learn how to do the job correctly. Control of the fire requires that all functions be properly performed. The fire cannot wait until later.

100. **(B)** Positive discipline includes instructing, training, correcting mistakes, and helping an employee to do a better job.

Answer Sheet for Practice Examination 3

1. Ⓐ Ⓑ Ⓒ Ⓓ	21. Ⓐ Ⓑ Ⓒ Ⓓ	41. Ⓐ Ⓑ Ⓒ Ⓓ	61. Ⓐ Ⓑ Ⓒ Ⓓ	81. Ⓐ Ⓑ Ⓒ Ⓓ
2. Ⓐ Ⓑ Ⓒ Ⓓ	22. Ⓐ Ⓑ Ⓒ Ⓓ	42. Ⓐ Ⓑ Ⓒ Ⓓ	62. Ⓐ Ⓑ Ⓒ Ⓓ	82. Ⓐ Ⓑ Ⓒ Ⓓ
3. Ⓐ Ⓑ Ⓒ Ⓓ	23. Ⓐ Ⓑ Ⓒ Ⓓ	45. Ⓐ Ⓑ Ⓒ Ⓓ	63. Ⓐ Ⓑ Ⓒ Ⓓ	83. Ⓐ Ⓑ Ⓒ Ⓓ
4. Ⓐ Ⓑ Ⓒ Ⓓ	24. Ⓐ Ⓑ Ⓒ Ⓓ	44. Ⓐ Ⓑ Ⓒ Ⓓ	64. Ⓐ Ⓑ Ⓒ Ⓓ	84. Ⓐ Ⓑ Ⓒ Ⓓ
5. Ⓐ Ⓑ Ⓒ Ⓓ	25. Ⓐ Ⓑ Ⓒ Ⓓ	45. Ⓐ Ⓑ Ⓒ Ⓓ	65. Ⓐ Ⓑ Ⓒ Ⓓ	85. Ⓐ Ⓑ Ⓒ Ⓓ
6. Ⓐ Ⓑ Ⓒ Ⓓ	26. Ⓐ Ⓑ Ⓒ Ⓓ	46. Ⓐ Ⓑ Ⓒ Ⓓ	66. Ⓐ Ⓑ Ⓒ Ⓓ	86. Ⓐ Ⓑ Ⓒ Ⓓ
7. Ⓐ Ⓑ Ⓒ Ⓓ	27. Ⓐ Ⓑ Ⓒ Ⓓ	47. Ⓐ Ⓑ Ⓒ Ⓓ	67. Ⓐ Ⓑ Ⓒ Ⓓ	87. Ⓐ Ⓑ Ⓒ Ⓓ
8. Ⓐ Ⓑ Ⓒ Ⓓ	28. Ⓐ Ⓑ Ⓒ Ⓓ	48. Ⓐ Ⓑ Ⓒ Ⓓ	68. Ⓐ Ⓑ Ⓒ Ⓓ	88. Ⓐ Ⓑ Ⓒ Ⓓ
9. Ⓐ Ⓑ Ⓒ Ⓓ	29. Ⓐ Ⓑ Ⓒ Ⓓ	49. Ⓐ Ⓑ Ⓒ Ⓓ	69. Ⓐ Ⓑ Ⓒ Ⓓ	89. Ⓐ Ⓑ Ⓒ Ⓓ
10. Ⓐ Ⓑ Ⓒ Ⓓ	30. Ⓐ Ⓑ Ⓒ Ⓓ	50. Ⓐ Ⓑ Ⓒ Ⓓ	70. Ⓐ Ⓑ Ⓒ Ⓓ	90. Ⓐ Ⓑ Ⓒ Ⓓ
11. Ⓐ Ⓑ Ⓒ Ⓓ	31. Ⓐ Ⓑ Ⓒ Ⓓ	51. Ⓐ Ⓑ Ⓒ Ⓓ	71. Ⓐ Ⓑ Ⓒ Ⓓ	91. Ⓐ Ⓑ Ⓒ Ⓓ
12. Ⓐ Ⓑ Ⓒ Ⓓ	32. Ⓐ Ⓑ Ⓒ Ⓓ	52. Ⓐ Ⓑ Ⓒ Ⓓ	72. Ⓐ Ⓑ Ⓒ Ⓓ	92. Ⓐ Ⓑ Ⓒ Ⓓ
13. Ⓐ Ⓑ Ⓒ Ⓓ	33. Ⓐ Ⓑ Ⓒ Ⓓ	55. Ⓐ Ⓑ Ⓒ Ⓓ	73. Ⓐ Ⓑ Ⓒ Ⓓ	93. Ⓐ Ⓑ Ⓒ Ⓓ
14. Ⓐ Ⓑ Ⓒ Ⓓ	34. Ⓐ Ⓑ Ⓒ Ⓓ	54. Ⓐ Ⓑ Ⓒ Ⓓ	74. Ⓐ Ⓑ Ⓒ Ⓓ	94. Ⓐ Ⓑ Ⓒ Ⓓ
15. Ⓐ Ⓑ Ⓒ Ⓓ	35. Ⓐ Ⓑ Ⓒ Ⓓ	55. Ⓐ Ⓑ Ⓒ Ⓓ	75. Ⓐ Ⓑ Ⓒ Ⓓ	95. Ⓐ Ⓑ Ⓒ Ⓓ
16. Ⓐ Ⓑ Ⓒ Ⓓ	36. Ⓐ Ⓑ Ⓒ Ⓓ	56. Ⓐ Ⓑ Ⓒ Ⓓ	76. Ⓐ Ⓑ Ⓒ Ⓓ	96. Ⓐ Ⓑ Ⓒ Ⓓ
17. Ⓐ Ⓑ Ⓒ Ⓓ	37. Ⓐ Ⓑ Ⓒ Ⓓ	57. Ⓐ Ⓑ Ⓒ Ⓓ	77. Ⓐ Ⓑ Ⓒ Ⓓ	97. Ⓐ Ⓑ Ⓒ Ⓓ
18. Ⓐ Ⓑ Ⓒ Ⓓ	38. Ⓐ Ⓑ Ⓒ Ⓓ	58. Ⓐ Ⓑ Ⓒ Ⓓ	78. Ⓐ Ⓑ Ⓒ Ⓓ	98. Ⓐ Ⓑ Ⓒ Ⓓ
19. Ⓐ Ⓑ Ⓒ Ⓓ	39. Ⓐ Ⓑ Ⓒ Ⓓ	59. Ⓐ Ⓑ Ⓒ Ⓓ	79. Ⓐ Ⓑ Ⓒ Ⓓ	99. Ⓐ Ⓑ Ⓒ Ⓓ
20. Ⓐ Ⓑ Ⓒ Ⓓ	40. Ⓐ Ⓑ Ⓒ Ⓓ	60. Ⓐ Ⓑ Ⓒ Ⓓ	80. Ⓐ Ⓑ Ⓒ Ⓓ	100. Ⓐ Ⓑ Ⓒ Ⓓ

TEAR HERE

PRACTICE EXAMINATION 3

The time allowed for the entire examination is 3½ hours.

DIRECTIONS: Each question has four suggested answers, lettered A, B, C, and D. Decide which one is the best answer. On the sample answer sheet, find the question number that corresponds to the answer you have selected and darken the area with a soft pencil.

Memory Question

The following passage describes an incident encountered by firefighters while engaging in their day-to-day activities. You will have five minutes to study this passage. Questions 1–15 are based on the details described in these paragraphs. Therefore, you should make every effort to memorize the highlights as they are described during the time allowed for memorization.

During the summer of 1981, a severe water shortage existed which resulted from an extended heat wave and drought. Hydrants opened by citizens in an effort to gain relief from the heat were depleting water resources rapidly, and emergency measures were undertaken by the city in order to conserve water.

Fire department personnel conducted a survey which revealed that 600 hydrants were opened daily by citizens and that an average of 1100 gallons of water per minute were being wasted at each open hydrant. Hydrant patrols were implemented using 75 engine companies on two-hour shifts to close down open hydrants.

One day, while on hydrant patrol, Engine 299 stopped to shut down an open hydrant which was being used by area residents in order to gain relief from the summer heat. While the firefighters were in the act of closing the hydrant, they were subjected to severe verbal abuse by the crowd that quickly formed around them. A potentially explosive situation was developing rapidly. The firefighters closed the hydrant and quickly left the scene to return to the firehouse.

They were in the firehouse for a very short time when an alarm was received to return to the same street for a reported fire in an occupied multiple-dwelling building. Engine 299 responded to the alarm. As they turned into the street, they were greeted with a barrage of rocks and bottles thrown from doorways and rooftops. The officer in command ordered the firefighters to take cover immediately and transmitted a radio message requesting police assistance. Police officers arrived and dispersed the crowd. The firefighters then checked the location of the reported fire and learned that there was no actual fire; it was a false alarm. They returned to the firehouse and a report of the incident was forwarded to the fire commissioner.

211

The following day, the deputy chief requested that a meeting be arranged between fire department officials and community leaders in an effort to discuss and resolve the problem. At the meeting, fire department representatives were to inform community leaders of the importance of having adequate water pressure available to extinguish fires and to convince residents that firefighters were there to help them. Residents would also be informed that sprinkler caps for fire hydrants were available at the firehouse. The use of these caps would result in the saving of much water. It would then be possible for the residents to use the hydrants to cool themselves, and sufficient water pressure would be maintained to fight fires.

Community leaders must be convinced that responding to false alarms can result in the destruction of property and the deaths of citizens since firefighters would not be able to respond to an actual fire if they were responding to a false alarm.

> *DIRECTIONS:* Questions 1–15 relate to the information given in the preceding passage and are to be answered in accordance with this information.

1. According to the passage, a severe water shortage existed in the city during the summer of 1981. This shortage was due to

   (A) broken water mains
   (B) open hydrants
   (C) an extended heat wave and drought
   (D) misuse of water by industry

2. The firefighters of Engine 299 stopped to shut down the open hydrant while

   (A) returning to quarters
   (B) responding to an alarm
   (C) on alarm box inspection duty
   (D) on hydrant patrol duty

3. From the information given in the passage, we can conclude that the main concern of the fire department was

   (A) to provide the area residents with sprinkler caps
   (B) to have water available for fire fighting
   (C) education of the residents in the area
   (D) to show the cooperation between the police and fire departments

4. If a lieutenant and five firefighters are on duty in each engine company, the total personnel hours spent by the fire department on hydrant patrol for one day was

   (A) 150 hours
   (B) 350 hours
   (C) 600 hours
   (D) 900 hours

5. The *most important* factor the fire officials wanted to convey to the community residents at the meeting was

   (A) to conserve water
   (B) to use sprinkler caps whenever possible
   (C) that false alarms can result in death
   (D) not to abuse firefighters while they are on duty

6. During extended periods of hot weather, hydrant patrol duty usually becomes a daily routine. Of the following, the action that should *not* be taken during those patrols is to

   (A) learn the location and the condition of the hydrants in the district
   (B) use force on area residents when encountering opposition to the firefighters
   (C) train chauffeurs to operate the fire apparatus
   (D) shut down open hydrants to insure adequate water pressure

7. The officer in command ordered the firefighters to take shelter from the barrage of rocks and bottles being thrown. The lieutenant's actions were

   (A) improper; the lieutenant should have first ordered a search for the fire
   (B) proper; however, the lieutenant should have tried to calm the crowd first
   (C) improper; the lieutenant should have requested more fire department units to respond to the scene
   (D) proper; the lieutenant's immediate responsibility was the safety and protection of the firefighters under attack

8. The firefighters shut down the hydrant and hastily left the scene in order to

   (A) respond to an alarm for a fire in a multiple dwelling
   (B) respond to a false alarm
   (C) obtain police assistance
   (D) avoid a physical confrontation with the crowd

9. Fire department engine companies were deployed to shut down open hydrants. The total number of hours on patrol duty by fire department companies was

   (A) 150 hours
   (B) 200 hours
   (C) 250 hours
   (D) 300 hours

10. It would be incorrect to state that the police were called to the scene

    (A) to shut down the open hydrant
    (B) to control the crowd
    (C) for protection of the firefighters
    (D) by the lieutenant in command of Engine 299

11. Open hydrants were depleting the water supply rapidly. In a two-hour period, an open hydrant was discharging approximately

    (A) 1100 gallons of water
    (B) 2200 gallons of water
    (C) 66,000 gallons of water
    (D) 132,000 gallons of water

12. According to the information contained in the passage, Engine 299 responded from quarters to

    (A) investigate a complaint of rubbish on a roof
    (B) a reported fire in an occupied private dwelling
    (C) a reported fire in a multiple dwelling
    (D) assist police at a false alarm

13. The average number of hydrants shut down each day by each engine company was

    (A) 4
    (B) 6
    (C) 8
    (D) 10

14. The officer in command of Engine 299 forwarded a report of the rock and bottle throwing incident. This report was forwarded to the

    (A) fire commissioner
    (B) deputy chief
    (C) community leaders
    (D) police department

15. It would be reasonable to believe that the main purpose of the meeting between fire officials and the community leaders was to

    (A) inform the residents of the importance of the firefighters
    (B) stress the importance of having adequate water pressure available for fire fighting
    (C) remind the community that the firefighter is their friend
    (D) make the community aware of the sprinkler caps

16. Each year many children die in fires they have started while playing with matches. Of the following measures, the one that would be most effective in preventing such tragedies is to

    (A) warn the children of the dangers involved
    (B) punish parents who are found guilty of neglecting their children
    (C) keep matches out of the reach of children
    (D) use only safety matches

17. Holding a fire drill during a performance in a Broadway theater would be ineffective mainly because

    (A) the audience is in the theater for pleasure and would resent interruption of the performance
    (B) the members of the audience are transients who do not return to this theater regularly
    (C) the members of the audience have easy access to many exits in the theater, permitting speedy and safe evacuation of the building
    (D) theaters are usually of fireproof construction and unlikely to catch fire

18. "The same care should be taken to avoid damage when extinguishing a fire in the home of a poor family as would be taken in the home of a wealthy family." The best justification for this statement is that the possessions of a poor family

    (A) probably are not covered by fire insurance
    (B) sometimes include items of great monetary value
    (C) may be very precious to the family
    (D) may contain valuable or irreplaceable documents such as birth certificates, military discharge certificates, etc.

19. Smoking in bed is considered to an unsafe practice because the smoker may

    (A) become drowsy and carelessly discard smoldering matches or cigarettes
    (B) fall asleep and burn his or her fingers or lips

    (C) fall asleep and set fire to his or her clothing, the bedding, or furniture

    (D) fall asleep and choke from the tobacco smoke and fumes

20. At a fire in a rooming house, the manager tells the officer in command that all the roomers have evacuated the building. The action which would be most proper for the officer to take in the situation is to

    (A) order a search of the building for possible victims

    (B) question the manager closely to make certain that everybody has in fact been evacuated

    (C) question the roomers to obtain confirmation that all occupants have been evacuated

    (D) obtain a list of the roomers and then verify that each one of them is outside of the building

21. An elevator in a large apartment house became stuck between floors and the fire department was called to remove the trapped passengers. Soon after arriving, the officer in command informed the passengers that the fire department was present and would start rescue operations immediately. The main reason for informing the passengers of their arrival was to reduce the chances that

    (A) another agency would receive credit for the rescue

    (B) the fire department would be criticized for being slow in responding

    (C) the passengers would not cooperate with the rescuers

    (D) the passengers would become panic-stricken

22. The election laws of many states require that general elections are to take place on the first Tuesday following the first Monday in November. The statement that is most accurate is that elections in those states

    (A) in some years may be held on November 1 and in other years on November 8

    (B) may not be held on either November 1 or November 8

    (C) in some years may be held on November 1 but never on November 8

    (D) in some years may be held on November 8 but never on November 1

23. At a smoky fire in a hotel, members of a ladder company are ordered to search all rooms and remove any persons they find remaining in the area threatened by fire or smoke. While performing this duty, a firefighter is prevented from entering a room by a ferocious dog standing in the doorway. In this situation, the best of the following actions for the firefighter to take is to

    (A) look into the room without entering and, if no person can be seen inside, assume that the room is empty

    (B) obtain assistance from other firefighters, subdue the dog, then enter and search the room

    (C) attempt to find the dog's owner so that the dog can be removed and the room entered and searched

    (D) call for assistance of the A.S.P.C.A. in removing the dog and then enter and search the room

24. Sparks given off by welding torches are a serious fire hazard. The best of the following methods of dealing with this hazard is to conduct welding operations

    (A) only in fireproof buildings protected by sprinkler systems

    (B) only out-of-doors on a day with little wind blowing

(C) only on materials certified to be noncombustible by recognized testing laboratories

(D) only after loose combustible materials have been cleared from the area and with a firefighter standing by with a hoseline

25. A firefighter on the way to report for work early one evening noticed a man loitering near a fire alarm box. The man appeared to be touching the alarm box handle but withdrew his hand as the firefighter approached. When asked by the firefighter to explain his actions, the man denied touching the handle. In this situation, the firefighter should

(A) summon a police officer and have the man arrested
(B) ignore the incident since no crime has been committed
(C) keep the man under surveillance until the man returns to his home
(D) locate the patrol officer on the beat and inform him or her of the incident

26. In New York, schools are required to hold at least 12 fire drills each school year, 8 of which must be conducted between September and December. Of the following, the best justification for holding most fire drills during the September 1–December 1 period is that

(A) most fires occur during that period
(B) pupils are trained in evacuation of their schools at the beginning of the school year
(C) the weather is milder during that period
(D) school attendance is higher at the beginning of the school year

27. Following a severe smoky fire in the lobby of a hotel, a dead body was found in a room on the eighth floor. The coroner who examined the body found that the victim died from the inhalation of toxic gases and smoke. The most probable explanation of the occurrence is that

(A) the coroner made an error in the findings
(B) the victim inhaled the gases and smoke in the lobby, fled to the eighth floor to escape the fire, and collapsed there
(C) the gases and smoke traveled through vertical openings to the upper stories where they were inhaled by the victim.
(D) a second fire occurred simultaneously in the upper stories of the hotel

28. The basic assumption of fire prevention educational programs is that people frequently

(A) must be forced into obeying fire laws
(B) are unaware of the dangers involved in some of their actions
(C) do not care whether or not their actions are dangerous
(D) assume that fire insurance protects them against all fire loss

29. The statement that is most accurate is that visibility in buildings on fire

(A) is not a serious problem mainly because fires give off sufficient light to enable firefighters to see clearly
(B) is a serious problem mainly because fires often knock out the electrical system of the building
(C) is not a serious problem mainly because most fires occur during daylight hours
(D) is a serious problem mainly because smoke conditions are often encountered

30. While operating at a five-alarm fire, a firefighter is approached by an obviously intoxicated man who claims to be a former firefighter and who offers to help put out the fire. In this situation, the best of the following courses of action for the firefighter is to

(A) give the man some easy task to perform
(B) refer the man to the officer in command of the fire
(C) decline the offer of help and ask the man to remain outside of the fire lines
(D) ask to see the man's credentials and, if he is a former firefighter as he claims, put him to work stretching hoselines

31. A civilian came into a firehouse to report that some adolescents had turned on a hydrant in the neighborhood. After getting the location of the hydrant, the firefighter on house watch asked the civilian for his name and was told "Smith." The firefighter then thanked the civilian and proceeded to process the report. The firefighter's actions in this situation were

(A) proper
(B) proper, except that the firefighter should not have wasted time asking for the civilian's name
(C) proper, except to find out whether the civilian's name really was Smith
(D) proper, except that the firefighter should have obtained the civilian's first name and address as well as his family name

32. Fire apparatus are not permitted to use their sirens when returning to quarters. Of the following, the main justification for this restriction is that

(A) the chances of being involved in traffic accidents are reduced
(B) there is no need for prompt return to quarters after a fire has been extinguished
(C) apparatus return to quarters by less crowded streets than are used when responding to alarms
(D) the officer in command is better able to exchange radio messages with the fire alarm dispatcher

33. A fire escape platform is attached to a building flush with the window sill leading onto it. The most important reason for this arrangement is that it reduces the chance that

(A) intruders will utilize the fire escapes for illegal purposes since they will be visible from inside the apartment
(B) tenants will place garbage or other obstructions on the platform since these objects will be visible from inside the apartment
(C) persons using the fire escape in an emergency will trip as they leave the apartment
(D) sagging or loosening of the supports will occur without coming to the attention of persons occupying the apartment

34. A civilian, on his way home from work one evening, hears an alarm ringing and sees water running out of a sprinkler discharge pipe on the side of a building. No smoke or other indications of fire are seen. There is a padlock on the entrance to the building, no fire alarm box is in sight, and a firehouse is located 1½ blocks away. In this situation, it would be most proper for the civilian to

(A) take no action since there is no evidence of fire or other emergency
(B) attempt to ascertain the name of the building's owner and notify him of the situation
(C) send an alarm from the closest fire alarm box
(D) go to the firehouse and inform the firefighter at the desk of the situation

35. Running a hoseline between two rungs of a ladder into a building on fire is generally considered to be a

(A) good practice, mainly because resting the hose on the rungs of the ladder reduces the load that must be carried by the firefighters
(B) poor practice, mainly because the maneuverability of the ladder is reduced
(C) good practice, mainly because the weight of the hoseline holds the ladder more securely against the side of the building
(D) poor practice, mainly because the ladder may be damaged by the additional weight of the hoseline filled with water

36. Suppose that you are the firefighter on housewatch duty when a woman rushes into the firehouse and reports that a teenage gang is assaulting a man on the street several blocks away. In this situation, the best of the following courses of action for you to take first is to

(A) notify the police department of the reported incident
(B) go to the scene of the disturbance and verify the woman's report
(C) suggest to the woman that she report the matter to the police department
(D) notify your fellow firefighters of the incident and all go to the aid of the man being assaulted

37. A firefighter on the way to work is standing on a subway platform waiting for a train. Suddenly, a man standing nearby collapses to the floor, showing no sign of breathing. In this situation, the firefighter should

(A) run to the nearest telephone and summon an ambulance
(B) designate a responsible person to look after the victim and then continue on the way to the firehouse
(C) administer first-aid measures to restore breathing
(D) take no action since the victim is obviously dead

38. Fire companies often practice methods of fire fighting and lifesaving in public places. The best justification for this practice is that it

(A) provides facilities for practicing essential operations which are not readily available in fire department installations
(B) impresses the taxpaying public that they are getting their money's worth
(C) attracts youngsters and interests them in fire fighting as a career
(D) makes certain that all company equipment is in the best possible working condition

39. The fire which would generally present the greatest danger to surrounding buildings from flying sparks or brands is a fire in a(an)

(A) warehouse storing paper products
(B) factory manufacturing plastics
(C) outdoor lumberyard
(D) open-type parking garage

40. A fire marshal, questioning firefighters at the scene of a suspicious fire, obtains some conflicting statements about details of the fire situation present upon their arrival. The most likely explanation of this conflict is that

(A) the firefighters have not been properly trained to carry out their part in arson investigations

(B) the fire marshal did not give the firefighters time to collect and organize their impressions of the fire scene

(C) witnesses to an event see the situation from their own point of view and seldom agree on all details

(D) details are of little importance if there is general agreement on major matters

41. A firefighter on duty at a theater who discovers standees obstructing aisles should immediately

(A) report the situation to his or her superior officer

(B) order the standees to move out of the aisles

(C) ask the theater manager to correct the situation

(D) issue a summons to the usher assigned to that area

42. At a fire on the fourth floor of an apartment house, the first engine company to arrive advanced a hoseline up the stairway to the third floor before charging the hose with water. The main reason that the firefighters delayed charging their line is that an empty line

(A) is less likely to whip about and injure firefighters

(B) is easier to carry

(C) won't leak water

(D) is less subject to damage

43. Suppose the owner of a burning tenement building complains that, although the fire is located on the first floor, firefighters are chopping holes in the roof. Of the following, the most appropriate reason you can give for their action is that the fire can be fought most effectively by permitting

(A) smoke and hot gases to escape

(B) firefighters to attack the fire from above

(C) firefighters to gain access to the building through the holes

(D) immediate inspection of the roof for extension of the fire

44. The fire department always endeavors to purchase the best apparatus and equipment and to maintain it in the best condition. The main justification for this policy is that

(A) public confidence in the department is increased

(B) failure of equipment at a fire may have serious consequences

(C) replacement of worn out parts is often difficult

(D) the dollar cost to the department is less in the long run

45. The statement about smoke which is most accurate is that smoke is

(A) irritating but not dangerous in itself

(B) irritating and dangerous only because it may reduce the oxygen content of the air breathed

(C) dangerous because it may reduce the oxygen content of the air breathed and often contains toxic gases

(D) dangerous because it supports combustion

46. Suppose that you are a firefighter making a routine inspection of a rubber goods factory. During the inspection, you discover some minor violations of the Fire Prevention Code. When you call these violations to the attention of the factory owner, he becomes annoyed and tells you that he is the personal friend of high officials in the fire department and the city government. Under these circumstances, the best course for you to follow is to

(A) summon a police officer to arrest the owner for attempting to intimidate a public official performing a duty
(B) make a very thorough inspection and serve summonses for every possible violation of the Fire Department Code
(C) ignore the owner's remarks and continue the inspection in your usual manner
(D) try to obtain from the owner the names and positions of his friends

47. It has been suggested that property owners should be charged a fee each time the fire department is called to extinguish a fire on their property. Of the following, the best reason for rejecting this proposal is that

(A) delay in calling the fire department may result
(B) many property owners don't occupy the property they own
(C) property owners may resent such a charge as they pay real estate taxes
(D) it may be difficult to determine on whose property a fire started

48. Standpipe systems of many bridges are the dry pipe type. A dry pipe system has no water in the pipes when not in use; when water is required, it is necessary first to pump water into the system. The main reason for using a dry standpipe system is to prevent

(A) corrosion of the pipes
(B) freezing of water in the pipes
(C) waste of water through leakage
(D) strain on the pumps

49. Assume that you are a firefighter on your way home after completing your tour of duty. Just as you are about to enter the subway, a man runs up to you and reports a fire in a house located five blocks away. You recognize the man as a mildly retarded but harmless person who frequently loiters around firehouses and at fires. Of the following, the best action for you to take is to

(A) run to the house to see if there really is a fire
(B) call in an alarm from a nearby telephone
(C) ignore the report because of the man's mental condition
(D) call the police and ask that a radio patrol car investigate the report

50. Sometimes a piece of apparatus is ready to leave the fire station before all members are completely dressed and equipped. In order to avoid delay, these firefighters finish dressing on the way to the fire. This practice is undesirable mainly because

(A) a poor impression is made on the public
(B) the firefighters are not able to size up the situation as they approach the fire
(C) the possibility of dropping equipment from the moving apparatus is increased
(D) the danger of injury to the firefighters is increased

DIRECTIONS: Answer questions 51–55 on the basis of the information found in the following passage.

1. The sizes of living rooms shall meet the following requirements:

a. In each apartment there shall be at least one living room containing at least 120 square feet of clear floor area, and every other living room, except a kitchen, shall contain at least 70 square feet of clear floor area.

b. Every living room which contains less than 80 square feet of clear floor area or which is located in the cellar or basement shall be at least 9 feet high and every other living room 8 feet high.

2. Apartments containing three or more rooms may have dining bays, which shall not exceed 55 square feet in floor surface area and shall not be deemed separate rooms or subject to the requirements for separate rooms. Every such dining bay shall be provided with at least one window containing an area at least one-eighth of the floor surface area of such dining bay.

51. The minimum volume of a living room, other than a kitchen, which meets the minimum requirements listed in the passage is one that measures

(A) 70 cubic feet
(B) 80 cubic feet
(C) 630 cubic feet
(D) 640 cubic feet

52. A builder proposes to construct an apartment house containing an apartment consisting of a kitchen which measures 10 feet by 6 feet, a room 12 feet by 12 feet, and a room 11 feet by 7 feet. This apartment

(A) does not comply with the requirements listed in the passage
(B) complies with the requirements listed in the passage provided that it is not located in the cellar or basement
(C) complies with the requirements listed in the passage provided that the height of the smaller rooms is at least 9 feet
(D) may or may not comply with the requirements listed in the passage depending upon the clear floor area of the kitchen

53. The definition of the term "living room" that is most in accord with its meaning in the passage is

(A) a sitting room or parlor
(B) the largest room in an apartment
(C) a room used for living purposes
(D) any room in an apartment containing 120 square feet of clear floor area.

54. Assume that one room in a four-room apartment measures 20 feet by 10 feet and contains a dining bay 8 feet by 6 feet. According to the passage, the dining bay must be provided with a window measuring at least

(A) 6 square feet
(B) 7 square feet
(C) 25 square feet
(D) 55 square feet

55. Kitchens, according to the preceding passage,

(A) are not considered "living rooms"
(B) are considered "living rooms" and must, therefore, meet the height and area requirements of the passage
(C) are considered "living rooms" but they need meet only the height or area requirements of the passage, not both
(D) are considered "living rooms" but they need not meet area requirements

DIRECTIONS: The following five questions, numbered 56–60, are based on the following regulations:

EMPLOYEE LEAVE REGULATIONS

"As a full-time permanent City employee under the Career and Salary Plan, firefighter Peter Smith earns an 'annual leave allowance.' This consists of a certain number of days off a year with pay, and may be used for vacation, personal business, or for observing religious holidays. As a newly appointed employee, during his first 8 years of City service he will earn an 'annual leave allowance' of 20 days off a year (an average of $1^2/_3$ days off a month). After he has finished 8 full years of working for the City, he will begin earning an additional 5 days off a year. His 'annual leave allowance' will then be 25 days a year and will remain at this amount for 7 full years. He will begin earning an additional 2 days off a year after he has completed a total of 15 years of City employment. Therefore, in his sixteenth year of working for the City, Smith will be earning 27 days off a year as his 'annual leave allowance' (an average of $2^1/_4$ days off a month).

A 'sick leave allowance' of 1 day a month is also given to firefighter Smith, but it can be used only in case of actual illness. When Smith returns to work after using 'sick leave allowance' he *must* have a doctor's note if the absence is for a total of more than 3 days, but he may also be required to show a doctor's note for absences of 1, 2, or 3 days."

56. According to the preceding passage, Mr. Smith's "annual leave allowance" consists of a certain number of days off a year which he

    (A) does not get paid for
    (B) gets paid for at time and a half
    (C) may use for personal business
    (D) may not use for observing religious holidays

57. According to the preceding passage, after Mr. Smith has been working for the City for 9 years his "annual leave allowance" will be

    (A) 20 days a year   (C) 27 days a year
    (B) 25 days a year   (D) 37 days a year

58. According to the preceding passage, Mr. Smith will begin earning an average of $2^1/_4$ days off a month as his "annual leave allowance" after he has worked for the City for

    (A) 7 full years   (C) 15 full years
    (B) 8 full years   (D) 17 full years

59. According to the preceding passage, Mr. Smith is given a "sick leave allowance" of

    (A) 1 day every 2 months   (C) $1^2/_3$ days per month
    (B) 1 day per month   (D) $2^1/_4$ days a month

60. According to the preceding passage, when he uses "sick leave allowance" Mr. Smith may be required to show a doctor's note

    (A) even if his absence is for only 1 day
    (B) only if his absence is for more than 2 days

(C) only if his absence is for more than 3 days
(D) only if his absence is for 3 days or more

DIRECTIONS: The following four questions, numbered 61 to 64, are based on the following paragraph.

A summons is an official statement ordering a person to appear in court. In traffic violation situations, summonses are used when arrests need not be made. The main reason for traffic summonses is to deter motorists from repeating the same traffic violation. Occasionally motorists may make unintentional driving errors, and sometimes they are unaware of correct driving regulations. In cases such as these, the policy is to have the Officer verbally inform the motorist of the violation and warn him or her against repeating it. The purpose of this practice is not to limit the number of summonses, but rather to prevent the issuing of summonses when the violation is not due to deliberate intent or to inexcusable negligence.

61. According to the preceding passage, the principal reason for issuing traffic summonses is to

(A) discourage motorists from violating these laws again
(B) increase the money collected by the City
(C) put traffic violators in prison
(D) have them serve as substitutes for police officers

62. The reason a verbal warning may sometimes be substituted for a summons is to

(A) limit the number of summonses
(B) distinguish between excusable and inexcusable violations
(C) provide harsher penalties for deliberate intent than for inexcusable negligence
(D) decrease the caseload in the courts

63. The author of the preceding passage feels that someone who violated a traffic regulation because he or she did *not* know about the regulation should be

(A) put under arrest
(B) fined less money
(C) given a summons
(D) told not to do it again

64. Using the distinctions made by the author of the preceding passage, the one of the following motorists to whom it would be most desirable to issue a summons is the one who exceeded the speed limit because he or she

(A) did not know the speed limit
(B) was late for an important business appointment
(C) speeded to avoid being hit by another car
(D) had a speedometer which was not working properly

DIRECTIONS: The following four questions, numbered 65 to 68, are based on the following paragraph.

Whenever a social group has become so efficiently organized that it has gained access to an adequate supply of food and has learned to distribute it among its members so well that wealth considerably exceeds immediate demands, it can be depended upon to utilize its surplus energy in an attempt to enlarge the sphere in which it is active. The structure of ant colonies renders them particularly prone to this sort of expansionist policy. With very few exceptions, ants of any given colony are hostile to those of any other colony, even of the same species, and this condition is bound to produce preliminary bickering among colonies which are closely associated.

65. According to the paragraph, a social group is wealthy when it

   (A) is efficiently organized
   (B) controls large territories
   (C) contains energetic members
   (D) produces and distributes food reserves

66. According to the paragraph, the structure of an ant colony is its

   (A) social organization
   (B) nest arrangement
   (C) territorial extent
   (D) food-gathering activities

67. It follows from the preceding paragraph that the *least* expansionist society would be one that has

   (A) great poverty generally
   (B) more than sufficient wealth to meet its immediate demands
   (C) great wealth generally
   (D) wide inequality between its richest and poorest members

68. According to the preceding paragraph, an ant is generally hostile *except* to other

   (A) insects
   (B) ants
   (C) ants of the same species
   (D) ants of the same colony

   DIRECTIONS: The following three questions, numbered 69 to 71, relate to the information given in the following paragraph and are to be answered in accordance with this information.

Fire records indicate that about ten percent of the total value of fire loss in the United States is due to fires caused by heating equipment, and that insufficient clearance to combustible materials is an important factor in a large percentage of these fires. The need for adequate clearance is emphasized in periods of severe cold weather, when heating equipment is run at full capacity, by a marked increase in the number of fires.

69. In a given year in the United States, 2,000,000 fires caused 10,000 deaths and destroyed $500,000,000 worth of property. We may conclude from the paragraph that heating equipment caused

   (A) 200,000 fires
   (B) fires resulting in the death of 1000 persons
   (C) fires resulting in the destruction of $50,000,000 worth of property

(D) 200,000 fires resulting in the death of 1000 persons and the destruction of $50,000,000 worth of property

70. The phrase "clearance to combustible materials," as used in the paragraph, means

(A) testing laboratory approval for heating equipment
(B) maintaining heating equipment in clean condition
(C) keeping sufficient space between heating equipment and inflammable material
(D) obtaining appropriate permits before installing the heating equipment

71. The month-to-month changes in the number of fires caused by heating equipment is most affected by the

(A) season of the year
(B) amount of clearance to combustible materials
(C) condition of the equipment
(D) long-term trend in fire statistics

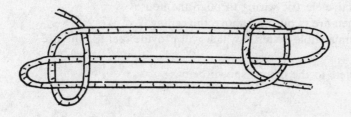

72. The knot shown in the above diagram would best be used to

(A) form a noose for hoisting tools
(B) wrap around fire victims to lower them to the ground
(C) shorten the effective length of a piece of rope
(D) lash two ladders together to extend their lengths

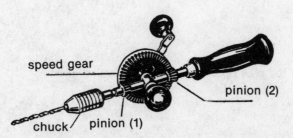

speed gear

pinion (2)

chuck   pinion (1)

73. The function of pinion gear (2) in the hand drill shown in the above diagram is to

(A) increase the speed of the chuck
(B) keep the speed gear from wobbling
(C) double the turning force on the chuck
(D) allow reverse rotation of the speed gear

74. Assume that two identical insulated jugs are filled with equal quantities of water from a water tap. A block of ice is placed in one jug and the same quantity of ice, chopped

into small cubes, is placed in the other jug. The statement that is most accurate is that the water in the jug containing the chopped ice, compared to the water in the other jug, will be chilled

(A) faster but to a substantially higher temperature
(B) faster and to approximately the same temperature
(C) slower but to a substantially higher temperature
(D) slower and to approximately the same temperature

75. The long pole and hook shown in the above sketch is called a pike pole. Firefighters sometimes push the point and hook through plaster ceilings and then pull the ceiling down. Of the following, the most likely reason for this practice is to

(A) let heat and smoke escape from the room
(B) trace defective electric wiring through the house
(C) see if hidden fire is burning above the ceiling
(D) remove combustible material which will provide fuel for the fire

76. Question 76 refers to the figures shown below.

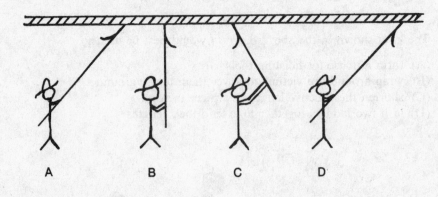

The firefighter using the pike pole in the best way is firefighter

(A) A
(B) B
(C) C
(D) D

77. As your company arrives at the scene of a fire in a large rooming house, a man is spied on the second-story fire escape, about to let himself down to the ground by means of a sheet which he has torn into strips, knotted, and attached to some object in the building. Two middle-aged women and a young boy are observed descending to the street by means of the fire escape. Flames and smoke are issuing from the third-floor windows. Of the following, the best characterization of this man's behavior is that it is

(A) intelligent, because the fire has cut off egress by means of the stairway

(B) not intelligent, because a firm anchor for the knotted sheet could be achieved by securing it to the fire escape

(C) intelligent, because the fire may heat the fire escape to the point where it is red hot

(D) not intelligent, because to descend the fire escape would be a quicker and safer means of escape

78. Suppose that the same quantity of water is placed in the cup and bowl pictured above and both are left on a table. The time required for the water to evaporate completely from the cup, as compared to the bowl, would be

(A) longer
(B) shorter
(C) equal
(D) longer or shorter depending upon the temperature and humidity in the room

79. Some tools are known as all-purpose tools because they may be used for a great variety of purposes; others are called special purpose tools because they are suitable only for a particular purpose. Generally, an all-purpose tool, as compared to a special tool for the same purpose, is

(A) cheaper
(B) less efficient
(C) safer to use
(D) simpler to operate

80. During a snowstorm, a passenger car with rear-wheel drive gets stuck in the snow. It is observed that the rear wheels are spinning in the snow and the front wheels are not turning. The statement which best explains why the car is not moving is that

(A) moving parts of the motor are frozen or blocked by the ice and snow
(B) the front wheels are not receiving power because of a defective or malfunctioning transmission
(C) the rear wheels are not obtaining sufficient traction because of the snow
(D) the distribution of the power to the front and rear wheels is not balanced

81. When the inside face of a rubber suction cup is pressed against a smooth wall, it usually remains in place because of the

(A) force of molecular attraction between the rubber and the wall
(B) pressure of the air on the rubber cup
(C) suction caused by the elasticity of the rubber
(D) static electricity generated by the friction between rubber and the wall

82. The self-contained breathing apparatus consists of a tank containing breathable air supplied to the user by means of flexible tubes. This type of breathing apparatus would be *least* effective in an atmosphere containing

(A) insufficient oxygen to sustain life

(B) gases irritating to the skin
(C) gases which cannot be filtered
(D) a combination of toxic gases

83. At a fire during very cold weather, a firefighter who was ordered to shut down a hose-line left it partially open so that a small amount of water continued to flow out of the nozzle. Leaving the nozzle partially open in this situation was a

(A) good practice, mainly because the hoseline can be put back into action more quickly
(B) poor practice, mainly because the escaping water will form puddles which will freeze on the street
(C) good practice, mainly because the water in the hoseline will not freeze
(D) poor practice, mainly because unnecessary water damage to property will result

84. Some sprinkler systems are supplied with water from a single source; others are supplied with water from two different sources. Generally, firefighters consider the two source system preferable to the single source system because the two source system

(A) is cheaper to install
(B) is less likely to be put out of operation
(C) is capable of delivering water at higher pressure
(D) goes into operation faster

85. The one of the following statements about electric fuses that is most valid is that they

(A) should never be replaced by coins
(B) may be replaced by coins for a short time if there are no fuses available
(C) may be replaced by coins provided that the electric company is notified
(D) may be replaced by coins provided that care is taken to avoid overloading the circuit

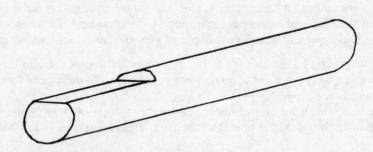

86. The function of the flat surface machined into the shaft in the diagram above is to

(A) prevent slippage of a pulley positioned at this end on the shaft
(B) provide a nonskid surface to hold the shaft steady as it is machined
(C) prevent the shaft from rolling about when it is placed on a flat workbench
(D) reveal subsurface defects in the shaft

DIRECTIONS: Answer questions 87 and 88 on the basis of the diagram below.
A, B, C, and D are four meshed gears forming a gear train. Gear A is the driver.
Gears A and D each have twice as many teeth as gear B, and gear C has four times

as many teeth as gear B. The diagram is schematic; the teeth go all around each gear.

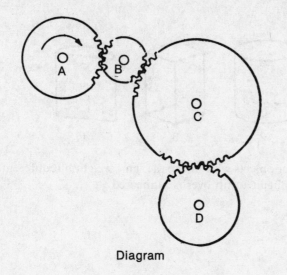

Diagram

87. Two gears which turn in the same direction are

(A) A and B
(B) B and C
(C) C and D
(D) B and D

88. The two gears which revolve at the same speed are gears

(A) A and C
(B) A and D
(C) B and C
(D) B and D

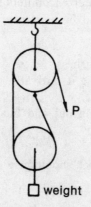

89. In the sketch of a block and fall above, if the end of the rope P is pulled so that it moves one foot, the distance the weight will be raised is

(A) ½ foot

(B) 1 foot
(C) 1¹/₂ feet
(D) 2 feet

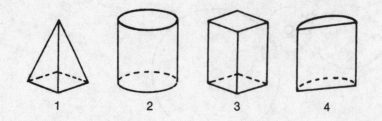

1   2   3   4

90. The sketches above show four objects which have the same weight but different shapes. The object which is most difficult to tip over is numbered

(A) 1
(B) 2
(C) 3
(D) 4

91. A firefighter, taking some clothing to a dry cleaner in his neighborhood, noticed that inflammable cleaning fluid was stored in a way which created a fire hazard. The firefighter called this to the attention of the proprietor, explaining the danger involved. This method of handling the situation was

(A) bad; the firefighter should not have interfered in a matter which was not his responsibility
(B) good; the proprietor would probably remove the hazard and be more careful in the future
(C) bad; the firefighter should have reported the situation to the fire inspector's office without saying anything to the proprietor
(D) good; since the firefighter was a customer, he should treat the proprietor more leniently than he would treat other violators

92. During the course of fighting a fire, a firefighter encounters an unconscious middle-aged man who is bleeding profusely. It would be most proper for the firefighter to administer first aid by initially

(A) attempting to bring him back to consciousness
(B) calling for medical assistance and awaiting its arrival
(C) administering artificial respiration
(D) calling for medical assistance and seeking out and blocking the source of the bleeding

93. The firefighter who encounters a victim of shock should give first aid to the victim by

(A) covering him with a wet cloth to keep him cool
(B) administering artificial respiration
(C) loosening his clothing and removing his shoes and stockings
(D) keeping the patient lying down and elevating extremities if there is no further injury

94. Modern acceptable first-aid technique recommends that, wherever possible, bleeding caused by punctures should be controlled by the application of a

(A) tourniquet
(B) pressure bandage
(C) coagulant
(D) figure eight bandage

95. A newly appointed firefighter becomes involved in an accident while on duty. Faced with this problem, it would be most appropriate for the firefighter's superior officer to first ascertain

(A) if the firefighter is accident-prone
(B) the cause of the accident
(C) if the orientation program stresses employee safety sufficiently
(D) if the firefighter has ever been involved in an accident previously which was covered by workmen's compensation

96. A veteran firefighter feels lightheaded soon after reporting for a tour of duty. It would be most proper for the firefighter to

(A) ask to be excused from ascending heights during this tour of duty
(B) say nothing because a replacement is not available
(C) promptly report this condition to his or her superior officer
(D) discuss the matter with fellow firefighters to see if they had ever had the same problem

97. A firefighter, after completing a special course of study on safety at work, is not convinced that a method taught during the course is the best way to avoid accidents. Upon returning to duty, the firefighter should

(A) perform the task as it was taught
(B) perform the task in the manner he or she is convinced is the best way
(C) attempt to gain the support of fellow firefighters in his or her method
(D) request a transfer to other duty where it will not be necessary to perform this task

98. A firefighter on inspection duty is offered a sum of money by a store owner to overlook a violation of the fire law. The best action to take would be to

(A) order the store owner to clear the violation without delay and report the incident to his or her superior officer
(B) say nothing to the store owner but return to the firehouse to seek the advice of fellow firefighters
(C) inform the store owner that firefighters do not accept bribes
(D) refuse the money but offer to assist the store owner to clear the violation during his or her off-duty hours

99. All newly appointed firefighters are put through an extensive training course before they are permitted to participate in the actual activity of fighting fires. The most probable reason for this is that

(A) it would be unwise for inept firefighters to be exposed to the scrutiny of the public
(B) the skills necessary for effective fire fighting are unique to the job
(C) the orientation training session is necessary to establish the officer–firefighter relationship

(D) newly appointed firefighters must be convinced of the dangers inherent in fire fighting.

100. The statement "Modern technology has greatly influenced fire fighting techniques" means most nearly that

(A) newly appointed firefighters are better equipped to fight fires than their predecessors because they are better educated

(B) the need for investigating whether or not changes should be made in procedures becomes important when new fire fighting equipment is acquired

(C) established successful fire fighting techniques should be retained whenever possible

(D) newly acquired equipment always calls for changes in established procedures

Answer Key for Practice Examination 3

1. C	21. D	41. C	61. A	81. B
2. D	22. D	42. B	62. B	82. B
3. B	23. B	43. A	63. D	83. C
4. D	24. D	44. B	64. B	84. B
5. C	25. D	45. C	65. D	85. A
6. B	26. B	46. C	66. A	86. A
7. D	27. C	47. A	67. A	87. D
8. D	28. B	48. B	68. D	88. B
9. A	29. D	49. B	69. C	89. A
10. A	30. C	50. D	70. C	90. A
11. D	31. D	51. C	71. A	91. B
12. C	32. A	52. C	72. C	92. D
13. C	33. C	53. C	73. B	93. D
14. A	34. D	54. A	74. B	94. B
15. B	35. B	55. D	75. C	95. B
16. C	36. A	56. C	76. D	96. C
17. B	37. C	57. B	77. D	97. A
18. C	38. A	58. C	78. A	98. A
19. C	39. C	59. B	79. B	99. B
20. A	40. C	60. A	80. C	100. D

Explanatory Answers for Practice Examination 3

1. **(C)** As stated in the first sentence of the first paragraph.

2. **(D)** As stated in the first sentence of the third paragraph.

3. **(B)** As stated in the second sentence of the fifth paragraph.

4. **(D)** 75 companies × 2 hours each = 150 hours × 6 persons per engine company = 900 personnel hours.

5. **(C)** See the last paragraph.

6. **(B)** If force is necessary to move people away from hydrants in order to shut the hydrants, the police department should be notified for assistance. The firefighter should not become involved in this action.

7. **(D)** To effectively fight a fire, firefighters should be in good physical condition. One of the lieutenant's responsibilities is to keep the firefighters from becoming injured.

8. **(D)** "A potentially explosive situation was developing rapidly." There was no reason for the firefighters to hang around waiting for trouble.

9. **(A)** 75 companies × 2 hours = 150 hours.

10. **(A)** The police were called by the lieutenant in command of Engine 299 to control the crowd and to protect firefighters, not to close the hydrant.

11. **(D)** 1100 gallons per minute $\times$ 120 minutes (2 hours) = 132,000 gallons.

12. **(C)** See the beginning of the fourth paragraph.

13. **(C)** 600 hydrants $\div$ 75 companies = 8 hydrants.

14. **(A)** As stated in the last sentence of the fourth paragraph.

15. **(B)** As stated in the second sentence of the fifth paragraph.

16. **(C)** Children are curious to try what they see adults doing, such as using matches or cigarette lighters, and working at stoves and fireplaces. Children who do not have access to matches will be unable to play with them.

17. **(B)** Since the theatergoers are transient the drill would be ineffective and disruptive to the performance. Fortunately, enforcement of strict fire codes for theaters results in very low fire incidence.

18. **(C)** The firefighter's job is to extinguish fires and protect life and property regardless of the social status of the occupant. This policy applies to every fire operation whether it is residential, commercial, or public building.

19. **(C)** The majority of bedroom and living-room couch fires originate when smoking material ignites upholstery and clothing. Usually the victim does not sense the fire until it has gained some headway. Carbon monoxide produced in a smoldering fire could paralyze the sleeping victim before the heat of the flame is felt.

20. **(A)** The officer cannot take anything for granted. A roomer may have had a visitor that the manager was not aware of, or a roomer may have returned to his or her room to retrieve some valuables.

21. **(D)** Informing the trapped passengers that the fire department has arrived and is actively working to free them should calm their fears and avert panic.

22. **(D)** If November 1 is on a Tuesday, November 8 would be the first Tuesday after the first Monday. The first Tuesday after the first Monday in November could be no earlier than November 2.

23. **(B)** Generally, a dog will attempt to get away from a heat and smoke condition, unless the owner is nearby. This is especially true if the animal is a Seeing Eye dog. You must search the room for the person the dog is trying to protect.

24. **(D)** The recommended procedure during welding or cutting operations is to have a person acting as a firewatch stand by with a hoseline or an extinguisher. An inspection should be made immediately after conclusion of operations and again one-half hour later.

25. **(D)** By notifying the police officer, the firefighter is fulfilling his responsibility as a concerned citizen. It is the responsibility of the police officer to then keep the man in question under surveillance. The firefighter must report to duty.

26. **(B)** Generally, there are many new students in a school at the beginning of the school year who are unfamiliar with the exits. By having the majority of drills in the early part of the school year, the children become familiar with the evacuation plan and the location of the exits.

27. **(C)** Toxic gases and smoke from a fire are light and rise rapidly through vertical openings in a building. They "mushroom" on the upper floors. This is why ventilation at the roof is essential.

28. **(B)** People frequently create unsafe situations. Poor housekeeping or poor safety habits could eventually result in a destructive fire. Educating people about the existing dangers can save them from injury, death, or destruction of their property.

29. **(D)** A top priority in fire fighting is ventilation of the building in order to remove smoke and toxic gases. Smoke removal will heighten visibility, save lives, reduce the chances of panic, and enable firefighters to perform more effectively.

30. **(C)** Only authorized, qualified firefighters are permitted to operate at a fire. Be firm. If the man insists upon "helping," have a police officer remove him from the scene.

31. **(D)** Generally it is a good policy to obtain as much information as you can from the complainant. It makes reporting easier and more complete. By having the full name and address of the complainant of record, contact could be made quite easily if necessary.

32. **(A)** When returning from an alarm, fire apparatus should adhere to traffic regulations like any other vehicle on the road. Sirens can cause motorists and pedestrians to become confused and nervous, thereby increasing the chance of an accident.

33. **(C)** People using a fire escape in an emergency are usually excited. Any obstruction in their path can present a hazard. By having the platform level with the sill, those using the fire escape would have better footing, reducing the chance of tripping.

34. **(D)** The alarm ringing and the discharge of water from the pipe are indications that one or more sprinkler heads have opened and there is water flow in the building. Since the location of the firehouse is known and nearby and the location of the alarm box is uncertain, the wise decision is to go to the firehouse and report the situation.

35. **(B)** Once a hoseline is run between the rungs of a ladder, the ladder is no longer maneuverable until the hoseline is shut down and disconnected.

36. **(A)** Notifying the police department is the quickest and most proper action to take. Radio communications enable the police to be at the scene rapidly. They are trained to handle this type of emergency. You must remain available to respond to fires.

37. **(C)** The immediate administration of first aid by the firefighter to restore breathing gives the victim a chance of surviving. The firefighter should also direct a passerby to call for medical assistance, either by telephone or at the change booth. If a transit officer is nearby, notify the officer.

38. **(A)** By conducting a drill or practice at the site of a possible emergency, the fire fighting forces become familiar with the various aspects of a particular type of building or

location. They get firsthand knowledge of the obstacles that may be encountered at an actual emergency.

39. **(C)** Lumberyards store a wide assortment of wood and wood products. Strong upward drafts develop at a lumberyard fire and carry sparks and large brands a considerable distance, especially if there is much wind.

40. **(C)** At a suspicious fire, the officer in charge should document pertinent facts as soon as possible. The passing of time generally results in a witness forgetting many of the details of what he or she saw originally. Documenting the facts will enable the firefighter to recall the incident.

41. **(C)** By asking the theater manager to correct the situation, the firefighter is making the manager aware of the existing violation. It is the manager's responsibility to correct the situation.

42. **(B)** An uncharged 50-foot length of 2 1/2-inch hose weighs approximately 40 pounds. Water weighs 8 1/3 pounds per gallon and there are 13 gallons in the 50-foot length. Obviously, the uncharged line is much lighter and much easier to carry to the point where water is needed.

43. **(A)** Smoke and hot gases from the fire will rise and "mushroom" at the roof or upper floors. If the roof is not vented, the hot gases will ignite and spread fire to the roof area.

44. **(B)** The fire equipment must be maintained to respond immediately in an emergency. Delays caused by apparatus breakdowns could result in loss of life or excessive property damage. That is why extensive preventative maintenance is performed on high-quality department apparatus.

45. **(C)** Smoke is a product of combustion; eliminate (D). The newspaper carries reports of death from smoke inhalation; eliminate (A). Smoke, as the product of combustion, not only reduces the oxygen content of the air but contains particles, carbon monoxide and whatever toxic gases might be produced by the substance that is burning.

46. **(C)** As long as the firefighter is performing his or her job properly, there is no reason to be intimidated by the owner.

47. **(A)** Charging a property owner a fee for calling the fire department is not a good policy. People might be unable or unwilling to afford the fee and attempt to extinguish the fire themselves. Experience has shown that delay in calling the fire department has resulted in fires getting out of control, causing loss of life and extensive property damage.

48. **(B)** The use of dry pipe standpipe systems where the pipes are exposed to freezing temperatures avoids having pipes which are blocked by ice or pipes that burst from the pressure of frozen water.

49. **(B)** Calling the fire alarm will bring firefighters to the reported location much faster than if the firefighter ran to the house to see if there really was a fire. Every report of fire must be taken seriously.

50. **(D)** The practice of firefighters dressing while responding to a fire is not wise. It is unsafe and should not be done. The few seconds' delay will insure that the firefighter is properly dressed, and this will enable him or her to hold on with both hands while the apparatus is in motion.

51. **(C)** A room must have at least 70 square feet of floor space but if it has less than 80 square feet must be at least 9 feet high. 70 square feet × 9 feet = 630 cubic feet.

52. **(C)** As stated in paragraph b. Every living room with less than 80 square feet of floor area shall be 9 feet high.

53. **(C)** When speaking of all rooms in apartments of three or more room as "living rooms," the passage must be referring to rooms used for living.

54. **(A)** The floor space of the dining bay is 6 × 8 = 48 square feet. One-eighth of 48 = 6 square feet.

55. **(D)** The words "every other living room except a kitchen" imply that a kitchen is a living room but that it is not subject to the size requirements of other living rooms.

56. **(C)** See the second sentence.

57. **(B)** From 8 to 15 years of service, the annual leave is 25 days.

58. **(C)** The sixteenth year, in which leave is earned at the rate of 2 1/4 days per month, comes after 15 full years.

59. **(B)** See the first sentence of the second paragraph.

60. **(A)** See the last sentence.

61. **(A)** The second sentence tells us that traffic summonses serve as a deterrent. People who receive summonses for violations tend not to repeat those violations.

62. **(B)** According to the last sentence, summonses are issued only for inexcusable violations.

63. **(D)** The purpose of the verbal warning is an educational one. The person who did not know that a particular action was a violation must be informed of the nature of the violation and must be told not to repeat the act.

64. **(B)** The person who is speeding because he or she is late for an appointment is speeding with deliberate intent. Such a person must be served with a summons.

65. **(D)** When supply exceeds current needs, the society is wealthy and can distribute its surplus.

66. **(A)** This information may be inferred from the first and second sentences.

67. **(A)** Poverty is the opposite of wealth. If there is poverty, there is no surplus and no expansionism.

68. **(D)** As stated in the third sentence.

69. **(C)** The question asks for fire loss. The first sentence of the paragraph states that ten percent of the fire loss in the United States was due to heating equipment. Ten percent of $500 million is $50 million ($50,000,000) or (C).

70. **(C)** Space between a heater and combustible material is necessary because heat radiation will vary inversely with the square of the distance separating the heat source from the object. Example: An object two feet away from the heater will receive one-fourth as much heat as an object one foot away. An object three feet away will receive one-ninth as much heat as the object one foot away.

71. **(A)** Month-to-month changes in fires related to heating equipment are based upon the extent of use of the equipment. The season determines the need for heating.

72. **(C)** The knot shown in the diagram is called a sheepshank and is used to shorten a length of rope without cutting it.

73. **(B)** The pinion gear is a small bevel or spur gear that meshes with the speed gear. The pinions "bind" the large gear to prevent it from wobbling.

74. **(B)** The small cubes of chopped ice have a greater surface area than the large block of ice. Therefore, more water will be exposed to the area of the chopped ice and it will chill faster to the same temperature.

75. **(C)** At a fire, flame and hot gases can enter cracks or holes in a ceiling or wall and burn undetected. If the ceilings are not opened and the fire extinguished, the fire will continue to burn and spread throughout the structure. The pike pole is used for this purpose.

76. **(D)** In figure D, notice that the hook is facing away from the firefighter and the pole is in front of his body. By working in this manner, the firefighter can see the area being opened and is out of the way of falling debris.

77. **(D)** There are people on the fire escape right now, so it obviously is not too hot. The fire escape would be quicker and safer than a sheet which may pull loose, tear, or burn. At moments of panic, people sometimes use poor judgment.

78. **(A)** Evaporation occurs when the molecules near the surface of the liquid escape into the air. The water in the bowl has a greater surface area exposed to the air which caused it to evaporate faster than the water in the cup.

79. **(B)** The special purpose tool is generally more efficient because it is designed for one specific purpose. The all-purpose tool can accomplish the intended job, but it usually requires more time and works less effectively. In fire fighting, time and efficiency are essential.

80. **(C)** In order for the rear wheels (which are the driving wheels) to drive or push the car, friction or resistance must be present between the tires and the road surface. The ice and snow create a slippery road condition, causing the tires to lose traction. The front wheels do not turn because power from the engine is transmitted to the rear wheels only.

81. **(B)** When the suction cup is pressed against the wall, the air in the cup is forced out and a vacuum or negative pressure remains. Atmospheric air with a pressure of 14.7

psi (pounds per square inch) pushes against the rubber cup and holds it against the wall.

82. **(B)** Many toxic gases and acids are absorbed directly through the skin. To operate safely and effectively in this type of atmosphere, self-contained suits made of rubber or other nonporous material should be used in conjunction with the self-contained breathing apparatus. The self-contained apparatus takes in no outside air so filtration is not a problem.

83. **(C)** During very cold weather, water will freeze much more quickly when it is idle than when it is moving. By keeping the water flowing, there is less chance of the water in the hoseline freezing.

84. **(B)** Some sprinkler systems require two sources of water. In the event that one source is placed out of service, the other is available to keep the system operating. Two source systems may be supplied by: a) two water mains, each on a different street; b) a street main and a gravity tank; c) a street main and a pressure tank.

85. **(A)** The function of a fuse is to protect against overload or overheating of a circuit. A burned-out fuse indicates that a dangerous condition on a circuit has been avoided. By replacing the fuse with a coin that cannot burn out, one bypasses the safety feature of the fuse and creates a fire hazard.

86. **(A)** The flat machined surface of the shaft is called a keyway. A corresponding opening in a gear or pulley accommodates the keyway. This prevents slippage of the gear or pulley on the shaft.

87. **(D)** Adjacent gears turn in opposite directions. Gear A turns in a clockwise direction, so Gear B, which is adjacent to it, will turn in a counterclockwise direction. Continuing, Gear C, adjacent to Gear B, will turn in a clockwise direction, and Gear D, adjacent to Gear C, will turn counterclockwise, the same as Gear B.

88. **(B)** Gears of the same size connected into the same system will turn at the same speed.

89. **(A)** The sketch is that of a fixed pulley and a moveable pulley. With this arrangement, the effort to move the weight will be one-half. However, the effort distance (length of the pull) will be twice the distance the weight travels. Therefore, a one-foot pull raised the weight one-half foot.

90. **(A)** With all the objects equal in weight, the pyramid in figure 1 is the most difficult to tip over. This is due to the fact that the bulk of the volume and weight is at the base.

91. **(B)** Surely the proprietor of the dry cleaning store does not want to create a dangerous situation for himself or for his property. He will be grateful for the information and will most likely change his manner of storing cleaning fluid. The firefighter's action falls into the category of friendly, helpful neighbor.

92. **(D)** Profuse bleeding can be life threatening. While waiting for medical assistance, the firefighter should initially attempt to control the bleeding by applying direct pressure to the wound with a sterile dressing if one is available. If bleeding persists, the firefighter should reach into the wound and pinch the blood vessel.

93. **(D)** Proper treatment for a person in shock includes assuring adequate breathing, controlling the bleeding, administering oxygen, elevating extremities, avoiding rough handling, preventing loss of body heat, keeping the victim lying down, and giving nothing by mouth.

94. **(B)** A pressure bandage will be sufficient. Generally, when a person suffers a puncture wound there is very little external bleeding. However, severe internal bleeding is possible.

95. **(B)** Before any corrective action can be taken, the cause of the accident must be known. By getting the facts, the officer is able to determine why the accident occurred. Knowing why the accident occurred can prevent a similar accident in the future.

96. **(C)** The superior officer, being aware of the firefighter's condition, can summon medical advice for the firefighter and, if necessary, relieve him or her from duty. Getting lightheaded or dizzy in a hazardous situation can result in serious injury to the firefighter or to a person he or she may be assisting.

97. **(A)** The information presented at the special course on safety was apparently compiled by using case studies, information gathered from accident reports, and firsthand experience. The firefighter could voice his or her opinion on the subject to the safety instructor or to his or her superior officers. However, the firefighter should perform the task as it was taught.

98. **(A)** A firefighter on inspection duty has a job to perform. That job is to note and report on violations and to follow through to be certain violations are corrected. Accepting a bribe to overlook violations would be a criminal act as well as dereliction of duty.

99. **(B)** An extensive training program is necessary because of the nature of the occupation of fire fighting. Fires and fire conditions vary with each situation. The tremendous hazards present when fighting fires are not found in any other occupation. The lives of the fire victims as well as those of fellow firefighters are at stake. In order for the firefighter to function effectively, he or she must be as well trained as possible before being permitted to take part in actual fire fighting.

100. **(D)** Changes in procedures are necessary due to the many technological advances that have been introduced into the field of fire fighting. These technological advances enable the firefighter to function more efficiently and with greater effectiveness. Some advances that have dictated changes are: improved self-contained breathing apparatus, handi-talkies, power saws, hydraulic cutting equipment, and sophisticated foams and fog nozzles.

Answer Sheet for Practice Examination 4

1. Ⓐ Ⓑ Ⓒ Ⓓ
2. Ⓐ Ⓑ Ⓒ Ⓓ
3. Ⓐ Ⓑ Ⓒ Ⓓ
4. Ⓐ Ⓑ Ⓒ Ⓓ
5. Ⓐ Ⓑ Ⓒ Ⓓ
6. Ⓐ Ⓑ Ⓒ Ⓓ
7. Ⓐ Ⓑ Ⓒ Ⓓ
8. Ⓐ Ⓑ Ⓒ Ⓓ
9. Ⓐ Ⓑ Ⓒ Ⓓ
10. Ⓐ Ⓑ Ⓒ Ⓓ
11. Ⓐ Ⓑ Ⓒ Ⓓ
12. Ⓐ Ⓑ Ⓒ Ⓓ
13. Ⓐ Ⓑ Ⓒ Ⓓ
14. Ⓐ Ⓑ Ⓒ Ⓓ
15. Ⓐ Ⓑ Ⓒ Ⓓ
16. Ⓐ Ⓑ Ⓒ Ⓓ
17. Ⓐ Ⓑ Ⓒ Ⓓ
18. Ⓐ Ⓑ Ⓒ Ⓓ
19. Ⓐ Ⓑ Ⓒ Ⓓ
20. Ⓐ Ⓑ Ⓒ Ⓓ

21. Ⓐ Ⓑ Ⓒ Ⓓ
22. Ⓐ Ⓑ Ⓒ Ⓓ
23. Ⓐ Ⓑ Ⓒ Ⓓ
24. Ⓐ Ⓑ Ⓒ Ⓓ
25. Ⓐ Ⓑ Ⓒ Ⓓ
26. Ⓐ Ⓑ Ⓒ Ⓓ
27. Ⓐ Ⓑ Ⓒ Ⓓ
28. Ⓐ Ⓑ Ⓒ Ⓓ
29. Ⓐ Ⓑ Ⓒ Ⓓ
30. Ⓐ Ⓑ Ⓒ Ⓓ
31. Ⓐ Ⓑ Ⓒ Ⓓ
32. Ⓐ Ⓑ Ⓒ Ⓓ
33. Ⓐ Ⓑ Ⓒ Ⓓ
34. Ⓐ Ⓑ Ⓒ Ⓓ
35. Ⓐ Ⓑ Ⓒ Ⓓ
36. Ⓐ Ⓑ Ⓒ Ⓓ
37. Ⓐ Ⓑ Ⓒ Ⓓ
38. Ⓐ Ⓑ Ⓒ Ⓓ
39. Ⓐ Ⓑ Ⓒ Ⓓ
40. Ⓐ Ⓑ Ⓒ Ⓓ

41. Ⓐ Ⓑ Ⓒ Ⓓ
42. Ⓐ Ⓑ Ⓒ Ⓓ
45. Ⓐ Ⓑ Ⓒ Ⓓ
44. Ⓐ Ⓑ Ⓒ Ⓓ
45. Ⓐ Ⓑ Ⓒ Ⓓ
46. Ⓐ Ⓑ Ⓒ Ⓓ
47. Ⓐ Ⓑ Ⓒ Ⓓ
48. Ⓐ Ⓑ Ⓒ Ⓓ
49. Ⓐ Ⓑ Ⓒ Ⓓ
50. Ⓐ Ⓑ Ⓒ Ⓓ
51. Ⓐ Ⓑ Ⓒ Ⓓ
52. Ⓐ Ⓑ Ⓒ Ⓓ
55. Ⓐ Ⓑ Ⓒ Ⓓ
54. Ⓐ Ⓑ Ⓒ Ⓓ
55. Ⓐ Ⓑ Ⓒ Ⓓ
56. Ⓐ Ⓑ Ⓒ Ⓓ
57. Ⓐ Ⓑ Ⓒ Ⓓ
58. Ⓐ Ⓑ Ⓒ Ⓓ
59. Ⓐ Ⓑ Ⓒ Ⓓ
60. Ⓐ Ⓑ Ⓒ Ⓓ

61. Ⓐ Ⓑ Ⓒ Ⓓ
62. Ⓐ Ⓑ Ⓒ Ⓓ
63. Ⓐ Ⓑ Ⓒ Ⓓ
64. Ⓐ Ⓑ Ⓒ Ⓓ
65. Ⓐ Ⓑ Ⓒ Ⓓ
66. Ⓐ Ⓑ Ⓒ Ⓓ
67. Ⓐ Ⓑ Ⓒ Ⓓ
68. Ⓐ Ⓑ Ⓒ Ⓓ
69. Ⓐ Ⓑ Ⓒ Ⓓ
70. Ⓐ Ⓑ Ⓒ Ⓓ
71. Ⓐ Ⓑ Ⓒ Ⓓ
72. Ⓐ Ⓑ Ⓒ Ⓓ
73. Ⓐ Ⓑ Ⓒ Ⓓ
74. Ⓐ Ⓑ Ⓒ Ⓓ
75. Ⓐ Ⓑ Ⓒ Ⓓ
76. Ⓐ Ⓑ Ⓒ Ⓓ
77. Ⓐ Ⓑ Ⓒ Ⓓ
78. Ⓐ Ⓑ Ⓒ Ⓓ
79. Ⓐ Ⓑ Ⓒ Ⓓ
80. Ⓐ Ⓑ Ⓒ Ⓓ

81. Ⓐ Ⓑ Ⓒ Ⓓ
82. Ⓐ Ⓑ Ⓒ Ⓓ
83. Ⓐ Ⓑ Ⓒ Ⓓ
84. Ⓐ Ⓑ Ⓒ Ⓓ
85. Ⓐ Ⓑ Ⓒ Ⓓ
86. Ⓐ Ⓑ Ⓒ Ⓓ
87. Ⓐ Ⓑ Ⓒ Ⓓ
88. Ⓐ Ⓑ Ⓒ Ⓓ
89. Ⓐ Ⓑ Ⓒ Ⓓ
90. Ⓐ Ⓑ Ⓒ Ⓓ
91. Ⓐ Ⓑ Ⓒ Ⓓ
92. Ⓐ Ⓑ Ⓒ Ⓓ
93. Ⓐ Ⓑ Ⓒ Ⓓ
94. Ⓐ Ⓑ Ⓒ Ⓓ
95. Ⓐ Ⓑ Ⓒ Ⓓ
96. Ⓐ Ⓑ Ⓒ Ⓓ
97. Ⓐ Ⓑ Ⓒ Ⓓ
98. Ⓐ Ⓑ Ⓒ Ⓓ
99. Ⓐ Ⓑ Ⓒ Ⓓ
100. Ⓐ Ⓑ Ⓒ Ⓓ

PRACTICE EXAMINATION 4

The time allowed for the entire examination is 3¹/₂ hours.

DIRECTIONS: Each question has four suggested answers, lettered A, B, C, and D. Decide which one is the best answer. On the sample answer sheet, find the question number that corresponds to the answer you have selected and darken the area with a soft pencil.

1. While performing a routine inspection of a factory building, a firefighter is asked a question by the plant manager about a matter which is under the control of the health department and about which the firefighter has little knowledge. In this situation, the best course of action for the firefighter to take is to

   (A) answer the question to the best of his or her knowledge
   (B) tell the manager that he or she is not permitted to answer the question because it does not relate to a fire department matter
   (C) tell the manager that it will be referred to the health department
   (D) suggest to the manager that he communicate with the health department about the matter

2. A firefighter on duty who answers a departmental telephone should give his or her name and rank

   (A) at the start of the conversation, as a matter of routine
   (B) only if asked for this information by the caller
   (C) only if the caller is a superior officer
   (D) only if the telephone message requires the firefighter to take some action

3. At a five-alarm fire in the West End, several companies from Northside were temporarily assigned to occupy the quarters and take over the duties of companies engaged in fighting the fire. The main reason for relocating the Northside companies was to

   (A) protect the firehouses from robbery or vandalism which might occur if they were left vacant for a long period of time
   (B) provide for speedy response to the fire if additional companies were required
   (C) give the Northside companies an opportunity to become familiar with the problems of the West End area
   (D) provide protection to the West End area in the event that other fires occurred

4. Two firefighters, on their way to report for duty early one morning, observe a fire in a building containing a supermarket on the street level and apartments on the upper stories. One firefighter runs into the building to spread the alarm to tenants. The other firefighter runs to a street alarm box two blocks away and sends an alarm. The latter firefighter should then

   (A) return to the building on fire and help evacuate the tenants
   (B) remain at the fire alarm box in order to direct the first fire company that arrives to the location of the fire

(C) look for a telephone in order to call his own fire company and explain that he and his companion will be late in reporting for duty

(D) look for a telephone in order to call the health department and request that an inspector be sent to the supermarket to examine the food involved in the fire

5. A man found an official fire department badge and gave it to his young son to use as a toy. The man's action was improper mainly because

(A) it is disrespectful to the fire department to use the badge in this manner
(B) the boy may injure himself playing with the badge
(C) an effort first should have been made to locate the owner of the badge before giving it to the boy
(D) the badge should have been returned to the fire department

6. In case of a fire in a mailbox, the fire department recommends that an extinguishing agent which smothers the fire, such as carbon tetrachloride, should be used. Of the following, the most likely reason for not recommending the use of water is that

(A) water is not effective on fires in small tightly enclosed spaces
(B) someone might have mailed chemicals that could explode in contact with water
(C) water may damage the mail untouched by fire so that it could not be delivered
(D) the smothering agent can be put on the fire faster than water can be

7. Of the following, the main difficulty in obtaining accurate information about the causes of fires is that

(A) firefighters are too busy putting out fires to have time for investigation of their causes
(B) most people have little knowledge of fire hazards
(C) fires destroy much of the evidence which would indicate the causes of the fires
(D) fire departments are more interested in fire prevention than in investigating fires that have already occurred

8. In an effort to discourage the sending of false alarms and to help apprehend those guilty of this practice, it is suggested that the handles of fire alarm boxes be covered with a dye which would stain the hand of a person sending an alarm and would not wash off for 24 hours. The dye would be visible only under an ultraviolet light. Of the following, the chief objection to such a device is that it would

(A) require funds that can be better used for other purposes
(B) have no effect on false alarms transmitted by telephone
(C) discourage some persons from sending alarms for real fires
(D) punish the innocent as well as the guilty

9. Automatic fire extinguishing sprinkler systems are sometimes not effective on fires accompanied by explosions chiefly because

(A) these fires do not generate enough heat to start sprinkler operation
(B) the pipes supplying the sprinklers are usually damaged by the explosion
(C) fires in explosive materials usually cannot be extinguished by water
(D) sprinkler heads are usually clogged by dust created by the explosion

10. When a fire occurs in the vicinity of a subway system, there is the possibility that water from the firefighters' hose streams will flood underground portions of the subway lines

through sidewalk gratings. Of the following methods of reducing this danger, the one that would generally be suitable is for the officer in command to order subordinates to

(A) use fewer hoselines and smaller quantities of water than they would ordinarily
(B) attack the fire from positions which are distant from the sidewalk gratings
(C) cover the sidewalk gratings with canvas tarpaulins
(D) advise the subway dispatcher to reroute the subway trains

11. When responding to alarms, fire department apparatus generally follows routes established in advance. The *least* valid justification for this practice is that

(A) motorists living in the area become familiar with these routes and tend to avoid them
(B) the likelihood of collision between two pieces of fire department apparatus is reduced
(C) the fastest response is generally obtained
(D) road construction, road blocks, detours, and similar conditions can be avoided

12. From a distance, an off-duty firefighter sees a group of teenage boys set fire to a newspaper and toss the flaming pages into the open window of a building which is being torn down. In this situation, the first action which should be taken by the firefighter is to

(A) send a fire alarm from the closest street alarm box
(B) chase the boys and attempt to catch one of them
(C) investigate whether a fire has been started
(D) call the police from the closest police alarm box or telephone

13. When responding to an alarm, officers are not to talk to chauffeurs driving the apparatus except to give orders or directions. Of the following, the best justification for this rule is that it

(A) gives the officer an opportunity to make preliminary plans for handling the fire problem
(B) enables the chauffeur to concentrate on driving the apparatus
(C) maintains the proper relationship between the ranks while on duty
(D) permits the officer to observe the chauffeur's skill, or lack thereof, in driving the apparatus.

14. The approved method of reporting a fire by telephone in most major cities is to dial

(A) central headquarters of the fire department
(B) 911
(C) the local fire station
(D) the telephone operator

15. Doors in theaters and other places of public assembly usually open outward. The main reason for this requirement is, in the event of fire, to

(A) provide the widest possible passageway for escape of the audience
(B) prevent panic-stricken audiences from jamming the doors in a closed position
(C) indicate to the audience the safe direction of travel
(D) prevent unauthorized persons from entering the building

16. Fire prevention inspections should be conducted at irregular hours or intervals. The best justification for this irregularity is that it permits the firefighters to

(A) make inspections when they have free time
(B) see the inspected establishments in their normal condition and not in their "dressed-up" condition
(C) avoid making inspections at times which would be inconvenient for the inspected establishments
(D) concentrate their inspectional activities on those establishments which present the greatest fire hazard

17. Some gas masks provide protection to the user by filtering out from the air certain harmful gases present in the atmosphere. A mask of this type would *not* be suitable in an atmosphere containing

(A) heavy black smoke
(B) a filterable gas under pressure
(C) insufficient oxygen to sustain life
(D) more than one filterable harmful gas

18. Firefighters are instructed never to turn on the gas supply to a house in which the gas was turned off by them during a fire. Of the following, the most important reason for this prohibition is that

(A) the fire may have made the gas meters inaccurate
(B) unburned gas may escape from the open gas jets
(C) the utility company's employees may object to firefighters performing their work
(D) firefighters should not do anything which is not directly related to extinguishing fires

19. A firefighter in uniform, performing inspectional duty, comes upon a group of young men assaulting a police officer. The firefighter goes to the aid of the police officer and, in the course of the struggle, receives some minor injuries. The action of the firefighter in this situation is

(A) proper, chiefly because members of the uniformed forces must stick together
(B) improper, chiefly because people in the neighborhood, as a result, might refuse to cooperate with the fire department's various programs
(C) proper, chiefly because all citizens have an obligation to assist police officers in the performance of their duty
(D) improper, chiefly because the fire department lost the services of the firefighter while the firefighter was recovering from his or her injuries

20. Members of the fire department may not make a speech on fire department matters without the approval of the fire commissioner. Requests for permission must be accompanied by a copy or summary of the speech. The main reason for this requirement is to

(A) determine whether the member is engaged in political activities which are forbidden
(B) reduce the chance that the public will be misinformed about fire department policies or procedures
(C) provide the department with a list of members who can serve in the department's speakers' bureau
(D) provide the department with information about the off-duty activities of members

21. For a firefighter to straddle a hoseline while holding the nozzle and directing water on fires is a

(A) good practice, mainly because better balance is obtained by the firefighter
(B) poor practice, mainly because the firefighter directing the hose may trip over it
(C) good practice, mainly because better control over the hoseline is obtained
(D) poor practice, mainly because the hose might whip about and injure the firefighter

22. Firefighters are required to wear steel reinforced innersoles inside their rubber boots. The main purpose of these innersoles is to

(A) make the boots more durable and long lasting
(B) protect the firefighter's feet from burns from smoldering objects or embers
(C) protect the firefighter's feet from injury from falling objects
(D) protect the firefighter's feet from nails or other sharp objects

23. Promoting good relations with the public is an important duty of every member of the fire department. Of the following, the best way for a firefighter to promote good public relations is to

(A) become active in civic and charitable organizations
(B) be well dressed, clean, and neat on all occasions
(C) write letters to newspapers explaining the reasons for departmental procedures
(D) perform his or her duties with efficiency, consideration, and courtesy

24. Firefighters are advised to avoid wearing rings on their fingers. The main reason for this advice is that the rings have a tendency to

(A) be damaged during operations at fires
(B) scratch persons receiving first-aid treatment
(C) catch on objects and injure the wearer
(D) scratch furniture and/or other valuable objects

25. Suppose that you are a firefighter on housewatch duty when a civilian enters the firehouse. He introduces himself as a British firefighter visiting the country to study American fire fighting methods. He asks you for permission to ride on the fire apparatus when it responds to alarms in order to observe operations firsthand. You know it is against departmental policy to permit civilians to ride apparatus without written permission from headquarters. In this situation you should

(A) refuse the request but suggest that he follow the apparatus in his own car when it responds to an alarm
(B) call headquarters and request permission to permit the visitor to ride the apparatus
(C) refuse the request and suggest that he apply to headquarters for permission
(D) refuse the request and suggest that he return the next time that the fire department holds an open house

26. When operating at a pier fire, firefighters usually avoid driving their apparatus onto the pier itself. The main reason for this precaution is to reduce the possibility that the apparatus will be

(A) delayed in returning to quarters
(B) driven off the end of the pier
(C) destroyed by a fire that spreads rapidly
(D) in the way of the firefighters

27. Pumpers purchased by the fire department are equipped with enclosed cabs. In the past, fire department apparatus was open with no cab or roof. The main advantage of the enclosed cab is that it provides

    (A) additional storage space for equipment
    (B) a place of shelter for firefighters operating in an area of radioactivity
    (C) protection for firefighters from weather conditions and injury
    (D) emergency first-aid and ambulance facilities

28. Heavy blizzards greatly increase the problems and work of the fire department. When such a situation occurs, the fire commissioner could reasonably be expected to

    (A) order members of the fire department to perform extra duty
    (B) limit parking on city streets
    (C) station firefighters at fire alarm boxes to prevent the sending of false alarms
    (D) prohibit the use of kerosene heaters

29. Regulations of the fire department require that, when placing hose on a fire wagon, care should be taken to avoid bending the hose at places where it had been bent previously. The most important reason for this requirement is that repeated bending of the hose at the same places will cause

    (A) kinks in the hose at those places
    (B) weakening of the hose at those places
    (C) discoloration of the hose at those places
    (D) dirt to accumulate and clog the hose at those places

30. While fighting a fire in an apartment when the occupants are not at home, a firefighter finds a sum of money in a closet. Under these circumstances, the firefighter should turn over the money to

    (A) a responsible neighbor
    (B) the desk sergeant of the nearest police station
    (C) the superintendent of the apartment house
    (D) his or her superior officer

31. "When it is necessary to remove a cornice, every effort should be made to pull it back on the roof." The most important reason for the direction in the above quotation is that pulling back the cornice on the roof rather than dropping it to the street below

    (A) requires less time
    (B) is safer for the people on the street
    (C) makes it possible to reuse the cornice
    (D) is less dangerous to firefighters working on the roof

32. After a fire has been extinguished, one firefighter often remains at the scene after the others have left. Of the following, the main reason for this practice is that this firefighter can

    (A) prevent looters from stealing valuables
    (B) watch for any rekindling of the fire
    (C) search the area for lost valuables
    (D) examine the premises for evidence of arson

33. Firefighters usually attempt to get as close as possible to the seat of a fire so that they can direct their hose streams with accuracy. However, intense heat sometimes keeps them at a distance. An unsatisfactory method of overcoming this problem is to

(A) have firefighters use some solid object, such as a wall, as a shield
(B) keep firefighters cool by wetting them with small streams of water
(C) use large high-pressure streams and operate at a great distance from the fire
(D) use a water spray to break down the heat waves coming from the fire

34. Suppose that you are an off-duty firefighter driving your car in the midtown area. As you cross an intersection, you hear sirens and, looking back, see fire apparatus approaching. In this situation, the best action for you to take is to

(A) attempt to clear a path for the fire apparatus by driving rapidly and sounding your horn
(B) drive to the next intersection and direct traffic until the apparatus has passed
(C) permit the apparatus to pass, then follow it closely to the fire, sounding your horn as you drive
(D) pull to the curb, permit the apparatus to pass, then continue on your way

35. Whenever a public performance is given in a theater involving the use of scenery or machinery, a firefighter is assigned to be present. The main reason for this assignment is that

(A) theaters are located in high property value districts
(B) the use of scenery and machinery increases the fire hazard
(C) theatrical districts have heavy traffic, making for slow response of apparatus
(D) emergency exits may be blocked by the scenery or machinery

DIRECTIONS: Questions 36–39 are based on the following paragraph.

A plastic does not consist of a single substance, but is a blended combination of several substances. In addition to the resin, it may contain various fillers, plasticizers, lubricants, and coloring material. Depending upon the type and quantity of substances added to the binder, the properties, including combustibility, may be altered considerably. The flammability of plastics depends upon the composition and, as with other materials, upon their physical size and condition. Thin sections, sharp edges, or powdered plastics will ignite and burn more readily than the same amount of identical material in heavy sections with smooth surfaces.

36. According to the paragraph, all plastics contain a

(A) resin
(B) resin and a filler
(C) resin, filler, and plasticizer
(D) resin, filler, plasticizer, lubricant, and coloring material

37. The conclusion that is best supported by the paragraph is that the flammability of plastics

(A) is generally high
(B) is generally moderate
(C) is generally low
(D) varies considerably

38. According to the paragraph, *plastics* can best be described as

(A) a trade name
(B) the name of a specific product
(C) the name of a group of products which have some similar and some dissimilar properties
(D) the name of any substance which can be shaped or molded during the production process

39. In a manufacturing process, large thick sheets of a particular plastic are cut, buffed, and formed into small tools. The statement that is most in accord with the information in the preceding paragraph is that

(A) the dust particles of the plastics are more flammable than the tools or the sheets
(B) the plastic tools are more flammable than the dust particles or the sheets
(C) the sheets of the plastic are more flammable than the dust particles or the tools
(D) there is insufficient information to determine the relative flammability of sheets, tools, and dust particles

DIRECTIONS: Questions 40–42 are based on the following paragraph.

To guard against overheating of electrical conductors in buildings, an overcurrent protective device is provided for each circuit. This device is designed to open the circuit and cut off the flow of current whenever the current exceeds a predetermined limit. The fuse, which is a common form of overcurrent protection, consists of a fusible metal element that, when heated by the current to a certain temperature, melts and opens the circuit.

40. According to the paragraph, a circuit which is *not* carrying an electric current is

(A) an open circuit
(B) a closed circuit
(C) a circuit protected by a fuse
(D) a circuit protected by an overcurrent protective device other than a fuse

41. As used in the paragraph, the best example of a conductor is

(A) a metal table which comes in contact with a source of electricity
(B) a storage battery generating electricity
(C) an electrical wire carrying an electrical current
(D) a dynamo converting mechanical energy into electrical energy

42. According to the paragraph, the maximum number of circuits that can be handled by a fuse box containing 6 fuses

(A) is 3
(B) is 6
(C) is 12
(D) cannot be determined from the information given in the paragraph

DIRECTIONS: Questions 43–45 are based on the following paragraph.

Unlined linen hose is essentially a fabric tube made of closely woven linen yarn. Due to the natural characteristics of linen, very shortly after water is introduced, the wet threads swell, closing the minute spaces between them and making the tube practically watertight. This type of hose tends to deteriorate rapidly if not thoroughly dried after use or if installed where it will be exposed to dampness or weather conditions. It is not ordinarily built to withstand frequent service or use where the fabric will be subjected to chafing from rough or sharp surfaces.

43. Seepage of water through an unlined linen hose is observed when the water is first turned on. From the preceding paragraph, we may conclude that the seepage

(A) indicates that the hose is defective
(B) does not indicate that the hose is defective provided that the seepage is proportionate to the water pressure
(C) does not indicate that the hose is defective provided that the seepage is greatly reduced when the hose becomes thoroughly wet
(D) does not indicate that the hose is defective provided that the seepage takes place only at the surface of the hose

44. Unlined linen hose is most suitable for use

(A) as a garden hose
(B) on Fire Department apparatus
(C) as emergency fire equipment in buildings
(D) in Fire Department training schools

45. The use of unlined linen hose would be *least* appropriate in

(A) an outdoor lumberyard
(B) a nonfireproof office building
(C) a department store
(D) a cosmetic manufacturing plant

DIRECTIONS: The following four questions, numbered 46–49, are based on the information given in the paragraph below and are to be answered in accordance with this information.

The principal value of inspection work is in the knowledge obtained relating to the various structural features of the building and the protective features provided. Knowledge of the location of stairways and elevators, the obstruction provided by merchandise, the danger from absorption of water by baled stock, the potential hazard of rupture of containers such as drums or cylinders, and the location of protective equipment are all essential features to be noted and later discussed in company school and officer's college.

46. According to the paragraph, the chief value of inspection work is to gather information which will aid in

(A) fixing responsibility for fires
(B) planning fire fighting operations
(C) training new firefighters
(D) obtaining compliances with the building code

47. The object which would be the most help in accomplishing the objective of the inspection as stated in the paragraph is a

(A) copy of the building code

(B) chemical analysis kit
(C) plan of the building
(D) list of the building's tenants

48. An example of a "structural feature" contained in the paragraph is the

(A) location of stairways and elevators
(B) obstruction provided by merchandise
(C) danger of absorption of water by baled stock
(D) hazard of rupture of containers such as drums or cylinders

49. Of the following, the best example of a "protective feature" as used in the preceding paragraph is

(A) a fire extinguisher
(B) a burglar alarm
(C) fire insurance
(D) a medical first-aid kit

50. Although the periodic inspection of buildings is required by law, a fire department can constructively utilize information gained during these inspections to assist in the

(A) training of its new firefighters
(B) substantiation of budget requests
(C) planning of fire fighting operations
(D) promotion of its community activities

51. A firefighter on inspection duty should behave in a manner that is likely to gain the support of the public that his or her agency serves. The firefighter can accomplish this by engaging in all but one of the following practices. The practice that will *not* accomplish this is

(A) being courteous and neat in appearance
(B) being prepared to give explanations of the Fire Prevention Code
(C) overlooking minor infractions of the Fire Prevention Code because correcting them would be expensive
(D) performing inspectional duties in a businesslike manner

52. A firefighter receives a call during the night while off duty from a neighbor who says his house is on fire. The most appropriate action for the firefighter to take *first* is to

(A) call for the assistance of the fire department if it has not already been called
(B) go to the neighbor's home to evaluate the extent of the fire
(C) call other off-duty firefighters who live close by to help fight the fire
(D) advise the neighbor that you will be over as soon as you are dressed

DIRECTIONS: The following four questions, numbered 53–56, relate to the information given in the paragraph below and are to be answered in accordance with this information.

Division commanders shall arrange and maintain a plan for the use of hose wagons to transport members in emergencies. Upon receipt of a call for members, the deputy chief of the division from which the firefighters are called shall have the designated hose wagon placed out of service and prepared for the transportation of members. Hose wagons shall be placed at central assembly points and members detailed instructed to report promptly to such locations equipped for fire duty. Hose wagons designated shall remain at regular assignments when not engaged in the transportation of members.

53. Preparation of the hose wagon for this special assignment of transporting of members would most likely involve

    (A) checking the gas and oil, air in tires, and mechanical operation of the apparatus
    (B) removal of hoselines to make room for the members being transported
    (C) gathering of equipment which will be needed by the members being transported
    (D) instructing the driver on the best route to be used

54. Hose wagons used for emergency transportation of members are placed out of service because they

    (A) are not available to respond to alarms in their own district
    (B) are more subject to mechanical breakdown while on emergency duty
    (C) are engaged in operations which are not the primary responsibility of their division
    (D) are considered reserve equipment

55. Of the following, the best example of the type of emergency referred to in the paragraph is

    (A) a firefighter injured at a fire and requiring transportation
    (B) a subway strike which prevents firefighters from reporting for duty
    (C) an unusually large number of false alarms occurring at one time
    (D) a need for additional manpower at a fire

56. A "central assembly point," as used in the paragraph, would most likely be a place

    (A) close to the place of the emergency
    (B) in the geographical center of the division
    (C) easily reached by the members assigned
    (D) readily accessible to the intersection of major highways

57. Modern firehouses have automatic hose dryers for drying rubber-lined hose after use. The controls are usually set to provide heat no higher than 25°F. above room temperature. Of the following, the most important reason for not using more heat is that the

    (A) wear and tear on the dryers would become excessive
    (B) hose dries more thoroughly if it dries slowly
    (C) hose would be too hot to handle
    (D) rubber lining would deteriorate too rapidly

58. The best explanation why smoke usually rises from a fire is that

    (A) cooler, heavier air displaces lighter, warm air
    (B) heat energy of the fire propels smoke upward
    (C) suction from the upper air pulls the smoke upward
    (D) burning matter is chemically changed into heat energy

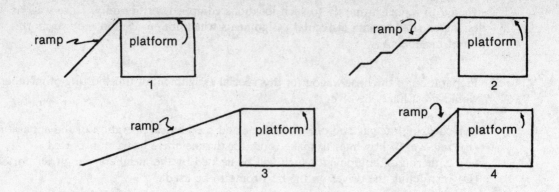

59. The diagram above shows various types of ramps leading to a loading platform. The ramp which would permit the load to be moved up to the platform with the least amount of force is

(A) 1
(B) 2
(C) 3
(D) 4

60. The practice of racing a car engine to warm it up in cold weather is generally

(A) good, mainly because repeated stalling of the engine and drain on the battery is avoided
(B) bad, mainly because too much gas is used to get the engine heated
(C) good, mainly because the engine becomes operational in the shortest period of time
(D) bad, mainly because proper lubrication is not established rapidly enough

61. Ice on sidewalks can often be melted by sprinkling salt on it. The melting of the ice results from

(A) a chemical reaction between the salt and ice that produces heat
(B) attraction of sun rays by the salt to the ice
(C) lowering of the freezing point of water by the salt
(D) heat of friction caused by persons walking on the salt

62. Only one of the following four statements relating to the temperature at which water will boil is correct. The correct statement is that

(A) water always boils at the same temperature regardless of pressure
(B) water heated slowly by a low flame will boil at a higher temperature than water heated quickly by a high flame
(C) a large quantity of water will boil at a higher temperature than a small quantity
(D) water heated at sea level will boil at a higher temperature than water heated on the top of a mountain

63. A substance which is a good conductor of heat is most likely to be a poor

(A) conductor of electricity
(B) insulator of heat
(C) vibrator of sound
(D) reflector of light

64. At a meeting concerned with fire prevention, it was said: "The fact that the fire loss has been maintained near to its previous levels is encouraging evidence that our increasing efforts over the years in the field of public education in fire protection have not been unavailing." Of the following, the most essential assumption that must be made if this statement is accepted is that

(A) further public education in fire protection is desirable
(B) fire loss has been computed upon the basis of real value rather than insured value
(C) reference is made here to losses due to fires caused by carelessness rather than sabotage
(D) there has been, in recent years, an increase in the potential fire hazard

65. A load is to be supported from a steel beam by a chain consisting of 20 links and a hook. If each link of the chain weighs 1 pound and can support a weight of 1000 pounds, and the hook weighs 5 pounds and can support a weight of 5000 pounds, the maximum load that can be supported from the hook is most nearly

(A) 25,000 pounds
(B) 5000 pounds
(C) 1000 pounds
(D) 975 pounds

66. Ice formation in water pipes often causes the bursting of the pipes because

(A) the additional weight of ice overloads the pipes
(B) water cannot pass the ice block and builds up great pressure on the pipes
(C) the cold causes contraction of the pipes and causes them to pull apart
(D) water expands upon freezing and builds up great pressure on the pipes

67. The suggestion has been made that groups of firefighters, without apparatus of any kind, be kept in reserve at a few centrally located points throughout the city. Of the following, the most valid justification for this proposal is that

(A) when second or third alarms are sent, the need is often for more firefighters rather than more apparatus
(B) the fire department is understaffed
(C) the fire districts in a city should be revised periodically to meet population trends
(D) discipline is as important as apparatus in extinguishing fires quickly

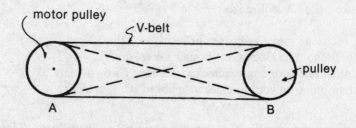

68. In the diagram above, crossing the V-belt as shown by the dotted lines will result in

(A) pulley A reversing direction
(B) no change in the direction of either pulley
(C) pulley B reversing direction
(D) stoppage of the motor

DIRECTIONS: Questions 69–71 are based on the following diagram. Assume that the teeth of the gears are continuous all the way around each gear.

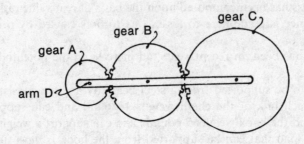

69. Fastening gear A to arm D at another point in addition to its shaft will result in

   (A) gear B rotating on its shaft in a direction opposite to gear A
   (B) gear C rotating on its shaft in a direction opposite to gear A
   (C) arm D rotating around the shaft of gear C
   (D) locking of all gears

70. If gear C is fastened to a supporting frame (not shown) so that it cannot rotate, and gear A turns clockwise on its shaft, then

   (A) gear B will turn counterclockwise and arm D will turn clockwise around the shaft of gear C
   (B) gear B will turn counterclockwise and arm D will turn counterclockwise around the shaft of gear C
   (C) gear B will turn clockwise and arm D will turn clockwise around the shaft of gear C
   (D) gear B will turn clockwise and arm D will turn counterclockwise around the shaft of gear C

71. If gear B is fastened to a supporting frame (not shown) so that it cannot rotate, and arm D rotates clockwise around the shaft of gear B, then for each complete revolution arm D makes, gear A will make

   (A) more than one turn clockwise about its own shaft
   (B) less than one turn clockwise about its own shaft
   (C) more than one turn counterclockwise about its own shaft
   (D) less than one turn counterclockwise about its own shaft

72. As a ship sails away from shore it appears to go below the horizon. The best explanation for this observation is that it results from the

   (A) rise and fall of tides
   (B) curvature of the earth's surface
   (C) refraction of light
   (D) effect of gravity on moving bodies

73. Of the following, the most important reason for lubricating moving parts of machinery is to

(A) reduce friction
(B) prevent rust formation
(C) increase inertia
(D) reduce the accumulation of dust and dirt on the parts

74. A canvas tarpaulin measures 6 feet by 9 feet. The largest circular area that can be covered completely by this tarpaulin is a circle with a diameter of

(A) 9 feet
(B) 8 feet
(C) 7 feet
(D) 6 feet

75. A firefighter caught a civilian attempting to reenter a burning building despite several warnings to stay outside of the fire lines. The civilian insisted frantically that he needed to save some very valuable documents from the fire. The firefighter then called a police officer to remove the civilian. The firefighter's action was

(A) wrong; it is bad public relations to order people about
(B) right; the firefighter is charged with the responsibility of protecting lives
(C) wrong; the firefighter should have explained to the civilian why he should not enter the building
(D) right; civilians must be excluded from the fire zone

DIRECTIONS: Questions 76–78 are to be answered on the basis of the diagram below. In this diagram, pulley A and pulley B are both firmly attached to shaft C so that both pulleys and the shaft can turn only as a single unit. The radius of pulley A is 4 inches and that of pulley B is 1 inch.

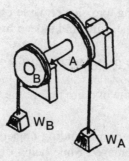

76. If the weight W_A weighs 20 pounds, the system will be in balance if the weight W_B weighs

(A) 5 pounds
(B) 10 pounds
(C) 20 pounds
(D) 80 pounds

77. If the rope on pulley A is pulled downward so that it unwinds, the rope on pulley B will

(A) wind slower
(B) wind faster
(C) unwind slower
(D) unwind faster

78. When pulley A makes one complete revolution, pulley B will make

(A) one quarter of a revolution
(B) one revolution
(C) four revolutions
(D) a number of revolutions which cannot be determined from the information given

79. For combustion to take place, the atmosphere must contain at least 12 percent oxygen. The maximum percent of oxygen in an atmosphere which can support combustion is

(A) 21 percent
(B) 25 percent
(C) 50 percent
(D) 100 percent

80. Firefighters, when not responding to alarms, are assigned to perform housekeeping chores on the equipment in the firehouse. The primary reason for these assignments is to

(A) keep the firefighters busy when not responding to alarms
(B) keep the equipment and quarters in the best possible condition
(C) save money by not hiring maintenance workers
(D) make firehouses competitive with each other

81. If a firefighter in the act of extinguishing a fire is forced to run through an area of dense smoke, it would be best for the firefighter to

(A) take deep slow breaths of air
(B) breathe deeply before starting and slowly while passing through the area
(C) take a shallow breath before starting and a deep breath while passing through the area
(D) refrain from breathing while passing through the area

82. Empty hoses are dragged to the site of a fire before the water supply is turned on. The reason for this is that

(A) full hoses are liable to catch on protruding objects
(B) full hoses are likely to be damaged while being dragged
(C) the source of the water may be too far removed from the fire
(D) empty hoses may be dragged more easily and quickly

83. An intense fire develops in a room in which carbon dioxide cylinders are stored. The principal hazard in this situation is that

(A) the CO_2 may catch fire
(B) toxic fumes may be released
(C) the cylinders may explode
(D) released CO_2 may intensify the fire

84. At a fire involving the roof of a five-story building, the firefighters trained their hose stream on the fire from a vacant lot across the street, aiming the stream at a point about 15 feet above the roof. In this situation, water in the stream would be traveling at the greatest speed

(A) as it leaves the hose nozzle
(B) at a point midway between the ground and the roof
(C) at the maximum height of the stream
(D) as it drops on the roof

85. The one of the following circumstances concerning a fire which indicates most strongly the possibility of arson is that

(A) there was heavy charring of wood around the point of origin of the fire
(B) three fires apparently broke out simultaneously in different parts of the building
(C) the heat was so intense that glass in the building became molten and fused
(D) when the firefighters arrived the smoke was very heavy

86. A firefighter searching an area after an explosion comes upon a man with a leg injury bleeding severely and in great pain. Under these circumstances, the first action the firefighter should take is to attempt to

(A) immobilize the man's leg
(B) cover the man with a blanket or coat
(C) stop the bleeding
(D) make the man comfortable

87. Spontaneous combustion may be the reason for a pile of oily rags catching fire. In general, spontaneous combustion is the direct result of

(A) application of flame
(B) falling sparks
(C) intense sunlight
(D) chemical action

88. Most fire departments advocate that firefighters become involved in community activities, such as coaching or sponsoring a Little League team. This course of action is likely to

(A) produce expert Little League coaching
(B) lead to better understanding of the activities of the fire department
(C) encourage the community members to assist in fighting fires when needed
(D) enlist community support of the firefighters' quest for higher salaries

89. There is a common bond among all firefighters functioning throughout the United States and even in some other parts of the world. There are many nationwide organizations of firefighters, and it is not uncommon for them to drop in at the local fire department when they are visiting other cities. The most likely reason for this is the

(A) similarity of techniques, equipment, and problems from one fire department to another
(B) need for firefighters to be strongly organized in order to achieve equitable pay
(C) extreme dangers involved in fighting fires
(D) usual competitiveness of police and fire departments

90. A firefighter stationed in a firehouse is approached by an irate citizen who declares that firefighters are well paid but spend most of their time lounging around the firehouse doing nothing. It would be most appropriate for the firefighter to respond by saying;

(A) "You had better forward your complaint to the mayor."
(B) "If you did my job for one week you would change your opinion."
(C) "How hard do you work for a living?"
(D) "The job of a firefighter is to always be ready to respond to an alarm."

91. Firefighters often visit schools to give talks on fire prevention to the children. The type of firefighter best suited for this duty would be

(A) one whose demeanor depicts authority but who has little experience in public speaking
(B) a newly appointed firefighter who is a recent college graduate and who has majored in speech
(C) a veteran firefighter who projects a strong image and who has addressed audiences in the past
(D) the firefighter who has achieved the highest rating on the eligible list

92. During an emergency situation, a firefighter may be the recipient of an order from his or her superior that is delivered in an abrupt manner. The recipient of this type of an order should assume that

(A) he or she is in disfavor with the superior officer
(B) the superior officer feels that he or she will not respond properly to any other type of order
(C) he or she is about to be brought up on charges
(D) emergency situations call for this type of order

93. Studies of accidents have indicated that more accidents occur at the end of a tour of duty than at its beginning. The most likely reason for this is

(A) carelessness on the part of the victims
(B) improper training in accident prevention
(C) lack of interest in safety
(D) mental and physical fatigue on the part of firefighters

94. Firefighters can best avoid accidents while doing heavy lifting by

(A) keeping the knees bent and the back straight
(B) keeping both knees and the back straight
(C) using the arms only
(D) using the method best suited to the individual

95. The *least* likely reason for including an accident prevention program in the training program for newly appointed firefighters is the

(A) nature of the work being performed
(B) number of reported accidents
(C) number of levels of command
(D) time lost as a result of accidents

96. A training officer assigned to teach safety to firefighters can best gain their cooperation by

(A) exercising the authority consistent with his or her rank
(B) waiting until a serious accident occurs
(C) making a distinct effort to become "one of the boys" during the training session
(D) indicating to the trainees how an accident can be costly both to the individual and to the department

97. A veteran firefighter observes a newly appointed firefighter performing a maintenance task in an unsafe manner. It would be most appropriate for the veteran firefighter to

(A) tell their superior officer of the indiscretion at once
(B) illustrate the proper manner for performing this task to the rookie firefighter
(C) not do anything because this is not an assigned responsibility
(D) tell another rookie firefighter to demonstrate the proper method because it is likely that there is a common bond between them

98. A well-executed accident report is extremely valuable principally because it

(A) isolates the cause of the accident so that future similar accidents can be avoided
(B) forms the basis of statistical studies
(C) complies with existing regulations
(D) informs everybody that an accident has taken place

99. It is essential that accident reports be submitted as soon after the accident has occurred as is practical. The most important reason for this course of action is

(A) witnesses tend to change their stories as time passes
(B) that accident reports should be considered trivial and therefore they should be gotten out of the way
(C) the details of the report will still be fresh in the mind of the reporter
(D) all reports should be submitted promptly even if they are incomplete

100. Following an accident involving a firefighter on the job, it is normal procedure for a fire department to spend considerable effort in an attempt to identify the cause of the accident. The most important reason for this procedure is

(A) no accident report is complete without the cause of the accident
(B) for statistical purposes
(C) so that the guilty firefighter may be disciplined
(D) that future similar accidents may be avoided

Answer Key for Practice Examination 4

1. D	21. D	41. C	61. C	81. D
2. A	22. D	42. D	62. D	82. D
3. D	23. D	43. C	63. B	83. C
4. B	24. C	44. C	64. A	84. A
5. D	25. C	45. A	65. D	85. B
6. C	26. C	46. B	66. D	86. C
7. C	27. C	47. C	67. A	87. D
8. C	28. A	48. A	68. C	88. B
9. B	29. B	49. A	69. D	89. A
10. C	30. D	50. C	70. B	90. D
11. A	31. B	51. C	71. A	91. C
12. C	32. B	52. A	72. B	92. D
13. B	33. B	53. B	73. A	93. D
14. B	34. D	54. A	74. D	94. A
15. B	35. B	55. D	75. B	95. C
16. B	36. A	56. C	76. D	96. D
17. C	37. D	57. D	77. A	97. B
18. B	38. C	58. A	78. B	98. A
19. C	39. A	59. C	79. D	99. C
20. B	40. A	60. D	80. B	100. D

Explanatory Answers
for Practice Examination 4

1. **(D)** The most effective course of action would be to suggest that the manager contact the department where he would be assured of receiving correct information.

2. **(A)** This immediately identifies the person receiving the call and is likely to save time and prevent misunderstandings.

3. **(D)** No area should remain without proper protection to cover possible fires. If a company is busy fighting one fire, coverage should be provided in the event that a second fire occurs in the same area.

4. **(B)** The responding fire company does not know the actual location of the fire. It will respond to the signal received from the box. When the fire apparatus arrives at the location of the box, the firefighter will be able to direct it to the actual location of the fire.

5. **(D)** The badge may fall into the wrong hands and may be used improperly. The badge belongs to the fire department. Any found property should be returned to the rightful owner, especially if the owner can be identified.

6. **(C)** Consideration should be given to the fact that as much mail as possible should be salvaged from a mailbox fire. Very often the contents of the mail are extremely important and valuable.

7. **(C)** One of the difficulties in determining the cause of a fire is that a fire is likely to destroy the evidence of what caused it.

8. **(C)** This device would discourage many people from using fire alarm boxes due to fear of possible incrimination.

9. **(B)** Water pipes leading to the sprinklers are often blown apart by an explosion.

10. **(C)** This course of action would prevent large amounts of water from entering the subway system.

11. **(A)** Note that the choice that is *not* true is the correct answer. Motorists do not avoid routes of travel because fire apparatus may use them.

12. **(C)** The first action to be taken is to determine if a fire has been started. The sooner a fire is discovered, the easier it will be to extinguish. This is generally true for all fires.

13. **(B)** Safe driving calls for total concentration. Firefighters should refrain from distracting the chauffeur while he or she is driving the apparatus.

14. **(B)** Most major cities have established a 911 emergency number which accepts all calls and sends assistance for police, fire, and medical emergencies.

15. **(B)** Doors that open outward provide the safest exit. Doors that open inward may jam closed as a panic-stricken crowd of people attempts to force its way through them.

16. **(B)** An unsafe condition could be covered up temporarily if one knows that the premises are scheduled to be inspected. Irregular inspection will prevent these temporary maskings of unsafe conditions.

17. **(C)** Oxygen is vital to life. A mask that filters out harmful gasses but does not provide oxygen would be useless in this situation.

18. **(B)** The gas should not be turned on after a fire until utility company employees determine that it is safe to do so.

19. **(C)** The firefighter was performing as a good citizen rather than specifically as a firefighter in this situation.

20. **(B)** Information concerning the operation of a fire department should be accurate so as to prevent misunderstandings with the general public.

21. **(D)** A firefighter may be knocked over by a hose he or she is straddling when the position of the hose is changed. It is always best for the firefighter to stand clear of the hose to prevent entanglement.

22. **(D)** A firefighter in the process of performing his or her duties is likely to step on sharp objects.

23. **(D)** Good public relations are best engendered from services extended.

24. **(C)** Rings always present a danger to people who are engaged in close contact with machinery or in the movement of heavy and irregularly-shaped objects.

25. **(C)** The fire department accepts liability when it permits a nonmember to ride on its equipment. The decisions to accept this liability should come from headquarters.

26. **(C)** If the pier is damaged during the fighting of the fire, the apparatus might fall from the pier into the water.

27. **(C)** These closed cabs provide greater protection for firefighters from inclement weather and hazardous conditions.

28. **(A)** Fires that occur during severe weather conditions are more difficult to extinguish. Therefore, increased manpower may be needed to combat a fire in a blizzard.

29. **(B)** Bends weaken the hose and should therefore be kept to a minimum.

30. **(D)** The decisions as to what to do with the money should come from a higher authority in the fire department.

31. **(B)** A section of a falling cornice may injure a pedestrian or a member of the fire department working on the street level.

32. **(B)** There is always a possibility that a fire may rekindle from the debris and ashes.

33. **(B)** Wetting the firefighters with small streams of water will accomplish nothing but making the firefighters wet.

34. **(D)** There is no reason for you to do anything but get out of the way of the oncoming fire apparatus. Driving rules require all vehicles to pull to the curb to allow fire apparatus to pass.

35. **(B)** Because of the use of scenery, which may be composed largely of wood and other highly combustible materials, and machinery, which may involve electrical wiring or other devices creating a potential fire hazard, a firefighter is assigned to be present to insure that these materials are handled correctly, and that in the event of a fire, the problem may be remedied as quickly as possible.

36. **(A)** The second sentence indicates this fact.

37. **(D)** This fact is indicated in the third and fourth sentences of the paragraph.

38. **(C)** As stated in the first, second, and third sentences.

39. **(A)** This may be derived from the last sentence of the paragraph.

40. **(A)** An open circuit does not carry electric current. When a fuse melts due to overheating, an open circuit condition results.

41. **(C)** An electrical wire carrying electrical current is a common example of a conductor.

42. **(D)** There is no information on the ratio of circuits to fuses in this paragraph.

43. **(C)** The second sentence indicates that the introduction of water will cause the threads to swell, thus reducing the space between threads and making the hose practically watertight.

44. **(C)** Unlined linen hose cannot withstand very hard wear; thus it is best used for emergencies.

45. **(A)** Unlined linen hose could withstand neither the punishment of being dragged over rough lumber nor exposure to weather conditions.

46. **(B)** As stated in the second sentence. All this knowledge helps in preparing to fight fires at the location.

47. **(C)** Inspections involve physical assessment of the property. A building plan would greatly assist in this endeavor.

48. **(A)** Structural features refer to the physical layout, doorways, stairways, etc., of the building rather than the location of any merchandise stored in the building.

49. **(A)** A protective feature, with reference to firefighting, would be an implement affording protection from fire damage. A fire extinguisher is one such device.

50. **(C)** Building inspections not only reveal violations of the law, but they also bring to light knowledge which would be helpful in extinguishing fires in those locations.

51. **(C)** It is not the firefighter's responsibility to decide which laws should be enforced and which should not be, regardless of expense. Nonprofessional behavior will not gain respect.

52. **(A)** The sooner the fire department arrives at the scene of a fire, the more likely it will be that damage to life and property will be kept to a minimum. A single firefighter would not be able to combat the fire alone without the assistance of other firefighters and the use of firefighting apparatus.

53. **(B)** The apparatus should already be in good working order. The hoselines will have to be removed to allow for the emergency transportation of firefighters.

54. **(A)** These wagons will be in use transporting firefighters, thus they will not be available to respond to other alarms.

55. **(D)** A call to a division for members would indicate a need for additional manpower at an extensive fire.

56. **(C)** This may be inferred from the overall purpose of the procedure: to get firefighters who may be scattered at different locations to the point of the emergency.

57. **(D)** Excessive heat is rubber's worst enemy.

58. **(A)** The cooler, heavier air, because it is dense, pushes the hot air upward.

59. **(C)** The incline of ramp 3 is not nearly as steep as that of ramp 1. Ramps 2 and 4 have irregularities in their inclines which would make it even more difficult to move a heavy load to the top of the platform.

60. **(D)** Oil pressure should be established before the engine is raced.

61. **(C)** Salt will affect ice in this manner.

62. **(D)** Water boils at 212°F. at sea level, at a lower temperature at lower air pressure. Cooking and baking instructions often vary for certain altitudes above sea level.

63. **(B)** It follows that a substance that carries heat well cannot be used as an insulator of heat.

64. **(A)** If some public education has helped to maintain fire losses at even level (costs only go up, so maintaining losses shows effect of education), then increasing public education should reduce the level of future losses.

65. **(D)** A chain is only as strong as its weakest link. All links in this chain are of equal strength. Each can support 1000 pounds. The weight of the hook reduces this amount slightly.

66. **(D)** Water expands as it freezes. If the expanding water is in an enclosed area, as in a pipe, the expansion must cause the pipe to burst.

67. **(A)** We first met this principle, the need for extra manpower, in the paragraph for questions 53–56.

68. **(C)** When it is desired that one pulley be turned in the opposite direction from its connecting pulley, drive belts must be crossed as shown in the diagram.

69. **(D)** With the center of gear A fastened at one point on the shaft, the gear is able to rotate at that point. If gear A is fastened at an additional point on the shaft, the gear becomes rigid and unable to rotate. It therefore cannot turn gear B and all gears will lock.

70. **(B)** By anchoring gear C, gear C then becomes the pivoting point and the remaining gears will then rotate around the pivot, which is gear C.

71. **(A)** Gear A, being smaller than gears B and C, must revolve at a greater rate of speed in order to travel the same distance. Therefore, gear A will make more than one rotation on its own shaft for every revolution that arm D makes.

72. **(B)** Because of the curvature of the earth's surface, one cannot see beyond the horizon.

73. **(A)** Lubrication with oil, grease, or any other suitable substance will reduce friction between moveable parts of machinery. It is not possible to operate most machinery well without lubrication. In addition, the machinery would wear out quickly because of the friction.

74. **(D)** Since the tarpaulin measures 6 feet by 9 feet, it will not be possible to completely cover any circular area which exceeds 6 feet in diameter.

75. **(B)** The firefighter had already given warnings, and, presumably some explanation to the civilian. The job of the firefighter is to save lives by fighting fires. This firefighter acted correctly in calling a police officer to remove the civilian from danger so that the firefighter could return to fighting the fire.

76. **(D)** There is a 1 to 4 ratio between pulleys B and A. In order to balance the system, the weight on the smaller pulley must be 4 times the weight on the larger pulley. The

weight on the large pulley is 20 pounds; therefore, 4 times 20 pounds equals 80 pounds, which is (D).

77. **(A)** Pulley A has a circumference that is 4 times greater than pulley B. With each revolution, pulley B will be traveling one-fourth the distance traveled by pulley A. This results in one-fourth the amount of rope being wound on the pulley. Therefore, pulley B would wind at a slower rate.

78. **(B)** Both pulleys are attached firmly to the shaft, therefore, both will make the same number of revolutions.

79. **(D)** Oxygen, though noncombustible, will support combustion. An atmosphere of 100 percent oxygen will cause burning material to accelerate with explosive rapidity.

80. **(B)** Housekeeping chores are part of the job of a firefighter. The motivated firefighter will take pride in the condition of the equipment used in the performance of his or her duties.

81. **(D)** By not breathing, the firefighter will expose his or her lungs to a minimum amount of irritation from the smoke.

82. **(D)** Full hoses weigh considerably more than empty hoses because of the weight of the water. Therefore, it is much easier to carry or to drag empty hoses to the site of a fire.

83. **(C)** The immediate danger involving the cylinders of CO_2 is that the contents of the cylinders may expand and explode from overheating. Not only will the explosion cause immediate danger to the premises, but personnel may be injured from the fragments of metal flying from the exploding cylinders.

84. **(A)** The water pressure will diminish gradually as the water gets farther from the nozzle of the hose.

85. **(B)** It would be most unlikely for three fires to break out simultaneously in the same building from accidental causes. Deliberate setting of the fires must be suspected.

86. **(C)** Whenever severe bleeding occurs, attention should be given to it immediately since such bleeding may quickly result in the loss of life.

87. **(D)** Spontaneous combustion is combustion (burning) that occurs from no external cause. Flame and falling sparks are obvious external causes. The heat of intense sunlight might act as a catalyst speeding up chemical activity in a pile of rags or other hazardous pile of debris, but the actual cause of the spontaneous combustion is a chemical reaction within the heap.

88. **(B)** When firefighters become involved in community activities, it is likely that the residents will become more aware of the purpose and the objectives of the fire department.

89. **(A)** Firefighters have the same responsibilities throughout the world—to extinguish fires, save lives, and minimize property damage. The equipment and techniques used are basically the same throughout the world, although larger cities may have more sophisticated fire fighting systems.

90. **(D)** There will always be citizens who are under the impression that firefighters do not work hard because they can often be seen at the firehouse instead of out fighting fires. It is very important that these citizens be made to understand that after housekeeping chores are done, the firefighter's prime responsibility is to make himself or herself ready to answer an alarm.

91. **(C)** Firefighting is a serious business and therefore a proper image of the fire department must be created in the minds of children. Of course, the firefighter selected must be skilled in public speaking and must be able to put forth a strong and serious image.

92. **(D)** Emergency situations call for immediate action. Very often courtesies are brushed over in the transmission of orders. The firefighter should realize that the abrupt manner in which he or she is addressed does not reflect on him or her personally.

93. **(D)** The tired worker is likely to become involved in accidents due to lack of alertness to dangerous situations.

94. **(A)** The weight of the object is evenly distributed with the knees bent and the back straight. No part of the body receives the bulk of the weight.

95. **(C)** The number of levels of command has no bearing on the necessity for such a program.

96. **(D)** The best approach to convince firefighters to take a course in safety and accident prevention seriously is by convincing them of how costly an accident can be to themselves as well as to the department.

97. **(B)** The newly appointed firefighter should be corrected at once before an accident occurs and before this method of working becomes a habit.

98. **(A)** Isolating the cause of an accident will alert firefighters to be aware of such situations in the future. This should reduce the chances of a similar accident occurring.

99. **(C)** Time plays tricks on an individual's memory. Therefore, the sooner the report is submitted, the more accurate is it likely to be.

100. **(D)** Once an accident occurs, there is little that can be done but to suffer its consequences. However, similar accidents may be avoided in the future if the cause of such an accident is identified.

Answer Sheet for Practice Examination 5

1. Ⓐ Ⓑ Ⓒ Ⓓ	21. Ⓐ Ⓑ Ⓒ Ⓓ	41. Ⓐ Ⓑ Ⓒ Ⓓ	61. Ⓐ Ⓑ Ⓒ Ⓓ	81. Ⓐ Ⓑ Ⓒ Ⓓ
2. Ⓐ Ⓑ Ⓒ Ⓓ	22. Ⓐ Ⓑ Ⓒ Ⓓ	42. Ⓐ Ⓑ Ⓒ Ⓓ	62. Ⓐ Ⓑ Ⓒ Ⓓ	82. Ⓐ Ⓑ Ⓒ Ⓓ
3. Ⓐ Ⓑ Ⓒ Ⓓ	23. Ⓐ Ⓑ Ⓒ Ⓓ	43. Ⓐ Ⓑ Ⓒ Ⓓ	63. Ⓐ Ⓑ Ⓒ Ⓓ	83. Ⓐ Ⓑ Ⓒ Ⓓ
4. Ⓐ Ⓑ Ⓒ Ⓓ	24. Ⓐ Ⓑ Ⓒ Ⓓ	44. Ⓐ Ⓑ Ⓒ Ⓓ	64. Ⓐ Ⓑ Ⓒ Ⓓ	84. Ⓐ Ⓑ Ⓒ Ⓓ
5. Ⓐ Ⓑ Ⓒ Ⓓ	25. Ⓐ Ⓑ Ⓒ Ⓓ	45. Ⓐ Ⓑ Ⓒ Ⓓ	65. Ⓐ Ⓑ Ⓒ Ⓓ	85. Ⓐ Ⓑ Ⓒ Ⓓ
6. Ⓐ Ⓑ Ⓒ Ⓓ	26. Ⓐ Ⓑ Ⓒ Ⓓ	46. Ⓐ Ⓑ Ⓒ Ⓓ	66. Ⓐ Ⓑ Ⓒ Ⓓ	86. Ⓐ Ⓑ Ⓒ Ⓓ
7. Ⓐ Ⓑ Ⓒ Ⓓ	27. Ⓐ Ⓑ Ⓒ Ⓓ	47. Ⓐ Ⓑ Ⓒ Ⓓ	67. Ⓐ Ⓑ Ⓒ Ⓓ	87. Ⓐ Ⓑ Ⓒ Ⓓ
8. Ⓐ Ⓑ Ⓒ Ⓓ	28. Ⓐ Ⓑ Ⓒ Ⓓ	48. Ⓐ Ⓑ Ⓒ Ⓓ	68. Ⓐ Ⓑ Ⓒ Ⓓ	88. Ⓐ Ⓑ Ⓒ Ⓓ
9. Ⓐ Ⓑ Ⓒ Ⓓ	29. Ⓐ Ⓑ Ⓒ Ⓓ	49. Ⓐ Ⓑ Ⓒ Ⓓ	69. Ⓐ Ⓑ Ⓒ Ⓓ	89. Ⓐ Ⓑ Ⓒ Ⓓ
10. Ⓐ Ⓑ Ⓒ Ⓓ	30. Ⓐ Ⓑ Ⓒ Ⓓ	50. Ⓐ Ⓑ Ⓒ Ⓓ	70. Ⓐ Ⓑ Ⓒ Ⓓ	90. Ⓐ Ⓑ Ⓒ Ⓓ
11. Ⓐ Ⓑ Ⓒ Ⓓ	31. Ⓐ Ⓑ Ⓒ Ⓓ	51. Ⓐ Ⓑ Ⓒ Ⓓ	71. Ⓐ Ⓑ Ⓒ Ⓓ	91. Ⓐ Ⓑ Ⓒ Ⓓ
12. Ⓐ Ⓑ Ⓒ Ⓓ	32. Ⓐ Ⓑ Ⓒ Ⓓ	52. Ⓐ Ⓑ Ⓒ Ⓓ	72. Ⓐ Ⓑ Ⓒ Ⓓ	92. Ⓐ Ⓑ Ⓒ Ⓓ
13. Ⓐ Ⓑ Ⓒ Ⓓ	33. Ⓐ Ⓑ Ⓒ Ⓓ	53. Ⓐ Ⓑ Ⓒ Ⓓ	73. Ⓐ Ⓑ Ⓒ Ⓓ	93. Ⓐ Ⓑ Ⓒ Ⓓ
14. Ⓐ Ⓑ Ⓒ Ⓓ	34. Ⓐ Ⓑ Ⓒ Ⓓ	54. Ⓐ Ⓑ Ⓒ Ⓓ	74. Ⓐ Ⓑ Ⓒ Ⓓ	94. Ⓐ Ⓑ Ⓒ Ⓓ
15. Ⓐ Ⓑ Ⓒ Ⓓ	35. Ⓐ Ⓑ Ⓒ Ⓓ	55. Ⓐ Ⓑ Ⓒ Ⓓ	75. Ⓐ Ⓑ Ⓒ Ⓓ	95. Ⓐ Ⓑ Ⓒ Ⓓ
16. Ⓐ Ⓑ Ⓒ Ⓓ	36. Ⓐ Ⓑ Ⓒ Ⓓ	56. Ⓐ Ⓑ Ⓒ Ⓓ	76. Ⓐ Ⓑ Ⓒ Ⓓ	96. Ⓐ Ⓑ Ⓒ Ⓓ
17. Ⓐ Ⓑ Ⓒ Ⓓ	37. Ⓐ Ⓑ Ⓒ Ⓓ	57. Ⓐ Ⓑ Ⓒ Ⓓ	77. Ⓐ Ⓑ Ⓒ Ⓓ	97. Ⓐ Ⓑ Ⓒ Ⓓ
18. Ⓐ Ⓑ Ⓒ Ⓓ	38. Ⓐ Ⓑ Ⓒ Ⓓ	58. Ⓐ Ⓑ Ⓒ Ⓓ	78. Ⓐ Ⓑ Ⓒ Ⓓ	98. Ⓐ Ⓑ Ⓒ Ⓓ
19. Ⓐ Ⓑ Ⓒ Ⓓ	39. Ⓐ Ⓑ Ⓒ Ⓓ	59. Ⓐ Ⓑ Ⓒ Ⓓ	79. Ⓐ Ⓑ Ⓒ Ⓓ	99. Ⓐ Ⓑ Ⓒ Ⓓ
20. Ⓐ Ⓑ Ⓒ Ⓓ	40. Ⓐ Ⓑ Ⓒ Ⓓ	60. Ⓐ Ⓑ Ⓒ Ⓓ	80. Ⓐ Ⓑ Ⓒ Ⓓ	100. Ⓐ Ⓑ Ⓒ Ⓓ

TEAR HERE

PRACTICE EXAMINATION 5

The time allowed for the entire examination is three hours and 55 minutes.

DIRECTIONS: Each question has four suggested answers, lettered A, B, C, and D. Decide which one is the best answer. On the sample answer sheet, find the question number that corresponds to the question you have just read and darken the circle with the letter of the answer you have selected. (Note: This is an examination measuring how well you read, think, and reason. It is not a test of your firefighting knowledge nor is it a course of instruction. Some questions spell out an order of action. You must accept the information as given in the question, even if you doubt its accuracy.)

Questions 1 through 18 are to be answered on the basis of the following sketch. You will have five minutes to study the sketch. Then you will have to answer the 18 questions without referring back to it.

Now turn the page and study the sketch for exactly five minutes. After five minutes turn the page and proceed to the 100 questions. You will have three hours and 50 minutes to answer the questions. Do not turn back to the sketch once you have begun the questions.

1. The vehicles involved in the motor vehicle accident are

    (A) a taxi and a bus
    (B) a taxi and a truck
    (C) a passenger car and a truck
    (D) a passenger car and a taxi

2. From studying the sketch, one could assume that the accident was caused by the

    (A) taxi
    (B) truck
    (C) bus
    (D) police car

3. The police officer who is not wearing a hat is

    (A) standing next to the vehicles involved in the accident
    (B) directing traffic at the intersection
    (C) helping a person into the ambulance
    (D) talking to a group of people on the corner

4. The police officer who is wearing a hat is

    (A) standing next to the tow truck
    (B) standing next to the vehicles involved in the accident
    (C) helping a person into the ambulance
    (D) directing traffic at the intersection

5. The fire is

    (A) at Tony's Pizza
    (B) at 26 Washington Avenue
    (C) above John's Drug Store
    (D) above Tony's Pizza

6. Which one of the following vehicles is *not* depicted in the sketch?

    (A) A motorcycle
    (B) A taxi
    (C) A garbage truck
    (D) A bus

7. The accident occurred in front of

    (A) Tony's Pizza
    (B) John's Drug Store
    (C) Public School #11
    (D) an apartment house

8. The quickest *legal* route by which a fire engine could travel to the scene of the fire is

    (A) west on Washington Avenue
    (B) north on Lincoln Street
    (C) east on Washington Avenue
    (D) south on Lincoln Street

9. The police car is parked

(A) on Washington Avenue
(B) on Lincoln Street
(C) on Washington Street
(D) in the middle of the intersection of Washington Avenue and Lincoln Street

10. The quickest, most convenient access to the fire scene is to approach from the

(A) west on Washington Avenue
(B) south on Washington Street
(C) north on Lincoln Street
(D) north on Washington Street

11. The person who is carrying a large portable radio is standing next to

(A) Tony's Pizza
(B) John's Drug Store
(C) the police car
(D) the tow truck

12. The ambulance shown in the sketch is from

(A) Jefferson Hospital
(B) Columbia Hospital
(C) Lincoln Hospital
(D) Washington Hospital

13. The type of establishment at 30 Washington Avenue is

(A) a gift shop
(B) a boxing studio
(C) a drug store
(D) not indicated

14. A woman with a baby carriage is

(A) in front of John's Drug Store
(B) on the corner of Washington Street and Lincoln Street
(C) in front of Tony's Pizza
(D) in front of Public School #11

15. A fire alarm box is

(A) located on the corner of Washington Avenue and Lincoln Street
(B) located in front of Tony's Pizza
(C) located in front of Public School #11
(D) not shown in the sketch

16. The person with the best view of the fire

(A) is looking out of a window
(B) has long hair
(C) is near the ambulance
(D) is driving a tow truck

17. Parking is prohibited on

(A) Washington Avenue between 8 A.M. and 7 P.M.
(B) Lincoln Street between 7 A.M. and 8 P.M.
(C) Washington Avenue between 8 A.M. and 7 P.M. on Sundays
(D) Washington Street between 7 A.M. and 8 P.M. weekdays

18. The fire hydrant closest to the fire is

(A) in front of Public School #11
(B) at the southeast corner of Washington Avenue
(C) at the northwest corner of Washington Avenue
(D) next to a trash basket

DIRECTIONS: Answer questions 19 and 20 by selecting the paragraph in which the sentences appear in the most logical order to make the best sense.

19. (A) You may admire Joe Montana, Louis Nizer, or a good homicide detective. But this does not mean that you can count on being successful or happy as a professional ball player, criminal lawyer, or famous detective. It is risky to choose an occupation just because you admire or are fond of someone who has chosen it. Sometimes young people make the mistake of picking a job just because a much-admired relative or friend likes that job.

(B) Sometimes young people make the mistake of picking a job just because a much-admired relative or friend likes that job. But this does not mean that you can count on being successful or happy as a professional ball player, criminal lawyer, or famous detective. You may admire Joe Montana, Louis Nizer, or a good homicide detective. It is risky to choose an occupation just because you admire or are fond of someone who has chosen it.

(C) Sometimes young people make the mistake of picking a job just because a much-admired relative or friend likes that job. It is risky to choose an occupation just because you admire or are fond of someone who has chosen it. But this does not mean that you can count on being successful or happy as a professional ball player, criminal lawyer, or famous detective. You may admire Joe Montana, Louis Nizer, or a good homicide detective.

(D) Sometimes young people make the mistake of picking a job just because a much-admired relative or friend likes that job. It is risky to choose an occupation just because you admire or are fond of someone who has chosen it. You may admire Joe Montana, Louis Nizer, or a good homicide detective. But this does not mean that you can count on being successful or happy as a professional ball player, criminal lawyer, or famous detective.

20. (A) Together they can fashion a program that will foster public understanding and enlist public assistance in combating crime. This important service makes private security a natural ally of the police and a formidable foe of the criminal. Both police and private security stand to gain, but more importantly, the public stands to gain the most. The business of the private-security sector is not only to sell safety and security but to educate people in the many ways they can protect themselves.

(B) The business of the private-security sector is not only to sell safety and security but to educate people in the many ways they can protect themselves. Together they can fashion a program that will foster public understanding and enlist public assistance in combating crime. This important service makes private security a natural ally of the police and a formidable foe of the criminal. Both police and private security stand to gain, but more importantly, the public stands to gain the most.

(C) The business of the private-security sector is not only to sell safety and security but to educate people in the many ways they can protect themselves. Both police and private security stand to gain, but more importantly, the public stands to gain the most. Together they can fashion a program that will foster public understanding and enlist public assistance in combating crime. This important service makes private security a natural ally of the police and a formidable foe of the criminal.

(D) The business of the private-security sector is not only to sell safety and security but to educate people in the many ways they can protect themselves. This important service makes private security a natural ally of the police and a formidable foe of the criminal. Together they can fashion a program that will foster public understanding and enlist public assistance in combating crime. Both police and private security stand to gain, but more importantly, the public stands to gain the most.

21. A firefighter may be required to administer first aid to people who are injured. It is of utmost importance that the firefighter be cognizant of certain conditions while performing this function. Of the following, which is the one condition he or she should *not* be concerned with?

(A) Are the injured person's legs twisted in a manner which may indicate that bones are broken?
(B) Is the injured party breathing?
(C) Is there an open medication bottle nearby?
(D) Is it an apparent case of attempted suicide?

22. Firefighters at the scene of a fire have just about succeeded in putting out the fire. Before it has been completely extinguished, a couple who live on the top floor request permission to enter their apartment to secure some of their belongings. It would be most proper for the firefighter who receives this request to

(A) permit the couple to enter their apartment since the fire is almost out
(B) suggest that the couple give him a written list and he will get whatever the couple wanted to remove
(C) permit one but not both persons to return to the apartment
(D) inform them that they will be permitted to enter only when the building is considered completely safe

DIRECTIONS: Answer questions 23 and 24 by choosing which implement in the lettered boxes is normally used with the implement or object on the left. Note: The tools are not drawn to scale.

23.

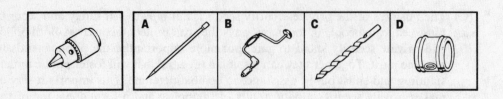

24.

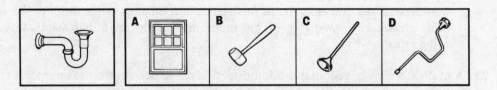

25. When responding to a fire in a nursing home, extra companies should be dispatched, even on a first alarm. The officer in charge of the first ladder company to arrive must size up the fire quickly to determine if there will be any need to evacuate residents. If evacuation will be extensive, the officer must immediately call for further reinforcements and must ask the nursing home administrator to direct firefighters to those areas in which residents will need assistance. Simultaneously with the evacuation of residents, steps must be taken to extinguish the fire.

A frantic nurse has reported a fire in the laundry room of the west wing of the Happy Hours Nursing Home at West 118th Street. The fire dispatcher has sent four pieces of apparatus to the scene. Captain Villegas has sized up the fire and has determined that the fire can be contained and extinguished without any threat to the residents. Now Captain Villegas should

(A) ask the administrator where the most disabled residents are located
(B) supervise the fire extinguishing operation
(C) call for reinforcements
(D) assist the firefighters in evacuating residents

26. Safety regulations require certain precautions to be taken when pulling fire vehicles out of the firehouse, regardless of the urgency of the fire. Sirens and flashing lights on both sides of the engine house doors must be activated before any vehicle is moved; doors may open at the same time. Then a firefighter runs outside to determine that there are no illegally parked cars or trucks blocking the entrance or turning area and that the street is not congested. Another firefighter controls pedestrian traffic, and one more stands in the street to detain oncoming traffic until the firefighting equipment has departed.

The alarm has sounded at Ladder Company 41, and the doors of the engine house are open. The siren is sounding and the lights are flashing. Firefighter Feeney has shouted and signaled that the coast is clear. Firefighter Moro is standing in the street to hold back oncoming cars. Firefighter Williams should

(A) drive out the first fire engine
(B) move forward with the first piece of equipment
(C) step outside and detain pedestrians until all danger has passed
(D) close the doors to the firehouse

27. If you are asked how to open a fire hydrant, the best way to give this information is to say:

(A) Turn the big nut to the left.
(B) Turn the big side nut counterclockwise.
(C) Turn the big side nut to the left.
(D) Turn the big side nut to your left.

28. There is a gas leak in the street, and firefighters must be certain that gas is cut off from each building before repairs can be completed. You will help the firefighters best by saying that the gas shut-off in your building is

(A) in the basement next to the electric meter
(B) in the far corner of the lower level
(C) downstairs and to your right
(D) in the west corner of the basement

DIRECTIONS: Questions 29 through 31 are based on the map on page 280. The flow of traffic is indicated by the arrows. If there is only one arrow shown, then traffic flows only in the direction indicated by the arrow. If there are two arrows shown, then traffic flows in both directions. You must follow the flow of traffic.

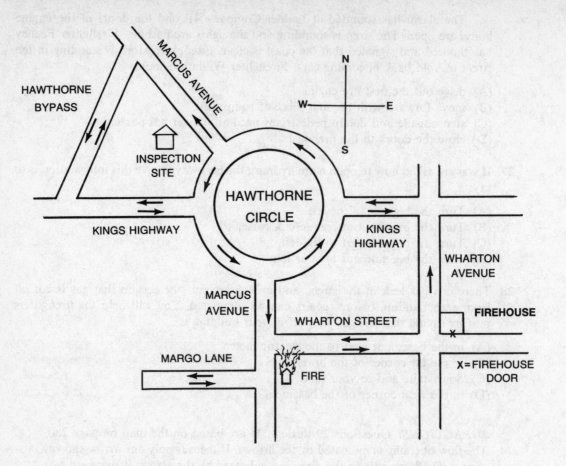

29. A single fire truck leaves the firehouse to make a scheduled fire safety inspection of a building on the northern spur of Marcus Avenue. The best way to get there is to travel

    (A) north on Wharton Avenue, turn left onto Kings Highway, go clockwise around Hawthorne Circle, and exit the circle onto Marcus Avenue just north of Kings Highway
    (B) west on Wharton Street, turn right onto Marcus Avenue, enter Hawthorne Circle going counterclockwise, pass one Kings Highway exit, and exit onto Marcus Avenue
    (C) west on Wharton Street and immediately north on Wharton Avenue, turn left to go west on Kings Highway, turn right onto Hawthorne Circle to the next intersecting road, and turn right onto Marcus Avenue
    (D) north on Wharton Street, west on Kings Highway into Hawthorne Circle, then counterclockwise on Hawthorne Circle directly onto Marcus Avenue

30. While the inspection is in progress, the alarm sounds for a fire on the southern spur of Marcus Avenue. The firefighters interrupt their inspection, get on the truck, and proceed to the fire by traveling

    (A) southeast on Marcus Avenue, then turning right onto Hawthorne Circle, continuing past Kings Highway, and turning right onto Marcus Avenue
    (B) northwest on Marcus Avenue, making a sharp left turn onto the Hawthorne Bypass, making another left onto Kings Highway, then a right onto Hawthorne Circle, and the next right to Marcus Avenue

(C) northwest on Marcus Avenue, making a sharp left turn onto the Hawthorne By-pass, making a right onto Kings Highway, then a right onto Hawthorne Circle, and the next right to Marcus Avenue

(D) southeast on Marcus Avenue to Hawthorne Circle, clockwise around Hawthorne Circle, past Kings Highway to the south exit for Marcus Avenue

31. Two engines respond to this same fire directly from the fire station. They travel

(A) south on Wharton Street and turn left onto Marcus Avenue

(B) north on Wharton Avenue, west on Kings Highway, counterclockwise on Hawthorne Circle past the northern spur of Marcus Avenue and the western extension of Kings Highway to the southern spur of Marcus Avenue

(C) west on Wharton Street and south on Marcus Avenue

(D) north on Wharton Avenue, west on Kings Highway, clockwise around Hawthorne Circle to Marcus Avenue

DIRECTIONS: Answer questions 32 and 33 by selecting the paragraph in which the sentences appear in the most logical order to make the best sense.

32. (A) They observed a vehicle accident with several injured persons lying on the freeway. The firefighters immediately requested that an ambulance be dispatched to the scene. Prompt action was taken to establish traffic control to protect the injured parties from further injury. Two firefighters returning to the station in the chief's car were traveling on a local freeway.

(B) Two firefighters returning to the station in the chief's car were traveling on a local freeway. They observed a vehicle accident with several injured persons lying on the freeway. The firefighters immediately requested that an ambulance be dispatched to the scene. Prompt action was taken to establish traffic control to protect the injured parties from further injury.

(C) Prompt action was taken to establish traffic control to protect the injured parties from further injury. The firefighters immediately requested that an ambulance be dispatched to the scene. They observed a vehicle accident with several injured persons lying on the freeway. Two firefighters returning to the station in the chief's car were traveling on a local freeway.

(D) Two firefighters returning to the station in the chief's car were traveling on a local freeway. The firefighters immediately requested that an ambulance be dispatched to the scene. They observed a vehicle accident with several injured persons lying on the freeway. Prompt action was taken to establish traffic control to protect the injured parties from further injury.

33. (A) As contradictory as it may sound, the slower a person reads, the less he absorbs. The faster he reads, the more he understands and retains. This is probably because heavier concentration is required for rapid reading, and concentration enables a reader to grasp important ideas in the reading material. The two basic elements in reading interpretation are speed and comprehension.

(B) The two basic elements in reading interpretation are speed and comprehension. As contradictory as it may sound, the slower a person reads, the less he absorbs. This is probably because heavier concentration is required for rapid reading, and concentration enables a reader to grasp important ideas in the reading material. The faster he reads, the more he understands and retains.

(C) As contradictory as it may sound, the slower a person reads, the less he absorbs. This is probably because heavier concentration is required for rapid reading, and concentration enables a reader to grasp important ideas in the reading material. The faster he reads, the more he understands and retains. The two basic elements in reading interpretation are speed and comprehension.

(D) The two basic elements in reading interpretation are speed and comprehension. As contradictory as it may sound, the slower a person reads, the less he absorbs. The faster he reads, the more he understands and retains. This is probably because heavier concentration is required for rapid reading, and concentration enables a reader to grasp important ideas in the reading material.

DIRECTIONS: Questions 34 through 36 are based on the storefront pictured below. In this drawing, the storefront is viewed from across the street.

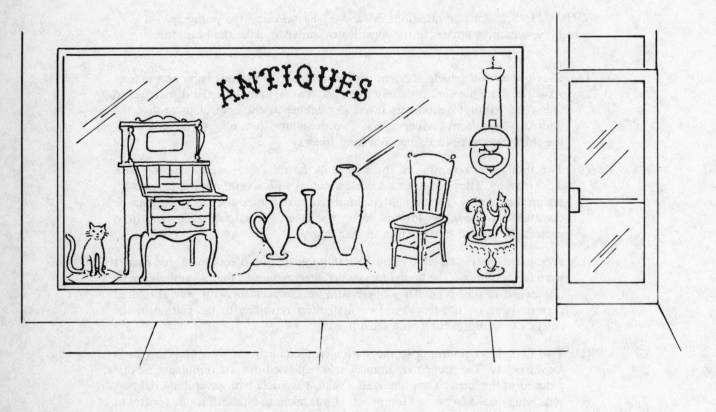

34. Which of the drawings below shows the same store window at the same time but as seen from inside the store?

35. Which of the drawings below shows the same store window at the same time but as seen from an aerial ladder 17 feet above the street?

36. Which of the drawings below shows the same store window at the same time but as seen through the open trap door in the ceiling at the front of the store?

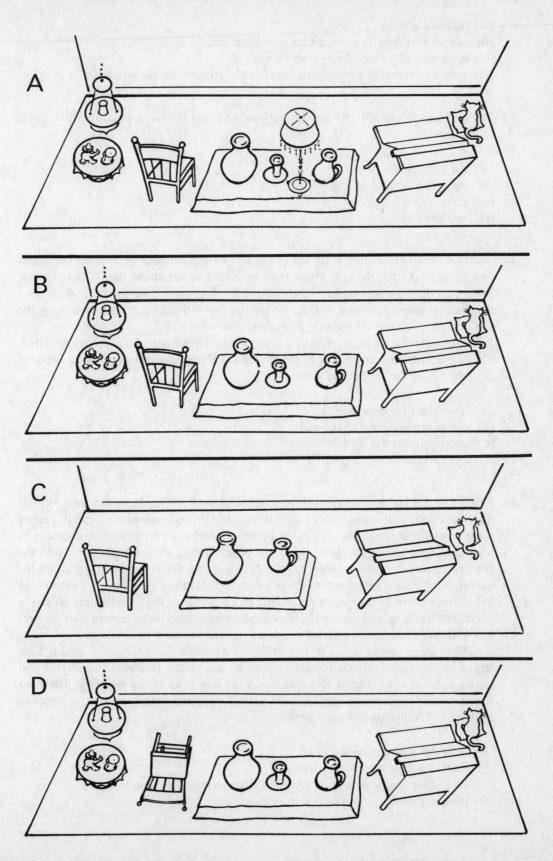

37. At a house fire, a woman is standing on the lawn screaming, "My baby, my baby!" The best instruction to give to the firefighters is to tell them to

(A) look for a baby
(B) search very quietly so that they can listen for crying
(C) systematically check every room for a crib
(D) start by searching rooms with "tot finder" stickers on the windows

38. The best way to explain why you strongly suspect that the cause of a fire was arson is to say that

(A) the floor was wet
(B) the place smelled of gasoline
(C) there was a funny smell
(D) you smelled gas

39. The purpose of ventilating a fire is to clear smoke and noxious gases from the premises. Since heat rises, the area above the fire should be ventilated first. Thus, the rule is to open the floor and walls directly above the fire; then open the roof; then open the floors between the floor directly above the fire and the roof; and then open the walls of the fire floor on the side away from the wind.

Firefighters have responded to a fire on the fourth floor of a six-story building. They have opened a large area of the fourth-floor ceiling and have broken the roof skylights. Now the firefighters should

(A) open the fifth-floor ceiling
(B) check the direction of the wind
(C) chop holes in the roof
(D) open windows on the fourth floor

40. Fire safety is a major concern in college dormitories. Early in the school year, dormitory managers must acquaint all students with fire procedures—alarm signals, proper areas in which to congregate, expected behavior—and must appoint floor captains to take head counts. The college is in charge of scheduling periodic fire drills, and the fire department provides oversight. At a fire drill, the fire department monitors the operation of alarms and observes the conduct of the drill. Firefighters are authorized to interrupt a fire drill to make comments or corrections. Then firefighters enter the dormitory building and conduct a door-to-door search to make certain that no person who was not on the roster of residents has remained in the building.

Firefighters attend a 2 A.M. fire drill in a women's dormitory at Stateside College. Fire alarms go off according to schedule, and sleepy residents—flashlights and towels in hand as instructed—congregate on the rear lawn of the building. The floor captains take head counts and account for all residents, either present or absence explained. The firefighters must now

(A) send the students back to bed
(B) escort each student to her room
(C) stick their heads through every open door to ask if anyone is there
(D) personally look into every room

41. An off-duty firefighter sprained an ankle while playing football. The next day the firefighter returned to work even though the ankle was quite painful. The firefighter felt that he was needed on the job. This course of action is considered improper mainly because the firefighter should have known that

(A) no one is indispensable
(B) medical attention was more necessary than reporting for duty
(C) because of this injury it might be impossible to perform his share of the work when required at a fire
(D) since the injury occurred off duty, the medical expenses should not be borne by the fire department

42. While fighting a fire, a firefighter searches an apartment looking for victims and comes across an unconscious woman on a bed. The firefighter would probably be of most assistance to the woman by *first*

(A) calling for a doctor
(B) giving the woman first aid until she regains consciousness
(C) moving the woman to a safe place
(D) remaining with the woman in case she regains consciousness

DIRECTIONS: Answer questions 43 and 44 by choosing which implement in the lettered boxes is normally used with the implement on the left. Note: The tools are not drawn to scale.

43.

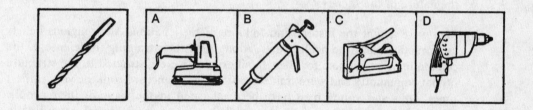

44.

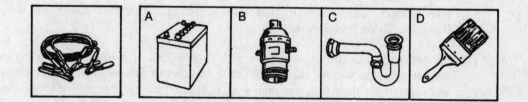

DIRECTIONS: Answer questions 45 and 46 by selecting the paragraph in which the sentences appear in the most logical order to make the best sense.

45. (A) A man told a passing firefighter that there was a bomb in the telephone booth outside the post office. The firefighter called the bomb squad. Visible at one end were some electrical tape and wires. The firefighter found a length of pipe wedged under the phone booth seat.

(B) The firefighter called the bomb squad. A man told a passing firefighter that there was a bomb in the telephone booth outside the post office. The firefighter found a length of pipe wedged under the phone booth seat. Visible at one end were some electrical tape and wires.

(C) The firefighter found a length of pipe wedged under the phone booth seat. Visible at one end were some electrical tape and wires. A man told a passing firefighter that there was a bomb in the telephone booth outside the post office. The firefighter called the bomb squad.

(D) A man told a passing firefighter that there was a bomb in the telephone booth outside the post office. The firefighter found a length of pipe wedged under the phone booth seat. Visible at one end were some electrical tape and wires. The firefighter called the bomb squad.

46. (A) From the very earliest times, boys and young men have been educated in the ways of their people. Men were taught the history and folklore of their nations, the religious beliefs and practices, and the healing arts. Along with their training in defense of the community and support for their families, men were instructed in the structure of that community and were trained to sit on its councils. Women were trained in "women's work"—housework or fieldwork according to the nature of the society.

(B) Men were taught the history and folklore of their nations, the religious beliefs and practices, and the healing arts. Along with their training in defense of the community and support for their families, men were instructed in the structure of that community and were trained to sit on its councils. From the very earliest times, boys and young men have been educated in the ways of their people. Women were trained in "women's work"—housework or fieldwork according to the nature of the society.

(C) Men were taught the history and folklore of their nations, the religious beliefs and practices, and the healing arts. Women were trained in "women's work"—housework or fieldwork according to the nature of the society. From the very earliest times, boys and young men have been educated in the ways of their people. Along with their training in defense of the community and support for their families, men were instructed in the structure of that community and were trained to sit on its councils.

(D) Women were trained in "women's work"—housework or fieldwork according to the nature of the society. Along with their training in defense of the community and support for their families, men were instructed in the structure of that community and were trained to sit on its councils. Men were taught the history and folklore of their nations, the religious beliefs and practices, and the healing arts. From the very earliest times, boys and young men have been educated in the ways of their people

47. Fire and its concomitant smoke and gases kill swiftly and disable even more quickly, so the search for potential victims must be swift and systematic within the bounds of safety for the firefighter. A firefighter preparing to enter a burning building must notify an officer before entering. The firefighter must put on protective gear, adjust a gas mask, and carry an ax and rope. The search should begin in the most heavily engulfed area that the firefighter can penetrate and proceed toward relative safety.

Firefighter Hu is about to enter a three-story frame house in which the third floor is already totally involved. Hu is wearing protective gear and a gas mask and is carrying an ax and rope. The next thing Hu must do is

(A) notify an officer of the plan to enter
(B) grab an oxygen pack
(C) climb directly to the third floor
(D) methodically search the second floor

48. A tunnel fire presents an especially dangerous situation. When notice is received of a fire in a tunnel, the dispatcher must ask the direction of the tube in which the fire is located; notify police in the jurisdiction of the entry end to block access at once; notify tunnel authorities to activate high-powered exhaust systems; dispatch firefighting equipment to the scene; alert ambulance crews to expect a call; and prepare wreckers but not send them to the scene until the fire has been extinguished and the victims evacuated.

A truck transporting live chickens has gone out of control in the westbound tube of the Bay Tunnel, has hit the tunnel wall, and has burst into flames. A police officer on duty in the tunnel has notified the fire department. Dispatcher Sikorsky has confirmed the precise location of the fire in the westbound tube. Now Sikorsky will

(A) tell police at the eastern end of the tunnel to block exit
(B) tell police at the western end of the tunnel to block access
(C) tell police at the eastern end of the tunnel to block access
(D) tell police at the western end of the tunnel to block exit

49. If you are on house watch and receive a call "Fire, Fire," the best way to get useful information is to ask:

(A) Where's the fire?
(B) What is the street address of the fire?
(C) Where are you?
(D) What is the name of the building on fire?

50. If you are asked when an alarm was received at the firehouse, your best answer would be:

(A) Four minutes before the first ladder company arrived at the scene.
(B) Just as the night shift was coming on.
(C) At 8:03 P.M.
(D) An hour and a half before the fire was declared under control.

DIRECTIONS: Use the map below to answer questions 51 through 53. The flow of traffic is indicated by arrows. If there is only one arrow shown, then traffic flows only in the direction indicated by the arrow. If there are two arrows shown, then traffic flows in both directions. You must follow the flow of traffic.

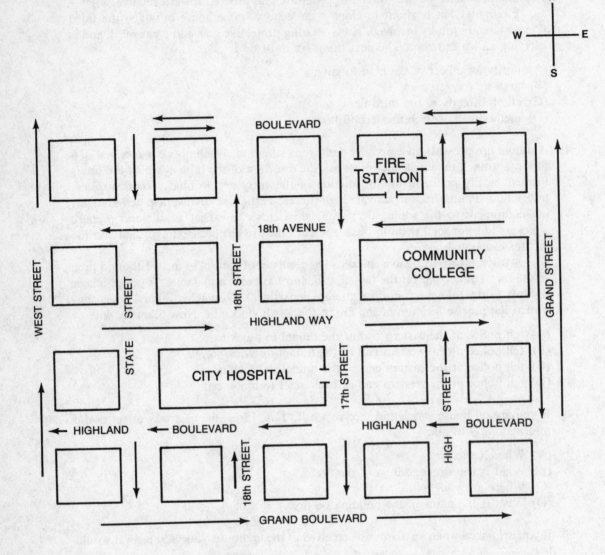

51. An alarm has just come into the firehouse from the alarm box at the corner of State Street and Grand Boulevard. The quickest route for the fire engines to take is

(A) out the east exit of the firehouse, north on High Street, south on Grand Street, west on Highland Boulevard to State Street, and south on State Street to the fire

(B) out the 18th Avenue exit of the fire station and west to 18th Street, north on 18th Street to Boulevard, east on Boulevard to Grand Street, south on Grand Street to Highland Boulevard, west on Highland Way to State Street, and south on State Street to the fire

(C) out the west exit of the fire station and up 17th Street to Boulevard, two blocks west on Boulevard to State Street, and all the way south on State Street to Grand Boulevard

(D) out the south exit of the fire station, west on 18th Avenue to State Street, then south on State Street to its corner with Grand Boulevard

52. A firefighter has been injured fighting this fire and must be transported by ambulance to City Hospital. The hospital emergency exit is located on 17th Street. The hospital is very nearby, but is not so easy to get to from the corner of State Street and Grand Boulevard. The best way to go is

    (A) Grand Boulevard to High Street, left onto High Street, left onto Highland Boulevard, and right onto 17th Street
    (B) Grand Boulevard to High Street, left onto High Street, right onto Highland Way, left onto Grand Street, left onto 18th Avenue, and left again onto 17th Street
    (C) east on Grand Boulevard to Highland Boulevard, west on Highland Boulevard to Highland Way, east on Highland Way, and south on 17th Street to the emergency room
    (D) east on Grand Boulevard to Grand Street, left onto Grand Street, then a left turn onto 18th Avenue, and south on 17th Street

53. A fire engine is about to leave the station house to make a routine inspection of a building located on the southwest corner of High Street and Highland Boulevard. The best route for the engine to take is to go

    (A) out the High Street exit of the fire station and take High Street straight to Highland Boulevard
    (B) out the east exit of the fire station and up to Boulevard, Boulevard to Grand Street, Grand Street to Highland Boulevard, and Highland Boulevard to High Street
    (C) out the west exit of the fire station and straight down 17th Street to Highland Boulevard, then Highland Boulevard to its corner with High Street
    (D) out the 18th Avenue exit of the firehouse to 17th Street, then south on 17th Street to Highland Way and the corner of High Street

54. A warehouse inspection should be conducted in a systematic manner so that no dangerous hazards are overlooked. A firefighter on inspection should follow the order of items to be checked as it appears on the inspection form. The form reads: Exit signs, visible and lighted? Exit doors, unlocked and unobstructed? Sprinkler heads, properly placed and uncorroded? Sprinkler shut-off accessible? Fire extinguishers, in place, fully charged, and inspected? Trash containerized?

    Firefighter Bligh has appeared at the ABC Warehouse and has asked the warehouse supervisor to accompany him on the inspection tour. Bligh notes that all exits are marked, but that the bulb is burned out in one sign. Then Bligh checks the exit doors, which are all unlocked, and suggests to the supervisor that the boxes partially obstructing one exit be moved out of the way. Bligh moves on to inspect the sprinkler system and asks the supervisor:

    (A) Where is the sprinkler shut-off?
    (B) When were the fire extinguishers last charged?
    (C) Where do you keep the trash?
    (D) Do you have extra light bulbs on hand?

55. All firefighters are thoroughly trained in first aid procedures including cardiopulmonary resuscitation (CPR). This training is necessary because fire victims are often injured, burned, or overcome by smoke. The firefighter who comes across an unresponsive victim must rapidly assess the person's condition and then must take appropriate first aid measures. The first step in assessing an apparently lifeless body is to check for a heartbeat by taking a pulse. If the heartbeat has just stopped, there is a possibility that the victim might be revived by CPR. Unless there is immediate danger to the life of the firefighter by remaining in the area, the firefighter should begin CPR right away. If there is a pulse but no sign of breathing, then mouth-to-mouth resuscitation is indicated. If the victim is injured but conscious, the firefighter must determine if helping the victim to safety is feasible. If not, the firefighter must go for assistance.

The fire in a three-family frame structure has been declared under control and firefighter Frantzman is searching the third floor for victims. Frantzman sees a leg sticking out from under a bed and pulls out a heavy-set, unconscious man. The man has a feeble pulse and is breathing shallowly. Visual check does not show any signs of obvious physical injury. Frantzman should now

(A) initiate mouth-to-mouth resuscitation
(B) drag and carry the man to safety
(C) get assistance
(D) help the man up and walk him out of the building

DIRECTIONS: Answer questions 56 and 57 by selecting the paragraph in which the sentences appear in the most logical order to make the best sense.

56. (A) A hysterical woman rushed up to a fire lieutenant and told him that her baby had been kidnapped. He asked her why she thought the baby had been kidnapped. The woman said that the baby had disappeared from in front of the supermarket next door while she was shopping. The officer suggested that she first look for the baby inside the supermarket.

(B) The woman said that the baby had disappeared from in front of the supermarket next door while she was shopping. A hysterical woman rushed up to a fire lieutenant and told him that her baby had been kidnapped. The officer suggested that she first look for the baby inside the supermarket. He asked her why she thought the baby had been kidnapped.

(C) The woman said that the baby had disappeared from in front of the supermarket next door while she was shopping. The officer suggested that she first look for the baby inside the supermarket. A hysterical woman rushed up to a fire lieutenant and told him that her baby had been kidnapped. He asked her why she thought the baby had been kidnapped.

(D) A hysterical woman rushed up to a fire lieutenant and told him that her baby had been kidnapped. The woman said that the baby had disappeared from in front of the supermarket next door while she was shopping. The officer suggested that she first look for the baby inside the supermarket. He asked her why she thought the baby had been kidnapped.

57. (A) Everyone entering the elevators for the upper floors of the postal facility must pass through a metal detector. The man was not permitted to take the exam. A sign in the lobby read, "All applicants for the postal police exam must check their weapons." As the man approached the elevator, the alarm began to sound.

(B) As the man approached the elevator, the alarm began to sound. The man was not permitted to take the exam. Everyone entering the elevators for the upper floors of the postal facility must pass through a metal detector. A sign in the lobby read, "All applicants for the postal police exam must check their weapons."

(C) A sign in the lobby read, "All applicants for the postal police exam must check their weapons." Everyone entering the elevators for the upper floors of the postal facility must pass through a metal detector. As the man approached the elevator, the alarm began to sound. The man was not permitted to take the exam.

(D) Everyone entering the elevators for the upper floors of the postal facility must pass through a metal detector. As the man approached the elevator, the alarm began to sound. The man was not permitted to take the exam. A sign in the lobby read, "All applicants for the postal police exam must check their weapons."

DIRECTIONS: Questions 58 through 60 are based on the boat collision scene pictured below. In this drawing, the scene is viewed from a helicopter coming to the rescue and just approaching the rear end of a barge from above a yacht it collided with.

58. Which of the drawings below shows the same scene at the same time as the original drawing, but from the viewpoint of a bystander standing on the shore?

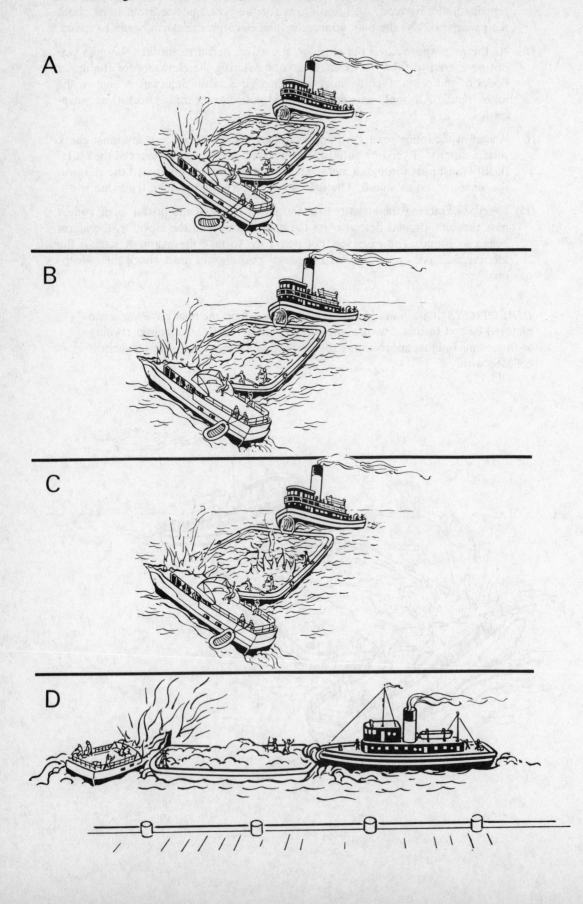

59. Which of the drawings below shows the same scene at the same time as the original drawing but as seen by firefighters about to launch a fireboat from the opposite side of the river?

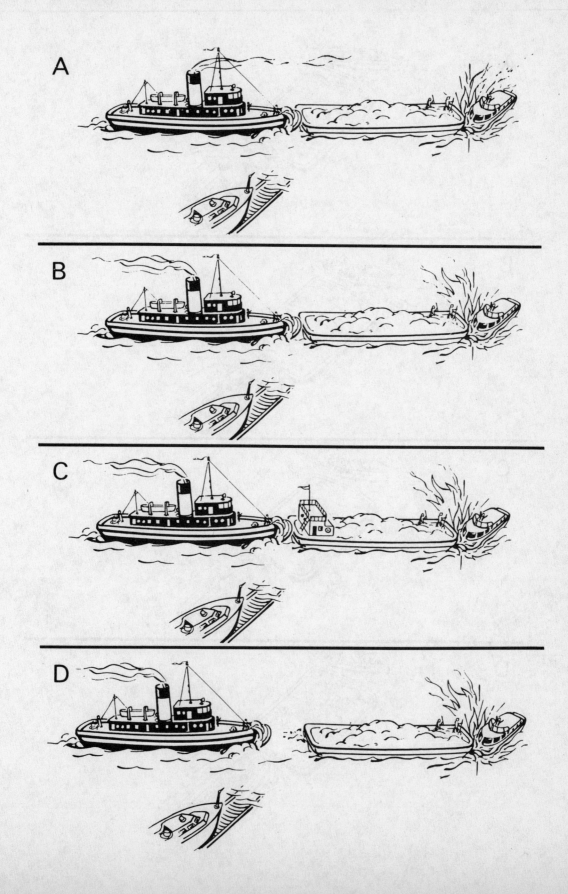

60. Which of the drawings below shows the same scene at the same time as the original drawing but as seen from above the tugboat?

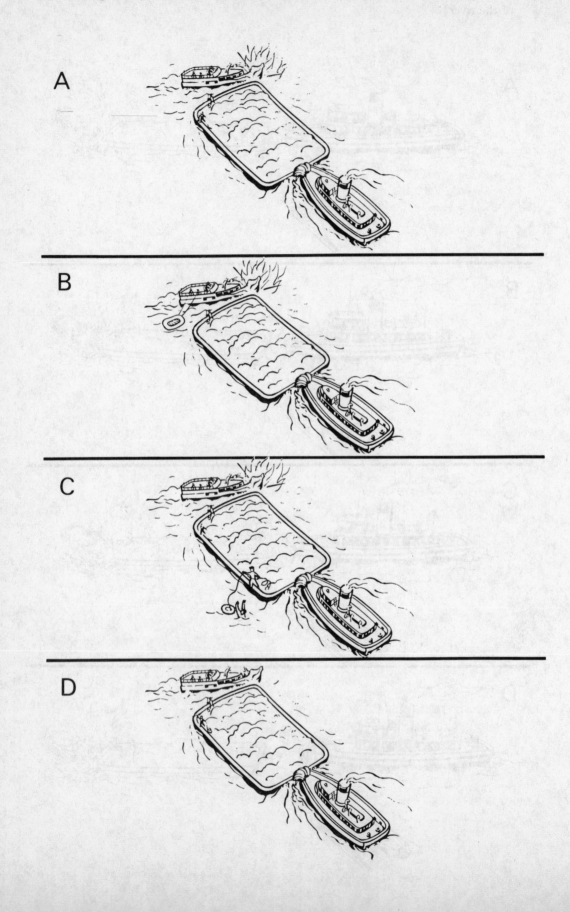

61. Firefighters responding to an alarm find a badly beaten unconscious man lying next to the alarm box. It would be appropriate for the firefighters to do all but which one of the following?

(A) summon the police
(B) give first aid to the man
(C) summon an ambulance
(D) question the homeowners in the vicinity

62. During the times when firefighters are not responding to alarms, they busy themselves checking the equipment and vehicles. The most likely reason for them to engage in this activity is to

(A) be busy when the public visits their firehouse
(B) provide free time to socialize
(C) promote neighborhood pride in the appearance of the firehouse
(D) assure that their equipment and vehicles are in good working order

DIRECTIONS: Answer questions 63 and 64 by choosing which implement in the lettered boxes is normally used with the implement on the left. Note: The tools are not drawn to scale.

63.

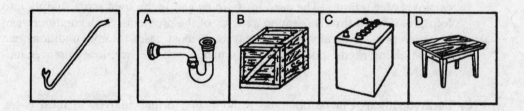

64.

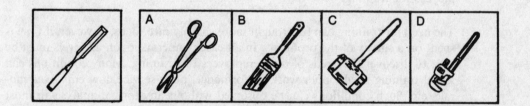

65. Firefighters at the scene of a major fire find they must get to the other side of a burning room. Gases fill the air and heavy smoke makes for poor visibility. Considering the firefighters' safety, their officer should command them to

(A) get to the other side in a hurry
(B) keep low
(C) run
(D) be careful

66. At the end of the day, each firefighter must file an activity report accounting for how his or her time was spent. A firefighter who spent the day inspecting business establishments should identify each clearly without giving more detail than necessary. Of the following, the best example of useful identification of premises is

(A) the dry cleaner on First Street between the post office and Brooks Drugstore
(B) the store at 21 First Street
(C) Harvey's Dry Cleaners at 21 First Street
(D) Harvey's Dry Cleaners on the east side of First Street between Owens Avenue and Miller Road

DIRECTIONS: Answer questions 67 and 68 by selecting the paragraph in which the sentences appear in the most logical order to make the best sense.

67. (A) Many conditions can signal training needs: poor work performance, poor service, low employee morale, etc. Such conditions can be observed without any formal analysis in most cases. The need for training can be brought more sharply into focus, however, if time is spent on a survey of the problems. In all cases, supervisors on all levels must be alert to discover the needs of their employees for training.

(B) In all cases, supervisors on all levels must be alert to discover the needs of their employees for training. The need for training can be brought more sharply into focus, however, if time is spent on a survey of the problems. Such conditions can be observed without any formal analysis in most cases. Many conditions can signal training needs: poor work performance, poor service, low employee morale, etc.

(C) Such conditions can be observed without any formal analysis in most cases. Many conditions can signal training needs: poor work performance, poor service, low employee morale, etc. The need for training can be brought more sharply into focus, however, if time is spent on a survey of the problems. In all cases, supervisors on all levels must be alert to discover the needs of their employees for training.

(D) The need for training can be brought more sharply into focus, however, if time is spent on a survey of the problems. In all cases, supervisors on all levels must be alert to discover the needs of their employees for training. Many conditions can signal training needs: poor work performance, poor service, low employee morale, etc. Such conditions can be observed without any formal analysis in most cases.

68. (A) She noticed a ticking suitcase under a workbench in the boiler room. The firefighter opened the suitcase and found pajamas and an alarm clock inside. The firefighter was searching the office building in response to a bomb threat. The area was dusty but the suitcase was clean of any dust.

(B) The firefighter was searching the office building in response to a bomb threat. The area was dusty but the suitcase was clean of any dust. She noticed a ticking suitcase under a workbench in the boiler room. The firefighter opened the suitcase and found pajamas and an alarm clock inside.

(C) The firefighter was searching the office building in response to a bomb threat. She noticed a ticking suitcase under a workbench in the boiler room. The area was dusty but the suitcase was clean of any dust. The firefighter opened the suitcase and found pajamas and an alarm clock inside.

(D) The area was dusty but the suitcase was clean of any dust. The firefighter opened the suitcase and found pajamas and an alarm clock inside. The firefighter was searching the office building in response to a bomb threat. She noticed a ticking suitcase under a workbench in the boiler room.

69. At the first alarm signaling a fire in its territory, a fire station dispatches two ladder companies and two engine companies along with one car in which a captain travels to the scene in order to size up the extent of the blaze. If the captain determines that the call-up is warranted, the fire is upgraded to two alarms, and two more engine companies and one more ladder company speed to the scene. A battalion chief joins them to further direct their efforts. If the battalion chief deems it necessary, the fire goes to three alarms. At this point, the nearest neighboring fire station sends all available equipment to assist in the firefighting effort.

   The fire at a small plastics factory has been actively burning for 40 minutes, and the wind has shifted so that it is now blowing toward the adjacent curtain factory. The battalion chief on the scene has decided that more equipment and manpower are needed. The battalion chief will now

(A) call for two more engine companies
(B) sound a second alarm
(C) go to the nearest neighboring fire station for assistance
(D) sound a third alarm

70. At the scene of an apartment house fire, the persons in the greatest danger from the fire must be rescued first. Thus, occupants of the room that is on fire must be removed first, followed by occupants of the remainder of the fire floor. Heat rises, so occupants of the floors above the fire floor are generally in greater danger than occupants of floors below the fire floor. Occupants of the floor directly above the fire floor have next priority for evacuation. While the general rule is that the floors above the fire present greater danger, the one exception is the floor directly below the fire. Since water is very heavy and fire tends to undermine the structure of a building, the fire floor may collapse onto the floor beneath. People on this floor directly below run the risk of being crushed by the floor and furniture from above. The floor below the fire, then, is the third to be emptied. So long as the fire is actively burning, upper floors have priority for further evacuation. Once the fire has been declared under control, evacuation should proceed on lower floors because of the continued danger from water damage.

   The fire on the fourth floor of a block-deep six-story apartment house has just been declared under control. Occupants of the fourth floor were all evacuated by way of the fire escape and ladders. Some fifth-floor occupants have come down the fire escape while others have made their way to the roof with some sixth-floor residents. Firefighters should now concentrate on evacuating

(A) the third floor
(B) the second floor
(C) the sixth floor
(D) the roof

DIRECTIONS: Questions 71 through 73 are based upon the airplane disaster pictured below. In this drawing, the scene is viewed from the driver's seat of a fire engine approaching the right side of the airplane near the tail.

71. Which of the drawings below shows the same scene at the same time but as seen from a rescue launch approaching the nose of the plane?

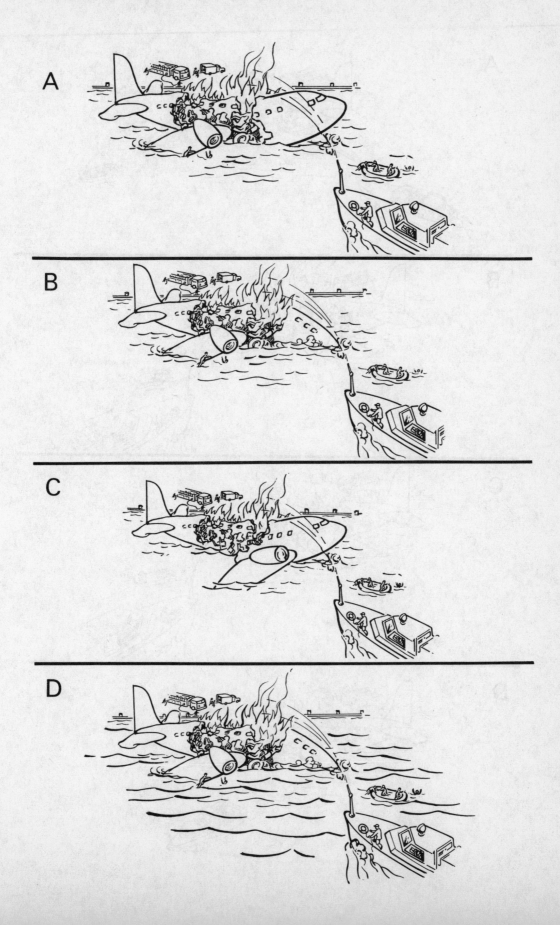

72. Which of the drawings below shows the same scene at the same time but as seen from a news helicopter hovering directly overhead?

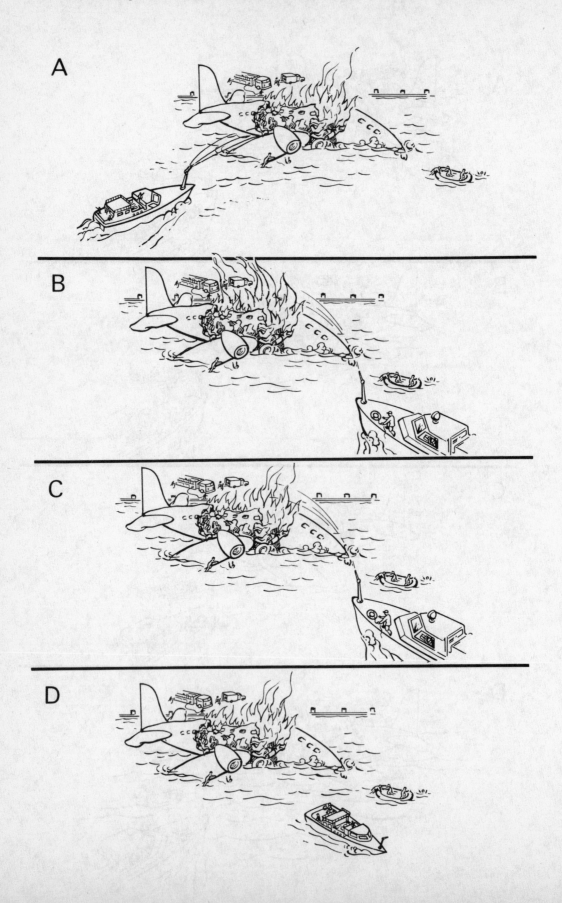

73. Which of the drawings below shows the same scene at the same time but as seen by a person running up from behind the tail?

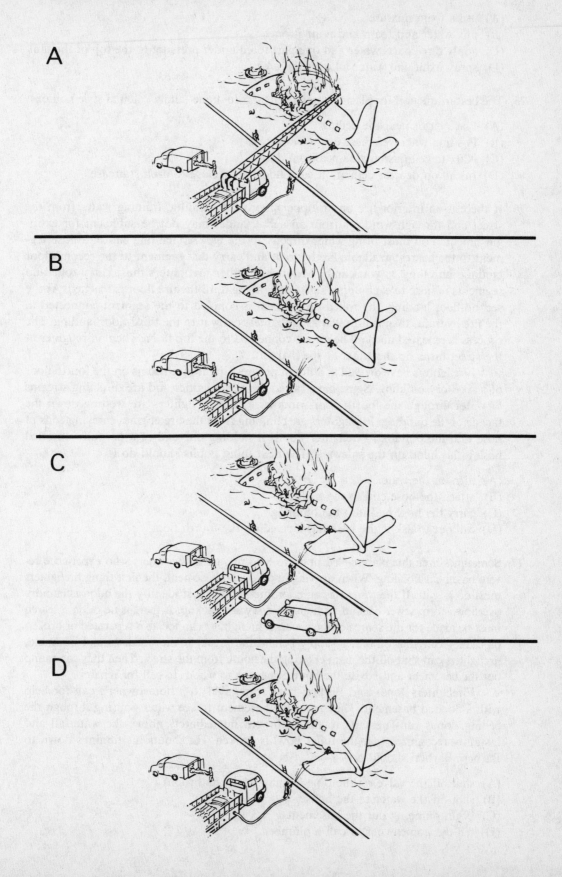

A

B

C

D

74. Fire is raging in a storage tank at an oil refinery. The officer in charge should instruct the firefighters to

(A) use a foam mixture
(B) mix water and foam and pour it on
(C) apply three parts water and one part foam under pressure to the top of the tank
(D) spray foam and water all over the area

75. The best instruction for placing smoke detectors in a one-family colonial-style house is:

(A) Place on or near the ceiling.
(B) Position wherever there could be smoke.
(C) Affix to ceilings outside bedroom doors.
(D) Install on or near ceiling on each floor fairly close to each staircase.

76. If there is an interior fire on an upper floor of a building, training water from the street and through windows from an aerial ladder may not be sufficient for extinguishment. You must bring water directly to the fire. Attach one end of a hose segment to the nearest available fire hydrant and carry that segment to the second-floor landing, unrolling as you climb. Meanwhile, other firefighters must carry rolled-up segments of hose to each upper staircase landing until the fire floor is reached. At the second-floor landing, the rolled-up segment is coupled to the segment connected to the fire hydrant, then is unrolled as firefighters carry it to the third-floor landing. The process is repeated until all hoses are connected to the fire floor. Then a firefighter at the street turns on the water at the hydrant.

Firefighters have arrived at a fire in progress in interior rooms on the fourth floor of a six-story building. Some have raised the aerial ladder and are directing streams of water through the fourth-floor windows. Other firefighters are wetting down the facade of the building. Firefighters are climbing from the fire engines carrying coils of hose into the building. Firefighter Poplis is holding one end of a partially unrolled hose in his hand on the sidewalk. The next thing Poplis should do is

(A) turn on the water
(B) attach the hose end to the hydrant
(C) carry the hose end into the building
(D) call out to ask if the hose has arrived on the fourth floor

77. Sometimes firefighters are called upon to help private homeowners who experience severe basement flooding. When arriving at a flooded basement, the first thing firefighters must do is sniff. If the water has an odor, firefighters must identify the odor as laundry overflow, storm sewer backup, or household sewage backup. If there is no odor, a search must be made for the source of the water. If firefighters can locate a separated or broken pipe, they can shut off water supply behind the break. If the break cannot be found, firefighters can shut off the main entering the house from the street. Then they can pump out the basement and advise the homeowners as to whom to call for repairs.

Firefighters Jonas and Barberi have responded to a homeowner's call for help with a flooded basement. They enter the basement to see water pouring through the ceiling. Jonas runs upstairs to the bathroom immediately above the waterfall and instantly recognizes that the toilet bowl is broken. He shouts his findings down to Barberi. Barberi should now

(A) shut off the valve on the pipe leading to the bathroom
(B) shut off the water to the house
(C) begin pumping out the basement
(D) tell the homeowner to call a plumber

DIRECTIONS: Answer questions 78 through 80 by selecting the paragraph in which the sentences appear in the most logical order to make the best sense.

78. (A) The superior officer should be particularly sensitive when personal issues are reached in the course of an interview. In situations like this, what most people want is a sympathetic, understanding listener rather than an advisor. They may ask for advice, but actually they only want a chance to express themselves. Even when advice-giving is successful, there is the danger that the subordinate may become dependent on the superior officer and run to the officer with every minor problem.

(B) In situations like this, what most people want is a sympathetic, understanding listener rather than an advisor. The superior officer should be particularly sensitive when personal issues are reached in the course of an interview. They may ask for advice, but actually they only want a chance to express themselves. Even when advice-giving is successful, there is the danger that the subordinate may become dependent on the superior officer and run to the officer with every minor problem.

(C) Even when advice-giving is successful, there is the danger that the subordinate may become dependent on the superior officer and run to the officer with every minor problem. In situations like this, what most people want is a sympathetic, understanding listener rather than an advisor. They may ask for advice, but actually they only want a chance to express themselves. The superior officer should be particularly sensitive when personal issues are reached in the course of an interview.

(D) In situations like this, what most people want is a sympathetic, understanding listener rather than an advisor. They may ask for advice, but actually they only want a chance to express themselves. Even when advice-giving is successful, there is the danger that the subordinate may become dependent on the superior officer and run to the officer with every minor problem. The superior officer should be particularly sensitive when personal issues are reached in the course of an interview.

79. (A) The accident had resulted in the death of the witness's brother, driver of one of the vehicles. The witness angrily replied that his brother was sober. The officer asked the witness if his brother had been drinking. A police officer on the scene was interviewing the witness to a serious car accident.

(B) The officer asked the witness if his brother had been drinking. The witness angrily replied that his brother was sober. A police officer on the scene was interviewing the witness to a serious car accident. The accident had resulted in the death of the witness's brother, driver of one of the vehicles.

(C) The officer asked the witness if his brother had been drinking. The accident had resulted in the death of the witness's brother, driver of one of the vehicles. A police officer on the scene was interviewing the witness to a serious car accident. The witness angrily replied that his brother was sober.

(D) A police officer on the scene was interviewing the witness to a serious car accident. The accident had resulted in the death of the witness's brother, driver of one of the vehicles. The officer asked the witness if his brother had been drinking. The witness angrily replied that his brother was sober.

80. (A) Mentally alert workers seldom fall victim to industrial accidents. They are naturally careful and attentive to their surroundings. It is believed by most experts in the field that there is a definite relationship between a worker's mental alertness and physical well-being and involvement in accidents. A worker who is bothered by a physical ailment or who is not feeling up to par while performing duties may be prone to accidents.

(B) It is believed by most experts in the field that there is a definite relationship between a worker's mental alertness and physical well-being and involvement in accidents. A worker who is bothered by a physical ailment or who is not feeling up to par while performing duties may be prone to accidents. Mentally alert workers seldom fall victim to industrial accidents. They are naturally careful and attentive to their surroundings.

(C) It is believed by most experts in the field that there is a definite relationship between a worker's mental alertness and physical well-being and involvement in accidents. They are naturally careful and attentive to their surroundings. A worker who is bothered by a physical ailment or who is not feeling up to par while performing duties may be prone to accidents. Mentally alert workers seldom fall victim to industrial accidents.

(D) A worker who is bothered by a physical ailment or who is not feeling up to par while performing duties may be prone to accidents. It is believed by most experts in the field that there is a definite relationship between a worker's mental alertness and physical well-being and involvement in accidents. Mentally alert workers seldom fall victim to industrial accidents. They are naturally careful and attentive to their surroundings.

81. While fighting a fire, a firefighter becomes trapped in a third-floor loft when the stairs catch fire. There is no fire escape. In this instance it would be best to

(A) call for a portable ladder to escape through a window
(B) wait for help to arrive
(C) go to the nearest window and jump
(D) go up the stairs to the next floor and await a rescue attempt

82. After a fire has been extinguished, a firefighter observes a newly appointed firefighter pulling down loose plaster from a ceiling without a protective helmet on. It would be most appropriate in this instance for the experienced firefighter to

(A) advise the superior officer concerning the rule violation
(B) say nothing to anyone
(C) explain to the new firefighter the danger involved and request that he or she wear a helmet while performing this function
(D) go to any extreme, using force if necessary, to convince the new firefighter to wear a helmet

DIRECTIONS: Answer questions 83 and 84 by choosing which implement in the lettered boxes is normally used with the object on the left. Note: The tools are not drawn to scale.

83.

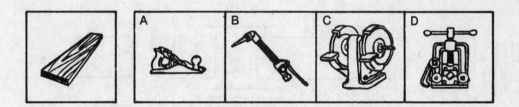

84.

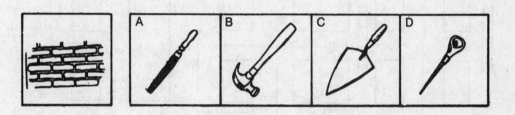

85. As you are chopping holes into warm partitions in a house fire that appears to be fully extinguished, the homeowner asks what you are doing. Your best answer is that you are

(A) chopping holes in the wall
(B) making certain that the fire is not spreading inside the walls
(C) following instructions
(D) finding out why the walls feel warm

86. If you are instructing a new firefighter in proper placement of a 16-foot extension ladder, you should tell the firefighter

(A) to lean the ladder on a firm wall
(B) to place the base four feet from the wall
(C) not to make the angle of the ladder too steep
(D) to place the base a distance from the wall equal to one-fourth of the ladder's height

DIRECTIONS: Use the map below to answer questions 87 through 89. The flow of traffic is indicated by the arrows. If there is only one arrow shown, then traffic flows only in the direction indicated by the arrow. If there are two arrows shown, then traffic flows in both directions. You must follow the flow of traffic.

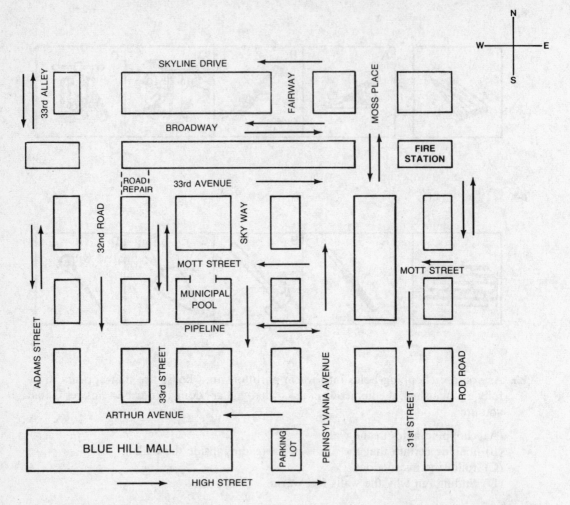

87. A patron at the municipal swimming pool is having an apparent heart attack. All emergency services crews are out on calls at this time, so the rescue team from the fire department must respond to the crisis. The best route for the rescue team to take from the fire station (which has exit doors at both east and west ends of the structure) is to

(A) exit by way of 31st Street, taking 33rd Avenue to Rod Road, making a right and taking Mott Street straight across to the pool

(B) exit the east end of the fire station, turn left onto Rod Road and then left onto Broadway, proceeding on Broadway to make a left onto 32nd Road, then turn right onto Mott Street to the pool

(C) exit onto Rod Road going north, then take Skyline Drive westbound to 32nd Road, go four blocks, turn left on Pipeline, then left onto Pennsylvania Avenue, and left from Pennsylvania Avenue onto Mott Street

(D) use the Moss Place exit of the fire station going north one-half block to Broadway, taking the long block of Broadway to a left onto 32nd Road and a right onto 33rd Avenue, followed by a turn south onto Sky Way and a right into Mott Street

88. Two firefighters are performing a routine fire safety inspection at a store in the Blue Hill Mall. Their fire engine is parked just outside the Sky Way entrance to the parking lot. Their next assignment is to check a malfunctioning street box on the southeast corner of 32nd Road and 33rd Avenue. The most efficient way for them to get there is by

(A) going south on Sky Way, east on High Street to the corner of Rod Road, then left onto Rod Road as far as Broadway, west on Broadway, and south on 32nd Road to its corner with 33rd Avenue

(B) going around the parking lot to Arthur Avenue, then north onto 33rd Street and west onto 33rd Avenue to the corner

(C) going three-quarters of the way around the parking lot onto Arthur Avenue going west, then taking Arthur Avenue to make a right turn onto Adams Street northbound, and a right onto 33rd Avenue

(D) continuing south on Sky Way to High Street, turning east for one block to make a left onto Pennsylvania Avenue, then right, left, and left again, so as to be headed west on Broadway, and left onto 32nd Road

89. Having repaired the street box, the firefighters are about to return to the fire station, but a road crew has just begun to repair a pothole on 33rd Avenue and has closed the street to all traffic between 33rd Street and 32nd Road. The firefighters will have to travel east on

(A) Broadway
(B) Pipeline
(C) Mott Street
(D) High Street

90. When arson is the suspected cause of a fire, the call must go out for witnesses. Each witness who comes forward must be interviewed to establish the credibility of the witness and the accuracy of the information. First, you must identify the witness by asking name, address, telephone number, place of business, and employer, if any. Then you must question the witness as to how he or she happened to be in a position to witness the arson and establish the precise spot and angle from which the witnessing took place. You must also ask the witness to tell you exactly what he or she saw happen and ask for a complete description of the alleged arsonist—sex, approximate age, size, race, and clothing. If the alleged arsonist was known to the witness, you can ask for name, address, etc.

As the last embers of a fire that totally destroyed a vacant storefront burn down, a somewhat disheveled elderly man approaches firefighter Kelliher and says, "I know who set that fire." Kelliher asks the man who he is, where he lives and works, and what he does for a living. Kelliher's next question should be:

(A) Why were you in this neighborhood this afternoon?
(B) So, who set the fire?
(C) Did you see the person strike a match?
(D) Was the arsonist a man or a woman?

91. The public is warned never to use an elevator during a fire. Next to every elevator is a sign that reads, "In case of fire, use the stairs." However, firefighters often use the elevators to transport themselves and their equipment to upper floors more quickly. When firefighters enter a burning elevator building, they must first ascertain that no elevators are currently in operation and that there are no passengers trapped in nonfunctional elevators. If there are passengers on an elevator stuck between floors, firefighters must first remove those passengers. They should begin with an attempt to restore power to the elevator and to bring it down to the ground floor. If the elevator cannot be moved by mechanical means, they should attempt to raise the elevator to the next higher floor by opening the elevator shaft and pulling manually on the cables. If this maneuver fails, firefighters must go to the floor immediately above the level on which the elevator is stuck, open the elevator door with the firefighter key, and descend to the roof of the cab with ladders. Then they can open the hatch of the elevator cab and enter the cab with their ladder. One firefighter must enter the cab to assist passengers with climbing up the ladder while two firefighters remain at the top, one to hold the ladder steady and the other to assist passengers as they alight. If there are no passengers trapped in elevators, firefighters must assess all safety factors. If water threatens the electrical system, firefighters must use the stairs.

Firefighters enter the 14-story building at 6 Livingston Plaza in response to a fire on the seventh floor. They check the elevator monitor board and note that of the eight elevators in the building, two are in the lobby, one is on the 14th floor, one is on the third floor, one is on the sixth floor, and one is on the 11th floor. One elevator is stuck between the fourth and fifth floors and one is stuck between the ninth and tenth floors. The building staff has shut off the power to the elevators. Using an intercom connected directly to the elevator cabs, firefighters learn that both of the stuck cabs are unoccupied. Next they should

(A) turn the power back on to get all cabs to the lobby
(B) attempt to manually hoist the stuck cabs to the floors above
(C) check the elevator electrical control box for water
(D) take the two elevators in the lobby to the seventh floor and fight the fire

92. At the scene of a building collapse, firefighters must search out victims and rescue them as quickly as possible. Search of a building collapse is especially hazardous because the area is often unstable. Debris may shift underfoot and further material may fall upon the victims and the rescuers. The procedure is to first rescue those victims who are easily accessible. Such victims usually can cooperate in their own rescue. Next to be rescued are victims who are visible but either severely disabled or trapped into immobility. Great caution must be taken in removing the next class of victims, those who can be heard crying out but who are buried under rubble. When these victims have been brought to safety, firefighters must try to communicate with other buried victims by calling out or by banging on exposed pipes. Attempts to call victims must be followed by total silence so that distant, weak calls or taps might be heard. The search for victims must continue even after known survivors have been removed because it is possible that unconscious survivors remain buried by the collapse.

A large building under construction has collapsed upon itself and has come down upon many construction workers. Firefighters arrive and immediately begin plucking workers from scaffolding and from cables to which they are clinging. Then they dig into the debris wherever they see flailing arms or legs or hear cries for help. Having carefully moved much rubble to extricate trapped workers, they call out "Is anybody else there?" Now they should

(A) go home, as everyone else must be dead
(B) stop and listen
(C) go on digging cautiously
(D) put out the fire

DIRECTIONS: Answer questions 93 and 94 by selecting the paragraph in which the sentences appear in the most logical order to make the best sense.

93. (A) In order to maintain public confidence in that organization, even the appearance of corruption must be avoided. It is not permissible for fire inspectors to accept a meal "on the house" or to receive any other such favors or gifts. They must pay for their meals just like everyone else. The possibility of corruption or corrupt practices exists within every large organization.

(B) It is not permissible for fire inspectors to accept a meal "on the house" or to receive any other such favors or gifts. In order to maintain public confidence in that organization, even the appearance of corruption must be avoided. They must pay for their meals just like everyone else. The possibility of corruption or corrupt practices exists within every large organization.

(C) In order to maintain public confidence in that organization, even the appearance of corruption must be avoided. The possibility of corruption or corrupt practices exists within every large organization. It is not permissible for fire inspectors to accept a meal "on the house" or to receive any other such favors or gifts. They must pay for their meals just like everyone else.

(D) The possibility of corruption or corrupt practices exists within every large organization. In order to maintain public confidence in that organization, even the appearance of corruption must be avoided. It is not permissible for fire inspectors to accept a meal "on the house" or to receive any other such favors or gifts. They must pay for their meals just like everyone else.

94. (A) Periodic fire drills are a very effective teaching device. Residents of a two-story house with only one staircase must have another way out of their second-floor bedrooms. Each member of the household must learn how to use the alternative route out. One solution is to purchase sturdy rope ladders in a hardware store and to bolt one securely to the underside of a windowsill in each bedroom.

(B) Residents of a two-story house with only one staircase must have another way out of their second-floor bedrooms. Periodic fire drills are a very effective teaching device. One solution is to purchase sturdy rope ladders in a hardware store and to bolt one securely to the underside of a windowsill in each bedroom. Each member of the household must learn how to use the alternative route out.

(C) Residents of a two-story house with only one staircase must have another way out of their second floor bedrooms. One solution is to purchase sturdy rope ladders in a hardware store and to bolt one securely to the underside of a windowsill in each bedroom. Each member of the household must learn how to use the alternative route out. Periodic fire drills are a very effective teaching device.

(D) Each member of the household must learn how to use the alternative route out. One solution is to purchase sturdy rope ladders in a hardware store and to bolt one securely to the underside of a windowsill in each bedroom. Periodic fire drills are a very effective teaching device. Residents of a two-story house with only one staircase must have another way out of their second floor bedrooms.

95. In the event of a house fire, it is crucial that all residents of the house be accounted for as soon as possible. In preparing your children for what to do if they must escape from a fire in their own home, the best instruction is to tell them to

(A) meet us under the tree directly across the street
(B) come to where mom is outside
(C) meet in a corner of the yard
(D) all get together right next to the fire hydrant

DIRECTIONS: Questions 96 through 98 are based on the forest fire pictured below. In this drawing, the scene is viewed by reinforcements running down a rise in the field toward the line of the fire.

96. Which of the drawings below shows the same scene at the same time but from the vantage point of a firefighting helicopter hovering above the trench?

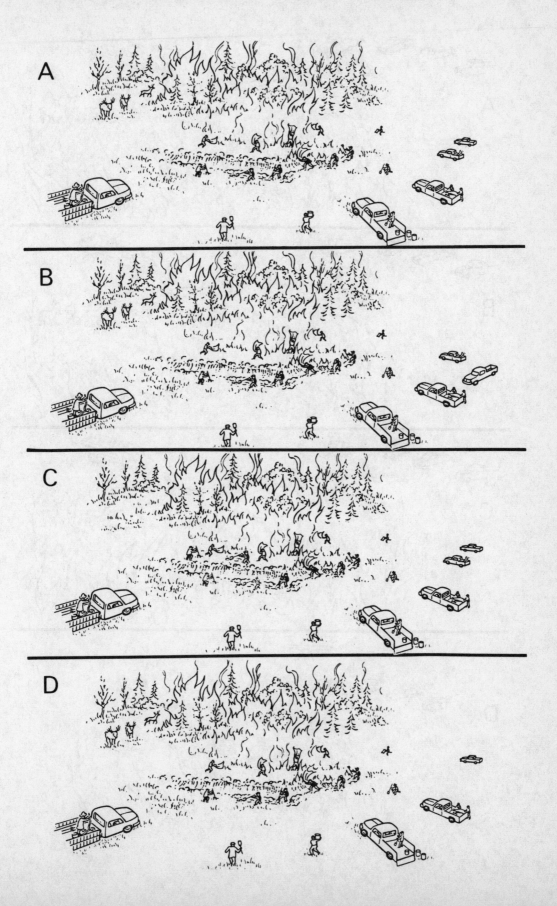

97. Which of the drawings below shows the same scene at the same time but as viewed from another helicopter approaching from over the flames?

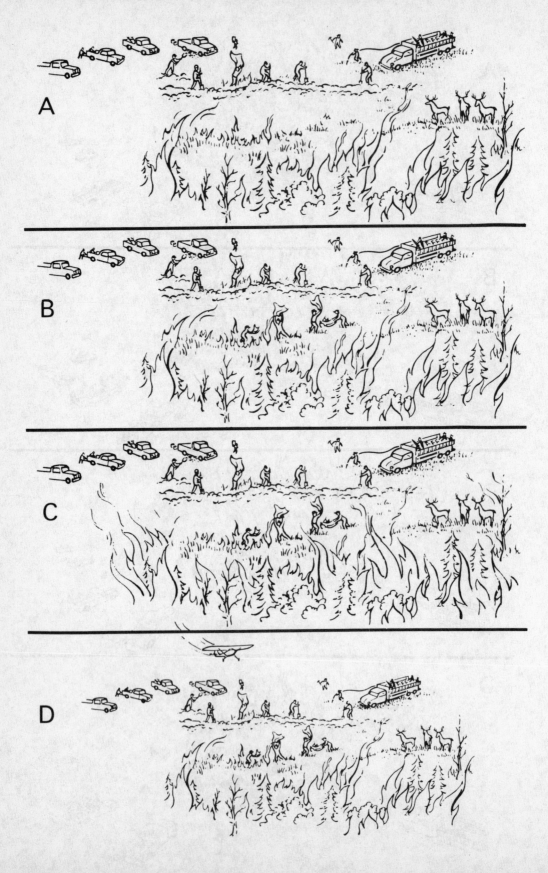

98. Which of the drawings below shows the same scene at the same time but as viewed from a pickup truck filled with supplies parked at one end of the trench?

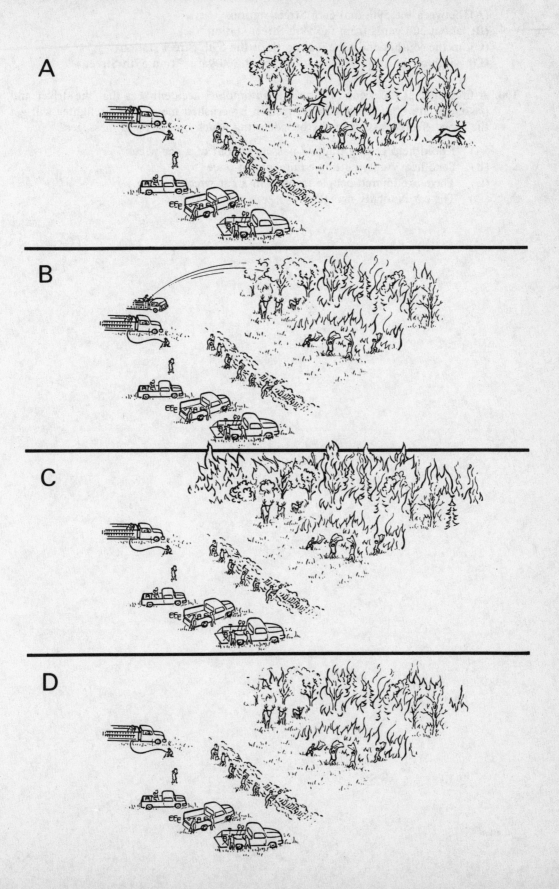

99. Firefighters must be called to a fire in a subway tunnel. The best instruction to give to help them locate the fire quickly and accurately is to tell them the fire is

(A) between the 59th and 66th Street stations
(B) about 300 yards from the 59th Street station
(C) in the southbound tube just north of the 59th Street station
(D) south of the 66th Street station, about 300 yards from 59th Street

100. A firefighter at the scene of a serious automobile accident sees that the driver and passenger are both thoroughly wedged in by crushed metal. The firefighter will get the most effective assistance by broadcasting which one of these messages?

(A) "I need help getting trapped passengers out of a car wreck."
(B) "Send heavy-duty cutting equipment at once."
(C) "There are injured people trapped in a car wreck."
(D) "The car is totally mangled."

Answer Key for Practice Examination 5

1. D	21. D	41. C	61. D	81. A
2. A	22. D	42. C	62. D	82. C
3. B	23. C	43. D	63. B	83. A
4. A	24. C	44. A	64. C	84. C
5. D	25. B	45. D	65. B	85. B
6. D	26. C	46. A	66. C	86. B
7. A	27. D	47. A	67. A	87. D
8. C	28. A	48. C	68. C	88. C
9. D	29. C	49. B	69. D	89. B
10. C	30. B	50. C	70. A	90. A
11. D	31. C	51. D	71. B	91. C
12. B	32. B	52. D	72. C	92. B
13. B	33. D	53. B	73. D	93. D
14. D	34. A	54. A	74. C	94. C
15. D	35. C	55. C	75. D	95. A
16. C	36. B	56. A	76. B	96. A
17. A	37. D	57. C	77. A	97. B
18. B	38. B	58. B	78. A	98. D
19. D	39. A	59. B	79. D	99. C
20. D	40. D	60. A	80. B	100. B

Explanatory Answers
for Practice Examination 5

Confirm the correct answers to questions 1 through 18 by looking back at the sketch. If you made many errors, give extra time to observation and memory training exercises.

19. **(D)** If you did not choose the correct answer to this question, reread the correct paragraph and notice the logical progression of the sentences and the even flow of the thoughts.

20. **(D)** See explanation above.

21. **(D)** The purpose of first aid is to provide immediate medical attention until professional help arrives. You need to know the cause of the injury or disability only to the extent that it affects the first aid treatment. Thus, overdose of pills dictates first aid steps, but the reasons for the overdose, bone breakage, or other disability are irrelevant to the first aid function.

22. **(D)** While firefighters do act to save property, their primary purpose is to save people. The people must be kept out until it is safe for them to enter. Likewise, there is no reason for a firefighter to put him- or herself in jeopardy solely in the interest of property.

23. **(C)** The drill chuck on the left is used to grip the twist drill in box C.

24. **(C)** The drain on the left is unclogged with the plunger in box C.

25. **(B)** Since the captain has determined that there is no danger to residents and that no one needs to be evacuated, the next step is to put the fire out. The captain directs this effort.

26. **(C)** The fire engine must cross the sidewalk before it enters and turns into the street. As the situation is described, traffic in the street has been halted, but pedestrians must still be cleared and restrained before the fire engine pulls out.

27. **(D)** There may be more than one big nut, so (A) is not the best answer. In giving a direction to turn, the question arises: Left or counterclockwise from the standpoint of the nut or from the standpoint of the person looking at it? The direction "to *your* left" makes this clear.

28. **(A)** Electric meters are large and very readily identified. There may be more than one lower level to a building. What is a "far" corner? And most certainly there are two west corners—northwest and southwest.

29. **(C)**

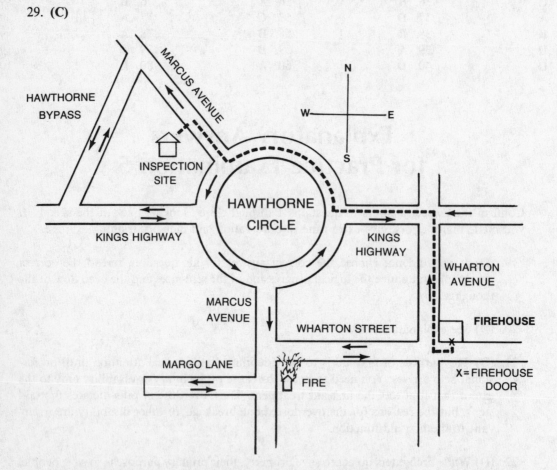

Try the other routes. (A) is wrong because you cannot exit directly onto Wharton Avenue; you must go west on Wharton Street, then turn right onto Wharton Avenue. Even if you could exit onto Wharton Avenue, Hawthorne Circle moves counterclockwise, not clockwise. (B) sends you north on the spur of Marcus Avenue that is one-way southbound. As for (D), Wharton Street will never get you to Hawthorne Circle.

30. **(B)**

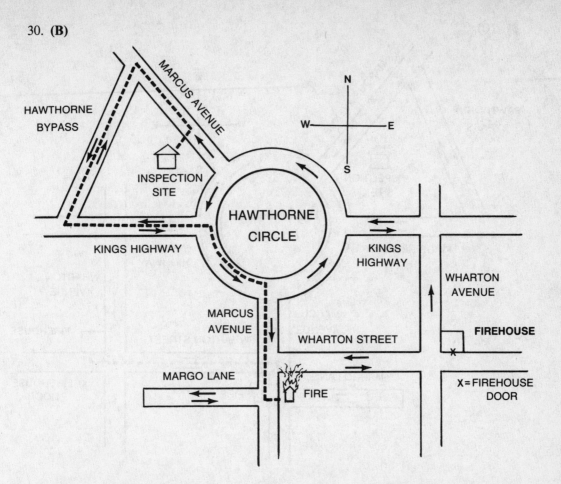

This spur of Marcus Avenue is one-way in a northwesterly direction, so (A) is wrong. As for (C), the firefighters must make a right turn from Kings Highway onto Hawthorne Circle, not a left. (D) presents the same problem as does (A).

31. **(C)**

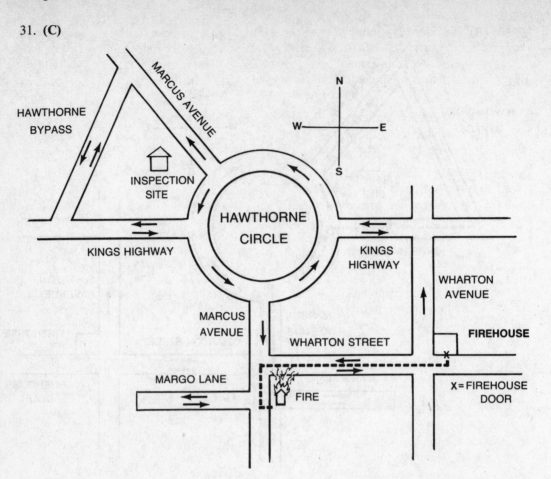

(A) cannot be correct because Wharton Street runs east-west. (B) is a legal route, but it is unnecessarily long and roundabout. (D) travels the wrong way around Hawthorne Circle.

32. **(B)** If your answer to question 32 does not agree with this, reread your choice and the correct paragraph once again. Subtle but real differences in continuity make the correct answer the best one.

33. **(D)** See explanation above.

34. **(A)** This is a precise mirror image of the original drawing. In (B) the cat has jumped to the desk; in (C) ANTIQUES appears as read from the outside instead of reversed as from within; in (D) a large vase is missing.

35. **(C)** This is the window as seen obliquely from 17 feet above the street. In (A) the chair is on the wrong side of the desk; in (B) flowers have suddenly appeared in a vase; in (D) there is a different desk.

36. **(B)** This is the view from above. In (A) a new lamp has appeared; in (C) the bric-a-brac has been removed; in (D) the chair has been knocked over.

37. **(D)** The purpose of "tot finder" stickers is to identify the rooms in which small children sleep and are most likely to be found. Firefighters should certainly start their search as directed by the stickers.

38. **(B)** The odor of gasoline in a structure other than a gasoline station or garage arouses a strong suspicion of arson. Gasoline does not belong inside the house, office, store, or school. A "funny smell" is too vague, and the floor can be wet from water. Gas is used for heating and cooking. A smell of gas should not arouse suspicion of arson.

39. **(A)** The ceiling of the fire floor has been opened and the roof skylights are broken, effectively opening the roof. The fifth floor is between the fire and the roof and is the next to be ventilated by opening its ceiling.

40. **(D)** According to the narrative, all the official residents are either outside on the lawn or away from the premises. However, this is a women's dormitory in the middle of the night. Unauthorized occupants are unlikely to announce themselves. In a real fire they would have to be evacuated. To make the fire drill authentic, firefighters must now enter every room and check for themselves.

41. **(C)** A firefighter who is not physically prepared to perform his or her functions can be the cause of much harm not only to him- or herself but also to fellow firefighters and to the public.

42. **(C)** The building is on fire, and evacuation of occupants is the first concern.

43. **(D)** The drill in box D uses the drill bit on the left to make holes.

44. **(A)** The jumper cables on the left connect the car battery in box A to a source of power, usually another car battery, in order to start a car with a dead battery.

45. **(D)** The correct answer to this question is represented by logically organized paragraphs with a natural sequence of events.

46. **(A)** See explanation above.

47. **(A)** The rule is clear. He must notify the officer in command of the fire before entering the building.

48. **(C)** The idea is to clear all vehicles from the tunnel, so exits must be kept open, not blocked. Access to the westbound tunnel is from the eastern end, so the next step in dealing with this emergency is to close off access at the eastern end of the westbound tunnel tube.

49. **(B)** The person calling to report a fire is likely to be very agitated. A specific question as to the location of the fire is likely to elicit the most specific answer. Even if the caller has the street number wrong, firefighters should have no trouble locating the fire once they arrive in the approximate neighborhood. A question like "Where's the fire?" may elicit an answer like "Next door." The answer to "Where are you?" might be "In the kitchen."

50. **(C)** Time of receipt of an alarm should be a matter of record. The question should be answered with the precise time.

51. **(D)**

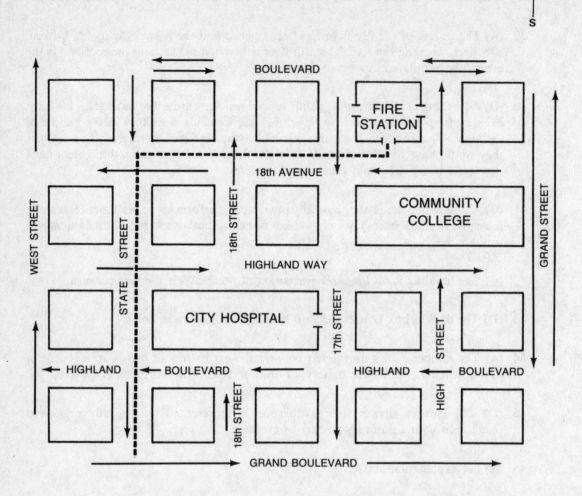

(A) cannot be correct because High Street does not intersect Grand Street. (B) confuses Highland Boulevard with Highland Way. (C) goes the wrong way on 17th Street.

52. **(D)**

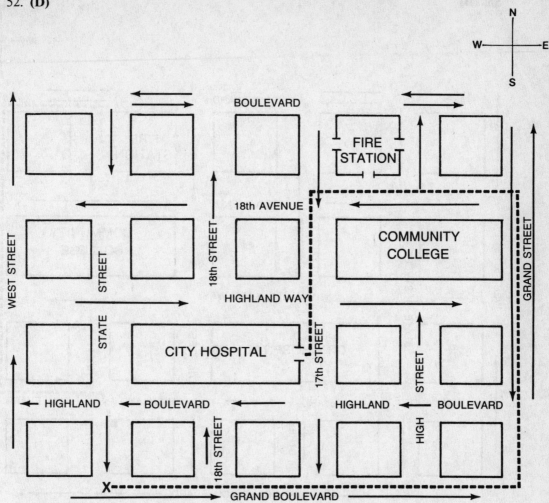

Route (A) takes you the wrong way on the one-way 17th Street. (B) is a legal route but a long one; the injured firefighter should get to the hospital as quickly as possible. As for (C), Grand Boulevard does not intersect Highland Boulevard.

53. **(B)**

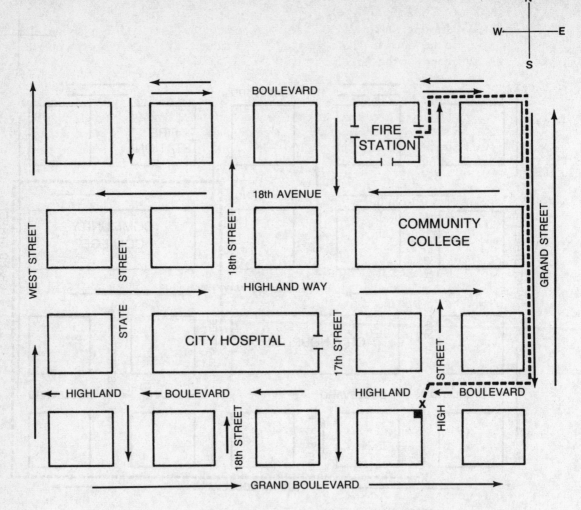

(A) is impossible for two reasons: High Street is one-way northbound, and Highland Boulevard is south of the fire station; and even if it were possible to go south on High Street, a straight run would be impossible because the Community College is two blocks wide and interrupts High Street. (C) is not correct because Highland Boulevard is one-way westbound, and High Street is east of 17th Street. (D) will not work because 17th Street alone will never get you to the corner of Highland Boulevard and High Street.

54. **(A)** The inspection must be systematic; the inspector must devote attention to one item after another. Since the topic of the inspection at this moment is the sprinkler system, the logical question concerns the location of the sprinkler shut-off. The question about light bulbs should already have been asked in connection with the exit lights. The other questions will arise later in the inspection.

55. **(C)** The man is unconscious; he cannot get up and walk out. The man is breathing on his own; he does not need artificial respiration. The fire is under control, so there is no point in dragging the man out. The man is heavy and is on the third floor; get help.

56. **(A)** If you chose any answer other than this, read the correct paragraph again. This paragraph sets the scene in the first sentence, then follows logically to a conclusion.

57. **(C)** See explanation above.

58. **(B)** This drawing is accurate. (A) is incorrect because the liferaft has been lowered all the way to the water. In (C) an extra fire has started on the barge. In (D) extra people have appeared on the barge.

59. **(B)** Obviously, from the opposite shore everything is seen from its other side. In (A) the fire and smoke are blowing in the wrong direction. At the same time, the wind direction should be the same; only the viewpoint should change. (C) is incorrect because a structure has suddenly appeared on the barge. (D) is wrong because the tug has backed off.

60. **(A)** The view is from the opposite end of the scene. In (B) the liferaft has been placed on the wrong side of the yacht. In (C) a person has appeared in the water and is being thrown a rope from the barge. (D) is incorrect because the fire has disappeared.

61. **(D)** In this instance, the job of the firefighters is clear. First, they must give first aid; then, they should send for an ambulance, which will bring professional medical assistance; and, finally, they should summon the police to investigate. The problem to be investigated is not a fire problem, so investigation is the job of the police, not of the firefighters.

62. **(D)** Between answering alarms, firefighters spend much of the time making sure their equipment is functioning properly. Ineffective, malfunctioning equipment leads to ineffective firefighting and danger to the firefighters.

63. **(B)** The crowbar on the left is used to open the crate in box B.

64. **(C)** The mallet in box C is used to tap the chisel on the left.

65. **(B)** Firefighters hardly need to be reminded to be careful or to move fast; but they might forget for the moment that heat, gases, and smoke all rise. If they are reminded to keep low, they will avoid the worst of the heat, will breathe somewhat clearer air, and will have greater visibility for greater safety in moving across the room.

66. **(C)** This response identifies the type of establishment and its precise location without giving any extraneous information.

67. **(A)** If you chose an incorrect answer for this question, read your wrong answer and the correct answer again. The first sentence of the paragraph must set the stage for the elaboration and explanation to follow. Pronouns and qualifiers like *however* and *such* must refer to nouns, activities, or events mentioned earlier in the paragraph.

68. **(C)** See explanation above.

69. **(D)** Since the battalion chief is already at the fire scene, the fire has obviously already gone to two alarms. According to the paragraph, the way to order additional equipment and manpower is to sound a third alarm.

70. **(A)** The fire that was on the fourth floor is under control, and occupants of that floor and the floor above have all reached safety. But the weight of water threatens those directly below on the third floor. They must be evacuated next.

71. **(B)** This is the view of the plane wreck as seen from a rescue launch approaching the nose of the plane. (A) is not correct because the nose is pointing too far up in the water. In the original, the plane has broken apart with nose down. (C) is not correct because the break in the plane is depicted behind the wings instead of in front. (D) is wrong because the waves are suddenly much too high.

72. **(C)** This is the scene from directly above. (A) is not correct because the firefighters with the hose are now working from a different angle. In (B) the wind has shifted, and the smoke and fire are now blowing in another direction. In (D) a rescue launch has turned and now is leaving the scene instead of approaching.

73. **(D)** This is the plane crash as seen from behind on the ground. In (A) an extra aerial ladder has suddenly appeared by the side of the plane near the break. The tail in (B) is not the tail of the original plane at all. (C) is not correct because the second ambulance was not in the original sketch.

74. **(C)** An oil refinery fire is not an everyday occurrence. Firefighters may not be familiar with all procedures. Therefore the officer in charge should explain the proportions of foam and water that should be mixed, how the mixture is applied, and where to apply it. The other answer choices are not specific enough for this circumstance.

75. **(D)** The area in which smoke detectors are most crucial is the bedroom, but most people who are placing smoke detectors are already aware of this. The important factor is that smoke rises. Smoke from a lower floor will be most quickly detected if the smoke detectors are located above open staircases. By adding the instruction "on every floor," the bedroom floors are automatically accounted for.

76. **(B)** The hose is partially unrolled, so someone is already carrying it into the building and up to the second floor landing. Clearly this is the first segment of hose, that which is to be connected to the hydrant. Firefighter Poplis should connect the hose end to the hydrant. The connections must be secure at the hydrant and on every floor before the water is turned on.

77. **(A)** Since the source of the water has been clearly identified, firefighters need only turn off water to that area. There is no need to stop the water supply to the entire house. With the water off they can pump out the basement and advise the home-owner to call a plumber to make repairs.

78. **(A)** The paragraph for this question requires careful reading to determine which pro-gression is most logical and which sequence is best.

79. **(D)** See explanation above.

80. **(B)** See explanation above.

81. **(A)** The stairs are on fire, so the firefighter cannot use them to go either down or up. The question does not describe a situation so desperate that the firefighter should risk jumping from the third floor, but just waiting would be foolhardy. The firefighter should call for a ladder and go out the window.

82. **(C)** This is the perfect time for an experienced firefighter to share knowledge of safe practices with a rookie.

83. **(A)** The plane in box A is used to taper and smooth the board on the left.

84. **(C)** The trowel in box C is used to spread mortar between the bricks when building a wall.

85. **(B)** The homeowner is entitled to know why you are chopping holes in the walls when the fire appears to be out, so explain that fire can travel undetected inside the walls, and since the walls feel warm to the touch, you must doublecheck.

86. **(B)** Even a new firefighter does not need to be told to lean the ladder on a firm wall, but he or she might not know that proper placement of the feet of the ladder is a distance from the wall equal to one-fourth of the height of the ladder. With experience this distance is easy to gauge, but we are advising a *new* firefighter. The height of the ladder is given as 16 feet, so tell the new firefighter to place the base four feet from the wall.

87. **(D)**

The answer cannot be (A) because there is no 31st Street exit and because Mott Street is not a through street. (B) ends up going east on a one-way westbound street. (C) would be legal and acceptable, but it is the long way round and so is not the answer to "Which is the best route?"

88. **(C)**

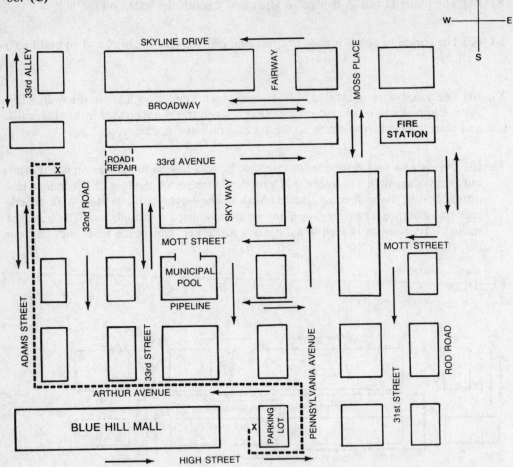

(A) is both possible and legal, but it is not the most efficient route. (B) is short and efficient, but it ends up going west on a one-way eastbound street. (D) is another possible route, but it is longer than (C) and involves many turns.

89. **(B)**

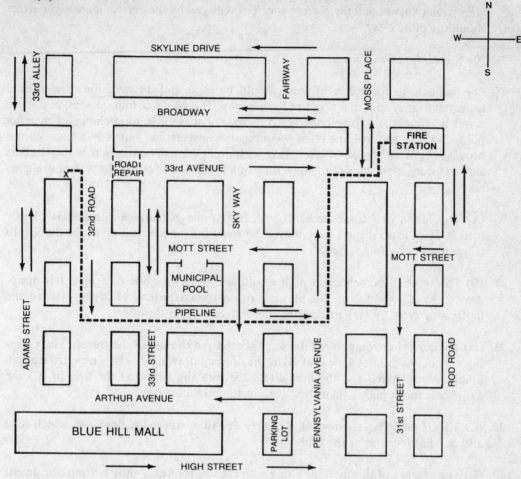

With 33rd Avenue closed, the firefighters have no way of getting to Broadway or, for that matter, to High Street from their current location. Mott Street is accessible, but they cannot go east on Mott Street.

90. **(A)** According to the procedure and order of questioning set out, the firefighter must next establish the witness's purpose in being at that location at the time of the fire.

91. **(C)** No one is trapped inside the cabs that are stopped between floors, so the firefighters need not concern themselves with moving the cabs. And there is no reason to call all the elevator cabs to the lobby. The firefighters must get to the seventh floor quickly to fight the fire, but before entering the elevator they must assure their own safety by checking the elevator shaft, control box, and wiring to rule out water damage.

92. **(B)** The next step according to the instructions given here is that the rescuers stop and listen for a response to their calls and taps. If they hear any response, they must search for the responding victim. If there is no response, they must continue in a cautious search for unconscious victims. The situation described in the narrative makes no mention of fire.

93. **(D)** If you chose the wrong paragraph as the answer to this question, reread both your wrong answer and the correct one. The reason why the correct answer is correct should be quite clear.

94. **(C)** See explanation above.

95. **(A)** Instructions for where to meet should be clear and unambiguous and should suggest a nearby spot that is out of harm's way. The yard has four corners, and mom could be almost anywhere outside, so neither of these is a good choice of meeting place. Next to the hydrant is an unambiguous instruction, but it is a poor choice because of the likelihood of interfering with firefighting. The tree across the street is easy to locate, close enough to allow for a quick count, and out of the way of firefighters.

96. **(A)** This is the view from the helicopter. In (B) one of the pickup trucks is slightly turned. In (C) two diggers have moved between the people beating at the flames. In (D) a truck has disappeared.

97. **(B)** This is what the helicopter pilot would see. In (A) the line of beaters has disappeared. In (C) the flames are blowing down trench instead of across toward the firefighting crew. In (D) an airplane has arrived to join the effort.

98. **(D)** The correct drawing shows the same scene from the end of the trench. The escaping deer in (A) did not appear at all in the original drawing. In (B) a new fire truck, a pumper, has arrived and the firefighters are spraying water on the fire. In (C) the flames are much higher than they were in the original.

99. **(C)** This is the only statement that clearly describes where the fire is—in which tube and at what location in the tube.

100. **(B)** The urgency of this message along with the call for heavy-duty cutting equipment makes clear that people are trapped in crushed metal. Choice (D) never mentions people and so does not imply urgency. (A) and (C) do not indicate what kind of help is needed.

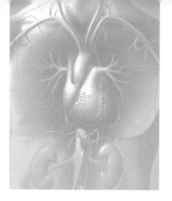

Essential Concepts in Cardiovascular Intervention

Published by Remedica Publishing
32–38 Osnaburgh Street, London, NW1 3ND, UK
Remedica Inc, Tri-State International Center, Building 25,
Suite 150, Lincolnshire, IL 60069, USA

Tel: +44 20 7388 7677
Fax: +44 20 7388 7457
Email: books@remedica.com
www.remedica.com/books

Publisher: Andrew Ward
In-house editor: Cath Harris

ISBN 1 901346 60 9

British Library Cataloguing in-Publication Data

A catalogue record for this book is available from the British Library.

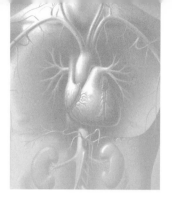

Essential Concepts in Cardiovascular Intervention

Deepak L Bhatt, Editor

REMEDICA
publishing

LONDON • CHICAGO

Editor

Deepak L Bhatt, MD, FACC, FSCAI, FESC
Director, Interventional Cardiovascular Fellowship
Associate Director, Cardiovascular Medicine Fellowship
Staff, Cardiac, Peripheral, and Cerebrovascular Intervention
Cleveland Clinic Foundation
Department of Cardiovascular Medicine
9500 Euclid Avenue, Desk F25
Cleveland, OH 44195, USA

Contributors

Ramtin Agah, MD
Assistant Professor of Internal Medicine
Division of Cardiology and the Program in
Human Molecular Biology and Genetics
University of Utah
Salt Lake City, UT 84112, USA

Youssef Al-Saghir, MD
Division of Cardiology and the Program in
Human Molecular Biology and Genetics
University of Utah
Salt Lake City, UT 84112, USA

Ivan P Casserly, MD
Cleveland Clinic Foundation
Department of Cardiovascular Medicine
9500 Euclid Avenue, Desk F25
Cleveland, OH 44195, USA

Roberto A Corpus, MD
University of Missouri-Kansas City
St Luke's Hospital, Mid America
Heart Institute, 4401 Wornall
Kansas City, MO 64111, USA

Shane Darrah, MD
Division of Cardiology
University of North Carolina
CB#7075, 130 Mason Farm Road
Chapel Hill, NC 27599, USA

Rupal Dumasia, MD
Division of Cardiology,
University of Michigan, TC B1 226
1500 E Medical Center Drive,
Ann Arbor, MI 48109–0311, USA

Stephen G Ellis, MD
Director, Sones Cardiac Catheterization
Laboratory, Cleveland Clinic Foundation
Department of Cardiovascular Medicine
9500 Euclid Avenue, Desk F25
Cleveland, OH 44195, USA

J Emilio Exaire, MD
Fellow, Interventional Cardiology
Cleveland Clinic Foundation
Department of Cardiovascular Medicine
9500 Euclid Avenue, Desk F25
Cleveland, OH 44195, USA

Cameron Haery, MD
Cleveland Clinic Foundation
Department of Cardiovascular Medicine
9500 Euclid Avenue, Desk F25
Cleveland, OH 44195, USA

Aamer H Jamali, MD
Cedars-Sinai Medical Center
UCLA School of Medicine
8631 West Third Street, Suite 415 East
Los Angeles, CA 90048, USA

Samir R Kapadia, MD
Interventional Cardiologist
Cleveland Clinic Foundation
9500 Euclid Avenue, Desk F25
Cleveland, OH 44195, USA

Rajendra R Makkar, MD
Co-director, Cardiovascular Intervention
Center; Co-director, Interventional Cardiology
Research Program
Cedars-Sinai Medical Center
UCLA School of Medicine
8631 West Third Street, Suite 415 East
Los Angeles, CA 90048, USA

Steven P Marso, MD
Assistant Professor
University of Missouri-Kansas City
St Luke's Hospital, Mid America
Heart Institute, 4401 Wornall
Kansas City, MO 64111, USA

Debabrata Mukherjee, MD, MS, FACC
Director, Peripheral Vascular
Interventions, Cardiology
Assistant Professor, Department
of Internal Medicine
University of Michigan Health System
University Hospital, TC B1 228
1500 E Medical Center Drive,
Ann Arbor, MI 48109–0311, USA

Ravish Sachar, MD
Cleveland Clinic Foundation
Department of Cardiovascular Medicine
9500 Euclid Avenue, Desk F25
Cleveland, OH 44195, USA

Jacqueline Saw, MD
Cleveland Clinic Foundation
Department of Cardiovascular Medicine
9500 Euclid Avenue, Desk F25
Cleveland, OH 44195, USA

Steven R Steinhubl, MD
Associate Professor of Medicine
Division of Cardiology
University of North Carolina
CB#7075, 130 Mason Farm Road
Chapel Hill, NC 27599–7075, USA

E Murat Tuzcu, MD
Director, Intravascular Ultrasound Laboratory
Cleveland Clinic Foundation
Department of Cardiovascular Medicine
9500 Euclid Avenue, Desk F25
Cleveland, OH 44195, USA

Jay S Varanasi, MD
Division of Cardiology
University of North Carolina
CB#7075, 130 Mason Farm Road
Chapel Hill, NC 27599–7075, USA

Jay S Yadav, MD
Director, Vascular Intervention
Cleveland Clinic Foundation
Department of Cardiovascular Medicine
9500 Euclid Avenue, Desk F25
Cleveland, OH 44195, USA

Khaled M Ziada, MD
Cleveland Clinic Foundation
Department of Cardiovascular Medicine
Contact at:
1047 Linden Lane
Lyndhurst, OH 44124, USA

Preface

Cardiovascular intervention has come of age. Percutaneous revascularization of coronary arteries has become the dominant mode of relieving cardiac ischemia, with the number of procedures surpassing coronary artery bypass surgeries. Along with this exponential growth in the number of interventions performed, so also has the complexity of coronary intervention increased. Patients with more advanced disease are treated with an increasingly sophisticated array of devices and concomitant pharmaceuticals. More recently, peripheral and cerebrovascular intervention has assumed a prominent role in revascularization strategies. In addition, noncoronary cardiac intervention is expanding the reach of the interventionalist into what was once the sole domain of the cardiac surgeon.

In keeping with this rapid expansion of knowledge in the field of cardiovascular intervention, this book provides a review of the major concepts that affect the daily practice of cardiovascular intervention. Using a Socratic format, the reader receives a detailed overview of the essentials of the field, as well as an update on the latest developments and future directions.

I am grateful to the authors for providing such excellent examples to illustrate optimal management of the mundane and expert treatment of the arcane. I hope that readers of sundry backgrounds at various stages in their careers will find this book informative and also entertaining, as they test their knowledge by tackling the questions that are posed.

Deepak L Bhatt, Cleveland Clinic Foundation

To my wife Shanthala, and to my sons Vinayak and Arjun,

for allowing me to stay late at the hospital performing

interventions and then giving me the time at home

to write a book about interventions.

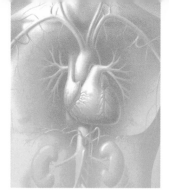

Contents

Chapter 1

Coronary anatomy

Jacqueline Saw and Deepak L Bhatt

Questions

1. All of the following statements are correct regarding the anomalous origin of the left main coronary artery from the right sinus of Valsalva or proximal right coronary artery (RCA), *except*:
   a. The incidence is approximately 0.02%
   b. This anomaly is considered benign if the left main coronary artery has a septal, anterior free wall or retroaortic course
   c. If this anomalous left main artery traverses between the aorta and pulmonary artery, angiography in the right anterior oblique (RAO) projection may show a posterior "dot" sign
   d. Septal perforating arteries may arise from the anomalous left main coronary artery if it has a septal course

2. Which of the following statements is *incorrect* regarding the condition in Figures 1a and 1b?

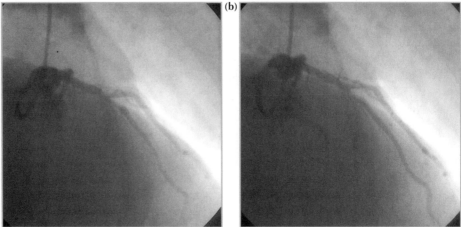

(a)   (b)

Figure 1. (a) LAD during systole; (b) LAD during diastole.

   a. Usually benign, with a good long-term prognosis
   b. More commonly involves the left anterior descending (LAD) artery than the circumflex artery and the RCA
   c. May have protective effects against atherosclerosis
   d. Administration of nitroglycerin diminishes the appearance of systolic narrowing
   e. May produce ischemic perfusion defects on nuclear imaging

3. All of the following statements are correct regarding coronary artery collaterals, *except*:
   a. Collateral circulation is dynamic, with collaterals that can promptly appear on angiography during vessel occlusion and rapidly diminish after blood flow resumes
   b. Kugel's artery is a collateral channel that supplies the LAD
   c. Collaterals can arise from intracoronary (ipsilateral) or intercoronary (contralateral) arteries, and can be antegrade (bridging) or retrograde
   d. Collateral arteries can have epicardial or endocardial routes
   e. Coronary collaterals can reduce angina and improve ventricular function

4. Which of the following is *true* regarding the significance of coronary artery stenosis?
   a. A 30% diameter stenosis significantly reduces coronary flow reserve
   b. A 50% diameter stenosis correlates to a 90% reduction in cross-sectional area
   c. A 70% diameter stenosis virtually abolishes the coronary flow reserve
   d. A 70% diameter stenosis correlates to a 99% reduction in cross-sectional area
   e. A 90% cross-sectional area stenosis reduces resting antegrade coronary flow

5. Which of the following statements *best* describes coronary artery dominance?
   a. The coronary artery that supplies the sinoatrial node is considered dominant
   b. The RCA is dominant in 55% of cases
   c. The circumflex artery is dominant in 85% of cases
   d. The circumflex artery is considered dominant if it supplies all of the posterolateral left ventricular branches
   e. Codominance with balanced circulation occurs when both the right and circumflex arteries supply the posterior descending artery

6. The left lateral projection is useful during coronary angiography for visualizing all of the following, *except*:
   a. The posterior descending artery
   b. The left internal mammary artery insertion into the mid LAD
   c. The mid and distal segments of the LAD
   d. The origin and mid segments of the RCA
   e. The ramus intermedius artery

7. Which of the following statements is *correct* regarding the left main coronary artery?
   a. It most frequently bifurcates into the LAD and circumflex arteries, but in 1% of patients it ends in a trifurcation, including a ramus intermedius artery
   b. The left main trunk is absent in 10% of patients, with the LAD and circumflex arteries arising from separate ostia
   c. The left anterior oblique (LAO) caudal and the shallow RAO (or anteroposterior [AP]) projections are excellent in identifying left main coronary artery disease
   d. Ionic contrast is preferred over nonionic contrast agents in patients with left main stenosis

8. All of the following statements are true regarding foreshortening in angiographic views, *except*:
   a. The LAO cranial projection foreshortens the circumflex artery, obtuse marginal branches, and the proximal LAD
   b. The LAO caudal projection foreshortens the proximal, mid, and distal LAD
   c. The RAO cranial projection foreshortens the circumflex artery
   d. The RAO caudal projection foreshortens the proximal circumflex and proximal obtuse marginal branches
   e. The AP cranial projection foreshortens the mid and distal LAD

9. All of the following statements are true regarding angiography of bypass grafts, *except*:
   a. The aortic anastomotic sites for bypass grafts to the circumflex branches are usually inferior to grafts supplying the LAD branches
   b. Bypass grafts usually run in a parallel course to the native vessel, just prior to the anastomotic site
   c. It is important to adequately visualize the origin, course, and anastomotic sites
   d. The best views for distal graft anastomotic sites are usually the same as the native vessels that are supplied
   e. Bypass grafts to the RCA or posterior descending artery can be adequately viewed in the RAO, LAO cranial, and lateral views

10. All of the following are recommended views for evaluating the anastomotic sites of bypass grafts, *except*:
    a. LAO cranial and RAO cranial views for grafts to the LAD
    b. LAO caudal and RAO caudal views for grafts to diagonal branches
    c. RAO cranial and RAO caudal views for grafts to obtuse marginals and posterolateral branches
    d. LAO cranial and LAO caudal views for grafts to the distal RCA
    e. LAO cranial and AP cranial views for grafts to the right posterior descending artery

11. Which of the following coronary anomalies is *not* benign?
    a. Circumflex artery origin from the right sinus of Valsalva
    b. Anomalous origin of the LAD from the posterior noncoronary sinus
    c. Absent circumflex artery
    d. Anomalous origin of the LAD from the pulmonary artery
    e. Small coronary fistulae

12. Which of the following *best* describes the abnormality shown in Figures 2a and 2b?

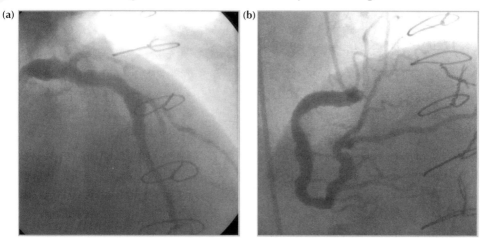

Figure 2. (a) LAD in PA cranial projection; (b) RCA in LAO cranial projection.

a. It is completely benign
b. It is most commonly caused by inflammation
c. The long-term prognosis in these patients is dismal
d. It predisposes to thrombus formation, vasospasm, and spontaneous dissection

13. Figures 3a and 3b are angiographic still-frames obtained with a right Amplatz catheter that is engaged in the right coronary ostium. Which of the following statements is *incorrect*?

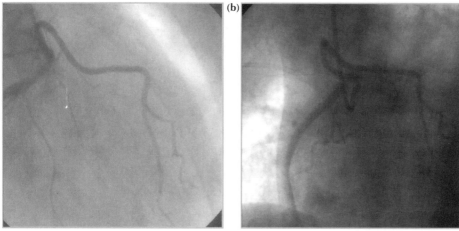

Figure 3. (a) RAO 30 projection; (b) LAO 50 projection.

a. This coronary anomaly is benign
b. This is a collateral branch from the proximal RCA to the mid LAD
c. This is an anomalous LAD that originates from the proximal RCA
d. This anomalous artery has a septal course, giving out early septal branches before turning down towards the apex

14. Which of the following statements is *incorrect* regarding the anomaly in Figures 4a and 4b?

(a) 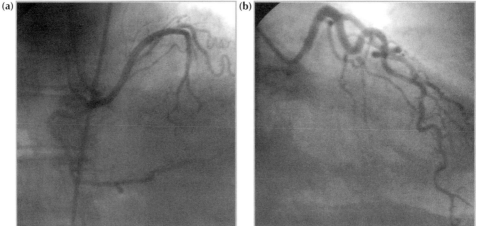 (b)

Figure 4. (a) JR 4 catheter: shallow RAO projection; (b) JL 4 catheter: AP caudal projection.

    a. The circumflex artery is occluded proximally
    b. The circumflex artery arises anomalously from the proximal RCA
    c. This is a benign variant of a coronary anomaly
    d. This type of anomalous artery almost always has a retroaortic course

15. The following statements are all true regarding the condition shown in Figures 5a and 5b, *except*:

(a) 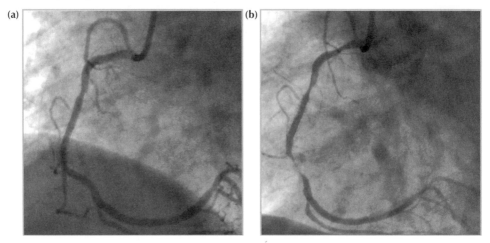 (b)

Figure 5. (a) Before provocative test; (b) after provocative test.

    a. Hyperventilation can stimulate the vasospasm shown in the figures
    b. Intracoronary adenosine can be used as a provocative test for vasospasm
    c. Ergonovine and methylergonovine stimulate α-adrenergic receptors in the smooth muscle of the coronary vasculature

   d. Vasodilating medications (such as calcium-channel blockers and nitrates) should be stopped for 24–48 hours prior to the provocative test for vasospasm

   e. The provocative test is considered positive if a focal spasm >70% in diameter is demonstrated, together with chest pain or electrocardiogram (ECG) changes

Answers

1. **c** The left main coronary artery may arise anomalously from the right sinus of Valsalva or the proximal RCA (seen in ~0.02% of coronary angiograms). This anomaly may result in sudden death if the left main artery has an interarterial course between the aorta and pulmonary artery. This has the appearance of an anterior "dot" sign on an RAO projection during angiography. Alternatively, the anomalous left main artery may follow one of three benign pathways: the intramuscular septal course, the anterior free wall of the right ventricle, or the retroaortic course (appearance of posterior "dot" sign on RAO projection). Septal perforating arteries may arise from the left main coronary artery if it has an intramuscular course through the septum.

[handwritten margin note: of slit like ostial stretching of orifice]

2. **d** **Figures 1a** and **1b** show myocardial bridging, which describes intramyocardial coronary artery segments that are compressed during ventricular systole. Angiographically, the arterial lumen appears compressed during systole, with relief from compression during diastole. The mid LAD is most commonly involved, but myocardial bridging may also involve the circumflex artery and the RCA. The administration of nitroglycerin paradoxically accentuates systolic narrowing, and may be useful as a provocative test during angiography. Myocardial bridging is mostly considered benign, although it may cause myocardial ischemia (perfusion defects on nuclear imaging have been demonstrated) that can potentially normalize after coronary stenting. The etiology of ischemia is unclear, especially since myocardial compression of the coronary artery occurs only during systole, whereas coronary flow predominantly occurs during diastole. The occurrence of atherosclerosis has also been noted to be infrequent in myocardial bridges.

3. **b** Coronary artery collaterals provide important alternate circulation when stenoses are present (complete or subtotal occlusions), improving myocardial flow, which can reduce ischemia and improve ventricular function. They can arise from the same artery as the stenosis (intracoronary or ipsilateral) or from other arteries (intercoronary or contralateral) in an antegrade or retrograde fashion, and can be epicardial or endocardial. Kugel's artery is a collateral channel that arises from either the proximal right or left coronary artery, traversing from the anterior atrial wall to the crux posteriorly to supply the posterior circulation (it may give collaterals to the sinoatrial and to the atrioventricular nodal arteries). Collateral circulation is dynamic. During percutaneous transluminal coronary angioplasty balloon occlusion, a prompt increase in collaterals is seen angiographically, with rapid decrease when the balloon is deflated. Collateral circulation can also be graded from 0 to 3 (0 = no collateral circulation, 1 = very weak opacification, 2 = slow filling and less dense opacification than the feeding vessel, 3 = rapid filling and dense opacification comparable with the feeding vessel).

4. **c** A coronary artery stenosis is considered hemodynamically significant when it reduces the arterial diameter by 50% (decreasing the cross-sectional area by 75%), since there is corresponding reduction in the coronary flow reserve. When the diameter stenosis is 70%, the cross-sectional area is reduced by 90%, which completely abolishes the coronary flow reserve. With a 90% diameter stenosis (about 99% reduction of cross-sectional area), the resting coronary flow will become compromised.

5. e "Coronary artery dominance" refers to the coronary artery that supplies the posterior diaphragmatic interventricular septum and left ventricle, giving rise to the posterior descending artery and the artery that supplies the atrioventricular node. The RCA is dominant 85% of the time, compared with 8% for the circumflex artery. Codominant circulation occurs in 7% of patients when both the RCA and circumflex artery supply the left ventricular diaphragmatic surface (both may supply the posterior descending artery; or the RCA supplies the posterior descending artery and then terminates, with the circumflex artery supplying all the posterolateral left ventricular branches). Coronary dominance is best evaluated from an LAO cranial projection.

6. a The left lateral projection is particularly useful to visualize the anastomosis of the internal mammary artery to the mid or distal LAD, since these segments are well delineated without overlap of the septal (above the LAD) or diagonal branches (below the LAD). This view is also useful for assessment of the RCA origin (especially if the take-off is more anterior), mid RCA (there is less motion of the RCA in this view compared with the RAO projection), proximal circumflex artery, and ramus intermedius branch. However, it is suboptimal for visualizing the posterior descending or posterolateral branches, which will be foreshortened, and the diagonal arteries, which frequently overlap. The origins of diagonal branches are usually best seen with cranial angulation in either the RAO or anteroposterior (AP) projections.

7. c The left main artery arises from the left coronary cusp, then travels between the pulmonary artery and the left atrial appendage in an anterior, inferior, and leftward direction. It ranges from 0 to 4 cm in length (typically 1–2 cm) and is absent in ~0.5% of the population (in which case the LAD and circumflex arteries will have either separate ostia or a common ostium). It usually bifurcates into the LAD and circumflex arteries, but in about one third of patients it ends in a trifurcation, which includes a ramus intermedius branch. The entire length of the left main artery is usually well visualized in the shallow RAO (or AP) and LAO caudal projections. Sometimes, the RAO caudal or AP caudal projections may also be useful. In patients with significant left main artery stenosis, it is important to limit the number of angiographic views, avoid strong contrast injections and frequent catheter engagements, watch for the presence of pressure damping, and utilize a nonionic or low-osmolar contrast agent (reducing hypotension and bradycardia).

8. e Angiographic x-ray images are viewed in only two dimensions. Thus, multiple projections are needed in various gantry positions to optimally compensate for limitations such as vessel-segment foreshortening (when arterial segments are viewed obliquely), overlap, and when visualizing complicated lesions (such as bifurcation and eccentric lesions). Recognizing the limitations of each view in terms of vessel-segment foreshortening is important, especially during percutaneous intervention, to ensure appropriate sizing of angioplasty balloons and stents (particularly when deploying drug-eluting stents). The LAO cranial projection, which delineates the mid and distal LAD well, unfortunately foreshortens the left main, proximal LAD, circumflex artery, and obtuse marginal branches. The LAO caudal projection, which is excellent for viewing the left main and circumflex (proximal, mid, and obtuse marginal branches) arteries, significantly foreshortens the LAD.

On the other hand, the RAO cranial projection shows the mid and distal LAD well, with good visualization of diagonal branches, but foreshortens the circumflex artery with overlap of the obtuse marginal branches. Similarly, the AP cranial projection foreshortens the circumflex artery and branches, but is excellent for visualizing the mid and distal LAD with their diagonal branches. The RAO caudal projection is usually very good for visualizing the mid and distal circumflex arteries with their obtuse marginal branches, but the ostial and proximal portions are foreshortened.

9. **a** It is important to visualize the entire length of bypass grafts, including the origin, body, and distal anastomotic sites. Grafts that supply the RCA or distal left dominant circumflex system usually arise from the lateral aortic wall above the right sinus of Valsalva. Grafts that supply the LAD or circumflex branches usually arise from the anterior aortic wall above the left sinus of Valsalva, typically in the following ascending take-off order: grafts to LAD, diagonal branches, circumflex branches (obtuse marginals and ramus intermedius). Bypass grafts generally travel parallel to the native coronary vessel, just prior to the anastomotic sites. The angiographic views that best depict the anastomosis sites tend to be those that best delineate the native vessels that the grafts supply. Thus, grafts to the RCA or its distal branches are well visualized in the RAO, LAO cranial, and lateral views. Grafts to the LAD are well seen in the LAO cranial, RAO cranial, and lateral views. Grafts to the circumflex branches are clearly seen in the RAO and LAO caudal views.

10. **b** The anastomotic sites of bypass grafts are often clearly seen in the views that show the supplied native vessels. Anastomoses of grafts to the LAD and diagonal branches are well visualized in the LAO cranial and the RAO cranial or caudal views. Anastomoses of grafts to obtuse marginal and posterolateral branches of the circumflex artery are well seen in the RAO caudal and cranial views. Anastomosis to the distal RCA is well seen in the LAO cranial or caudal views. Anastomosis to the right posterior descending artery is usually clearly seen in the AP or LAO cranial views.

11. **d** Coronary anomalies are seen in ~1% of angiograms performed for suspected coronary disease. The majority (80%) of these anomalies are benign (eg, high anterior take-off of RCA, separate LAD and circumflex ostia, circumflex origin from the RCA or right sinus of Valsalva, anomalous origins of coronary arteries from the posterior sinus, small coronary fistulae, absence of the circumflex artery), and only 20% have serious consequences (eg, myocardial ischemia, bacterial endocarditis, congestive heart failure, ventricular fibrillation, sudden death). These potentially lethal anomalies include origin of the coronary artery from the opposite aortic sinus (if the course is interarterial between the aorta and pulmonary artery, which may be diagnosed with the anterior "dot" sign on the RAO projection), origin of the coronary artery from the pulmonary artery, a single coronary artery, and large coronary fistulae.

12. **d** **Figures 2a** and **2b** demonstrate coronary artery ectasia. This abnormality is defined as diffuse, irregular dilatation involving most of the length of an epicardial coronary artery. The diameter is up to 1.5-times that of normal segments. (Coronary aneurysm is defined as dilatation >1.5-times that of normal segments.) Coronary ectasia occurs in ~1%–5% of patients with coronary obstructive disease. The main underlying etiology

is atherosclerosis (~50%–90% of patients have coexisting obstructive coronary artery disease), but other causes include inflammation (eg, Kawasaki disease, polyarteritis nodosa, Takayasu's arteritis, scleroderma, Ehlers–Danlos syndrome, syphilis) and congenital anomalies. Even in the absence of overt coronary stenosis, coronary artery ectasia may predispose to thrombus formation, vasospasm, and spontaneous dissection. However, the long-term prognosis for patients with coronary ectasia is good. The optimal treatment has not been defined, but potential treatment strategies include aspirin, coumadin, and calcium-channel blockade (to prevent spasm).

13. b **Figures 3a** and **3b** demonstrate an anomalous LAD that originates from the proximal RCA. Angiography of the left coronary system from the left sinus of Valsalva (see **Figure 6**) shows a large circumflex arterial territory with small LAD territory vessels. The incidence of anomalous LAD from either the right sinus of Valsalva or the proximal RCA occurs in ~0.03% of the population. This anomalous artery usually follows either an anterior free-wall course (anterior surface of the right ventricle) or a septal course (intramuscular through the septum along the floor of the right ventricular outflow tract), prior to being directed towards the apex after the mid septum. Both of these courses are benign.

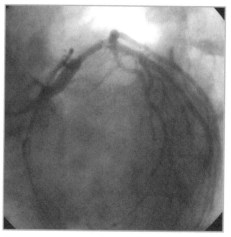

Figure 6. LAO caudal projection.

14. a **Figures 4a** and **4b** demonstrate an anomalous circumflex artery that originates from the proximal RCA. This anomalous artery can arise from either the right sinus of Valsalva or the proximal RCA. It almost always has a retroaortic course, initially traveling posterior to the aortic root. Injection from the left coronary ostium shows a long left main segment, and may be mistaken for an occluded circumflex artery. This is a benign variant, which occurs in 0.37% of the population.

15. b **Figures 5a** and **5b** demonstrate focal epicardial coronary artery spasm, which can occur in the presence or absence of significant coronary artery disease. This condition can cause myocardial ischemia, which typically presents as variant or Prinzmetal angina, with transient ECG changes of ST-segment elevation. Confirmation with a provocative test during coronary angiography can be performed with ergonovine, methylergonovine,

acetylcholine (but not adenosine, which causes vasodilatation), and hyperventilation. However, long-acting nitrates and calcium-channel blockers should be discontinued for 24–48 hours prior to testing. Ergonovine is no longer available in the US, but intracoronary methylergonovine can be administered up to a maximum dose of 50 μg. Both of these agents stimulate the α-adrenergic and serotonin receptors, causing constriction of coronary vascular smooth muscle. Mild generalized coronary narrowing (<15%) can occur in response to these vasoconstrictors, even in normal individuals. Thus, the provocative test is considered positive only when severe focal spasm (>70%) occurs, in association with symptoms or ECG changes.

> BAV Associations:
① Coarctation of Ao.
② Separate LAD & Circ ostia
④ ⑫ dominance.

Chapter 2

Hemodynamics

J Emilio Exaire and Samir R Kapadia

Questions

1. The following are all suggested indications for intra-aortic balloon pump (IABP)-supported percutaneous coronary intervention (PCI), *except:*
    a. Cardiogenic shock
    b. Large at-risk territory with ongoing ischemia
    c. Use of atherectomy devices
    d. Left ventricular (LV) dysfunction
    e. Intervention in a long, tortuous, calcified coronary artery (American College of Cardiology [ACC] type C lesion)

2. Of the following PCI complications, which is *best* treated with IABP?
    a. National Institutes of Health (NIH) type C coronary dissection
    b. Refractory no reflow
    c. Threatened vessel closure
    d. Elastic recoil
    e. NIH type B coronary dissection

3. Right heart catheterization (RHC) is indicated for all of the following, *except:*
    a. Cardiogenic shock
    b. Suspected coronary artery perforation
    c. Carotid stenting
    d. PCI in a hemodynamically unstable patient
    e. PCI in a patient with severe LV dysfunction

4. All of the following statements are true regarding "damping" and "ventricularization" of the coronary pressure waveform, *except:*
    a. It may be the result of an unfavorable position of the catheter tip against an arterial wall
    b. It reflects the actual ventricular pressure
    c. The waveform is derived from a reduction in both systolic and diastolic pressures
    d. It may happen if the catheter is not coaxial to the ostium of the left main coronary artery
    e. Right coronary artery dampening can be due to ostial disease

5. Which of the following hemodynamic patterns identifies an *adverse* prognosis in patients with cardiogenic shock?
    a. Cardiac index >2 L/min/m^2 and pulmonary arterial wedge pressure (PAWP) <20 mm Hg
    b. Cardiac index >5 L/min/m^2 and PAWP <20 mm Hg
    c. Cardiac index <1.5 L/min/m^2 and PAWP >20 mm Hg
    d. Cardiac index >3 L/min/m^2 and PAWP <20 mm Hg
    e. Cardiac index <5 L/min/m^2 and PAWP >10 mm Hg

6. Figure 1 shows an IABP curve. What physiologic consequence would you *expect?*

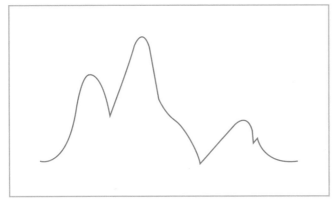

Figure 1.

    a. Significant afterload reduction
    b. Decreased myocardial oxygen (MO_2) demand
    c. Increased coronary blood flow
    d. Impeded LV ejection and increased afterload
    e. Increased ejection fraction by 30%

7. The IABP curve in Figure 2 *corresponds to*:

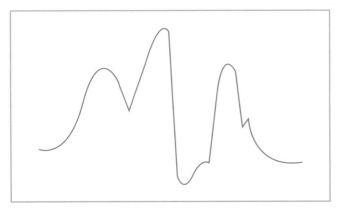

Figure 2.

    a. Early inflation
    b. Early deflation
    c. Late inflation
    d. Late deflation
    e. A normal IABP curve

8. Point 1 in the following IABP curve (Figure 3) *represents*:

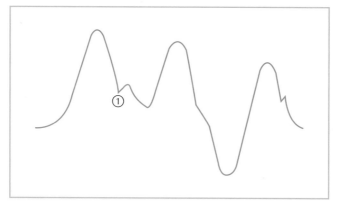

Figure 3.

   a. Initiation of diastolic augmentation
   b. Assisted systole
   c. Dicrotic notch
   d. Unassisted systole
   e. Assisted diastole

9. The IABP curve in Figure 4 has all of the following physiologic consequences, *except*:

Figure 4.

   a. Potential premature closure of the aortic valve
   b. Potential increase in LV end diastolic pressure (LVEDP)
   c. Increase in afterload
   d. Decreased MO_2 demand
   e. Potential aortic regurgitation

10. What is the *most* common IABP placement complication?
a. Thrombosis
b. Vascular injury and limb ischemia
c. Stroke
d. Acute renal failure
e. Spinal cord necrosis

11. Which of the following is *not* an absolute contraindication for IABP placement?
a. Mild aortic insufficiency
b. Severe peripheral vascular disease
c. Acute stroke
d. Bleeding diathesis
e. Thrombocytopenia ($<$50,000 cells/μL)

12. When significant coronary artery stenosis is *present*, placement of an IABP:
a. Increases the flow velocity distal to the stenosis
b. Decreases the flow velocity distal to the stenosis
c. Increases the flow velocity proximal to the stenosis, but fails to increase the flow velocity distal to the stenosis
d. Increases the flow velocity throughout the entire coronary vessel
e. Has no impact on coronary perfusion

13. A 75-year-old patient presents to the emergency room complaining of ongoing chest pain. On auscultation, a new systolic murmur is heard at the apex. Electrocardiography shows ST-segment elevation in II, III, and aVF. A bedside echocardiogram reveals moderate hypokinesis of the inferior wall, LV ejection fraction of 30%, dilated and hypokinetic right ventricle (RV), and severe mitral regurgitation. If an RHC were placed, you would expect all of the following, *except*:
a. Elevated mean right atrial pressure
b. Large V waves
c. Delayed RV relaxation
d. Rapid rise in diastolic pressure to an elevated plateau
e. High cardiac output

Answers

1. **e** ACC type C lesions, although considered complicated for intervention, do not necessarily need to be supported by an IABP. There is a lack of randomized controlled trials addressing IABP-supported PCI in cardiogenic shock patients. However, the weight of evidence from observational databases suggests that there is a decrease in mortality for IABP-supported PCI in this particular subset of patients, and the American College of Cardiology/American Heart Association guidelines recommend early consideration of IABP placement. In the SHOCK (Should we Emergently Revascularize Occluded Coronaries in Cardiogenic Shock) trial, IABP was used in 87% of patients [1]. In this study, IABP was strongly encouraged, whether patients had medical stabilization or early revascularization.

One of the suggested uses of IABP-supported PCI is during unprotected left main PCI. Since the at-risk territory exceeds 50% of the myocardium, its use is recommended. Patients with borderline hemodynamics and ongoing ischemia benefit from the insertion of an IABP just prior to coronary instrumentation. Distal embolization frequently occurs with the use of atherectomy devices, resulting in slow flow or no reflow. Use of prophylactic IABP may decrease the severity and occurrence of these phenomena. Patients with extremely depressed LV function may not tolerate balloon inflation or stent deployment. IABP increases the cardiac output by 30% and improves the coronary perfusion pressure, reducing the risk of the intervention.

2. **b** No reflow is a feared complication following PCI. It is associated with myocardial ischemia, myocardial infarction (MI), and increased mortality. Refractory no reflow usually results in ST-segment elevation MI. IABP placement may ease this condition by increasing diastolic coronary flow and decreasing systolic pressure.

NIH type B and C dissections are not flow limiting and are best treated by stenting. Threatened vessel closure is defined as a combination of >50% residual lesion, dissection, angina, or ST-segment changes. The vast majority of these cases are successfully treated by stenting. Elastic recoil is a frequent phenomenon after plain old balloon angioplasty (POBA). It is usually not flow limiting, and is successfully treated by stent deployment.

3. **c** Routine use of RHC for carotid stenting is not indicated. Its use is reserved for patients with concomitant moderate to severe aortic stenosis, LV dysfunction, or left main trunk disease. Patients presenting with hypotension may respond to fluid challenge. RHC may not be needed if the condition improves. If the blood pressure does not return to normal, RHC is warranted to determine the etiology of the shock and facilitate medical treatment.

Most coronary perforations in the contemporary PCI setting are related to wire technique, resulting in insidious-onset tamponade. RHC is extremely useful to guide therapy and monitor whether blood is accumulating in the pericardium. If the perforation is secondary to balloon or stent inflation, resulting in free flow into the pericardium, RHC may be deferred until pericardiocentesis is performed.

4. **b** The coronary pressure waveform obtained with damping or ventricularization is derived from the reduction in both systolic and diastolic pressures. Some of the conditions that may cause ventricularization include an unfavorable position of the catheter tip against the arterial wall, matching diameters of the coronary artery and catheter, or a deeply seated subselective position of the catheter tip against an arterial wall. Right coronary artery dampening or ventricularization may occur secondary to coronary spasm, subelective conus branch engagement, ostial disease, or smaller caliber arteries.

5. **c** In the large, multicenter GUSTO (Global Use of Strategies to Open Coronary Arteries) trial, cardiogenic shock occurred in 7.2% of patients, but accounted for 58% of all deaths in the trial. The estimated incidence of cardiogenic shock in the US is 50,000 patients/year. In the SHOCK registry, in-hospital mortality was approximately 60% in patients with post-MI cardiogenic shock. Significant hemodynamic predictors for adverse outcome were cardiac index <1.5 L/min/m^2 or PAWP >20 mm Hg.

6. **d** The trace corresponds to late deflation. Afterload reduction is essentially absent, there is an increase in MO$_2$ consumption, and IABP may impede LV ejection.

7. **b** Early deflation is characterized by a sharp drop following diastolic augmentation. The physiologic consequences include suboptimal coronary perfusion, suboptimal afterload reduction, and increased MO$_2$ demand.

8. **c** The trace corresponds to late inflation. This IABP curve is characterized by inflation after the normal dicrotic notch. The physiologic consequence is inadequate augmentation of diastolic pressure, resulting in suboptimal coronary artery perfusion.

9. **d** Early inflation is characterized by augmentation encroaching onto the systolic curve. The physiologic consequences include potential premature closure of the aortic valve and increased LVEDP, afterload, and MO$_2$ demand. It may also cause aortic regurgitation. A normal IABP inflation/deflation tracing is shown in **Figure 5**:

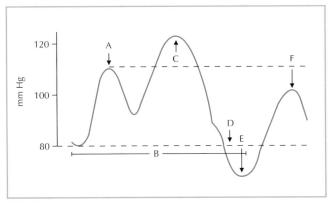

Figure 5. A normal intra-aortic balloon pump inflation/deflation tracing. **A**: unassisted systolic pressure; **B**: one complete cardiac cycle; **C**: diastolic augmentation; **D**: unassisted aortic end diastolic pressure; **E**: assisted aortic end diastolic pressure; **F**: reduced systolic pressure.

10. b Although all of the options are complications related to IABP placement, by far the most common is vascular injury and limb ischemia. Only 1% of these cases result in irreversible limb ischemia. Risk factors associated with this complication include hypertension, diabetes, female sex, and pre-existing peripheral vascular disease.

11. a IABP may be safely used in patients with mild aortic insufficiency. However, moderate to severe aortic insufficiency constitutes an absolute contraindication for IABP placement. Augmenting diastolic pressure in this setting can cause LV dilatation and worsening cardiac performance. Severe peripheral vascular disease may impede the use of an IABP or precipitate limb ischemia. The need for anticoagulation when using an IABP contraindicates its use in acute stroke, bleeding diathesis, or in patients with severe thrombocytopenia.

12. c Kern and colleagues showed that IABP augments coronary flow velocity proximally in a coronary, but not if it is measured distally to high-grade stenosis [2]. This reverts after successful PCI.

13. e The case corresponds to an RV infarction with concomitant mitral regurgitation. Typically, the hemodynamic evaluation reveals an increase of right heart filling pressures, equalization of right- and left-sided diastolic pressures, and low cardiac output despite preserved LV function. Acute RV dilatation also shifts the interventricular septum towards the LV, impairing LV compliance and limiting its filling. Constrictive or restrictive physiology is not uncommon due to the pericardial constraint applied to the dilated RV. This results in both systolic and diastolic dysfunction, and increased right atrial pressure. Large V waves result from ischemic mitral regurgitation.

References

1.   Hochman JS, Sleeper LA, Godfrey E et al. SHould we emergently revascularize Occluded Coronaries for cardiogenic shocK: an international randomized trial of emergency PTCA/CABG-trial design. The SHOCK Trial Study Group. *Am Heart J* 1999;137:313–21.
2.   Kern MJ, Aguirre F, Bach R et al. Augmentation of coronary blood flow by intra-aortic balloon pumping in patients after coronary angioplasty. *Circulation* 1993;87:500–11.

Chapter 3

Vascular biology of atherosclerosis

Youssef Al-Saghir and Ramtin Agah

Questions

1. The trilaminar structure of a normal artery includes all of the following layers, *except*:
   a. Adventitia
   b. Serosa
   c. Intima
   d. Media

2. Atheroma typically form focally, despite the presence of equal concentrations of blood-borne risk factors, including lipoproteins, throughout the vasculature. The hypotheses invoked to explain the focality of lesion formation include all of the following, *except*:
   a. Multicentric origin hypothesis
   b. Hydrodynamic hypothesis
   c. Monoclonal hypothesis
   d. Focal genetic mutation hypothesis

3. The *first* ultrastructurally definable event in the initiation of atherogenesis is:
   a. Leukocyte recruitment
   b. Intracellular lipid accumulation
   c. Extracellular accumulation of small lipoprotein particles in the intima
   d. Local mineralization

4. The *second* ultrastructurally definable event in the initiation of atherogenesis is:
   a. Leukocyte recruitment and accumulation
   b. Intracellular lipid accumulation
   c. Accumulation of small lipoprotein particles in the intima
   d. Local mineralization

5. Chemoattraction and proliferation of leukocytes in response to oxidized lipoprotein and other stimuli involves *which* of the following proteins?
   a. Monocyte chemoattractant protein (MCP)-1
   b. Macrophage colony-stimulating factor (M-CSF)
   c. Interleukin (IL)-8
   d. Oxidized low-density lipoprotein (oxLDL)
   e. All of the above

6. Once monocytes are recruited to the arterial intima, they can imbibe lipids to become foam cells. Each of the following cell receptors has a major role in the formation of foam cells, *except*:
   a. Scavenger receptor A
   b. LDL receptor
   c. Scavenger receptor B
   d. Macrosialin receptor

7. Which of the following is *true* regarding fatty streaks?
   a. They are present only in adults
   b. They are mainly formed of leukocytes
   c. Withdrawal of an atherogenic diet or the use of lipid-lowering drugs can partially or totally reverse the lesion
   d. They are ubiquitous in vascular beds

8. Which of the following is *not* involved in evolution of the atheroma into a more complex plaque?
   a. Smooth muscle cell migration
   b. Smooth muscle cell replication
   c. Red blood cell recruitment
   d. All of the above

9. What is the *role* of metalloproteinases in atherogenesis?
   a. Breakdown of the extracellular matrix (ECM)
   b. Biosynthesis of the ECM
   c. Biosynthesis of collagen I and III
   d. Biosynthesis of versican

10. Positive remodeling *refers to*:
    a. Migration of the plaque cellular components
    b. Apoptosis and accumulation of cellular debris
    c. Distention of the elastic lamina
    d. Outward growth of the plaque

11. Microanatomical features of vulnerable atherosclerotic plaques include all of the following, *except*:
    a. Decreased collagen synthesis and increased collagen catabolism
    b. Apoptosis of endothelial cells
    c. "Dropout" of smooth muscle cells
    d. Accumulation of large lipid pools
    e. Accumulation of basophils

Answers

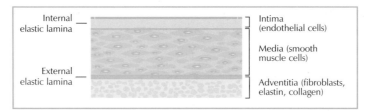

Figure 1. The arterial lumen.

1. **b** Normal arteries are composed of a trilaminar structure, encompassing the intima, media, and adventitia (see **Figure 1**).

The intima is the innermost layer. It is formed of an endothelial cell monolayer that rests upon a basement membrane containing nonfibrillar collagen types (the internal elastic lamina). The endothelial cell monolayer is involved in many highly regulated mechanisms that are of importance in vascular homeostasis. By the second decade of life, the intima develops into a more complex structure that contains arterial smooth muscle cells and fibrillar forms of interstitial collagen. This process is known as "intimal thickening".

The media of elastic arteries is composed of concentric layers of smooth muscle cells, interleaved with layers of elastin-rich extracellular matrix (ECM).

The adventitia contains collagen fibrils in a looser array than the intima or the media. The vasa vasorum and nerve endings reside in this outermost layer. The adventitia also contains fibroblasts and mast cells.

2. **d** Two hypotheses are invoked in order to explain this phenomenon.

The monoclonal hypothesis of atherogenesis (or the multicentric hypothesis) proposes that atheroma arise as benign leiomyomas of the vessel walls. This hypothesis is supported by the monotypia of various molecular markers, such as glucose-6-phosphate dehydrogenase (G6PD) isoforms, in individual atheroma; specifically, females who are heterozygous for the X-linked enzyme G6PD (the A and B isoenzyme) predominantly demonstrate only one subtype of G6PD within the plaque.

The hydrodynamic hypothesis suggests that hydrodynamic forces and local flow disturbances form the basis of early lesion formation. This is supported by the predilection of lesions to form at the proximal region of arteries, bifurcations, and vascular sites with local flow disturbance. Locally disturbed flow induces alterations that may promote the early steps of atherogenesis, such as decreased expression of vasoactive agents (eg, nitric oxide).

Laminar flow at sites that do not tend to develop early lesions may elicit antiatherogenic mechanisms. *In vitro* data support the notion that laminar shear stress promotes the expression of genes that may protect against atherosclerosis. These include genes that encode the enzymes superoxide dismutase (which helps to reduce oxidative stress by catabolizing the reactive and injurious superoxide anion) and endothelial nitric oxide synthase (which produces nitric oxide, an anti-inflammatory and vasodilating agent).

3. c In the presence of an atherogenic diet, the first ultrastructural change in the artery wall is the extracellular accumulation of small lipoprotein particles in the intima. Some studies suggest that permeability of the endothelial monolayer increases at sites of lesion formation, allowing these lipid particles to penetrate.

Proteoglycans in the media allow extracellular attachment of lipid particles and retention of these particles within the vessel wall. The proteoglycan-bound particles appear to have increased susceptibility to oxidative modification. Oxidation of these lipoproteins in the artery wall is considered the key step in initiation of the atherosclerotic lesion. The underlying mechanism is not clearly understood.

4. a Normal endothelial cells generally resist adhesive interaction with leukocytes. Most recruitment and trafficking of leukocytes occurs in the postcapillary venules, not in the arteries. However, early in the atherosclerotic process, adhesion molecules (cell surface proteins, primarily vascular cell adhesion molecule-1) are selectively expressed by endothelial cells overlying early lesions. T cells and monocytes can bind to these adhesion molecules, and thereby interact with endothelial cells. Furthermore, endothelial cells also express chemokines (chemoattractant cytokines such as monocyte chemoattractant protein [MCP]-1), which allows recruitment of leukocytes to the endothelium.

Leukocytes adhere and diapedese between endothelial cell junctions to enter the intima. Here, they accumulate lipids and secrete proinflammatory cytokines and matrix-degrading proteins, leading to dynamic growth and remodeling within the vessel wall.

5. e Monocytes and T cells are the principal leukocytes found in atherosclerotic lesions. Monocytes are the dominant cell type involved in early lesion formation (known as a fatty streak), while T cells appear at a later stage.

Migration of monocytes into the artery wall is, in part, directly stimulated by oxLDL. OxLDL also stimulates the production of other chemoattractants by endothelial cells, mainly MCP-1 and its receptor (chemokine receptor [CCR]-2). The role of hypercholesterolemia in this response is directly supported by an increased chemoattractant response to MCP-1 of monocytes derived from hypercholesterolemic patients. IL-8, which is also present in human atherosclerotic lesions, has been shown to have a role in monocyte trafficking. Once within the artery wall, M-CSF, leads to proliferation of monocytes. Paradoxically, the injury response by inflammatory cells in early atherogenesis may be maladaptive. In mice, inhibition of MCP-1, M-CSF, and CCR-2 (via gene knockout experiments) significantly reduces both the formation and the progression of atherosclerotic lesions.

Less is known about the role of T cells and their recruitment and "activation" in atherosclerotic lesions; however, interferon γ, a cytokine known to be present in atheromatous plaques, appears to play a key role in the T-cell response.

6. **b** The LDL receptor, which is expressed by most cells, is highly regulated by cholesterol. As soon as the cell collects enough LDL, an elaborate transcriptional control mechanism halts expression of the receptor. This mechanism regulates intracellular LDL levels in most cells. However, the LDL receptor appears to have a minimal role in extracellular lipid uptake by macrophages in atherosclerotic lesions. Instead, "scavenger" receptors appear to mediate the excessive lipid uptake that is characteristic of foam cell formation.

   These surface receptors bind modified, not native, lipoproteins (ie, oxLDL) and participate in their internalization. Scavenger receptors A, B, and macrosialin receptors have all been shown to participate in the formation of foam cells; macrosialin receptors are known to have a predilection for binding oxLDL.

7. **c** Fatty streaks are formed by the accumulation of foam cells, mostly of the monocyte type. Withdrawal of an atherogenic diet or the use of lipid-lowering drugs can partially or totally reverse the lesion. Fatty streaks are seen in children and adults, and are mainly present in the arteries.

8. **c** Evolution of the atheroma into a more complex plaque involves smooth muscle cell migration and proliferation. Although some smooth muscle cells arrive in the arterial intima early in life, the majority migrate from the underlying media; platelet-derived growth factor (PDGF), a potent chemoattractant, is probably involved. Even though the rate of smooth muscle cell division is low (approximately 1% of the smooth muscle cells are replicating at any one time), considerable cell accumulation may occur over decades of lesion evolution. In addition, smooth muscle cell proliferation may occur as bursts of cell replication.

   As plaque progression continues, smooth muscle cells may undergo apoptosis, potentially contributing to plaque complications such as rupture. In particular, certain T-cell populations can express fas ligand on their surface. This engages the fas receptor on the surface of smooth muscle cells and, in conjunction with soluble proinflammatory cytokines, leads to the death of the smooth muscle cell.

   Therefore, the smooth muscle cell population in the growing atherosclerotic plaque results from a "tug-of-war" between cell replication and cell death. The balance between smooth muscle cells and inflammatory cells within the plaque may dictate its propensity for rupture, differentiating stable from unstable plaques.

9. **a** ECM makes up most of the volume of the advanced atherosclerotic plaque. The main extracellular macromolecules include interstitial collagens and proteoglycans; elastin fibers may also accumulate within the plaque.

Similarly to the accumulation of smooth muscle cells, ECM secretion depends on a balance – in this case, between the biosynthesis and breakdown of ECM molecules. Biosynthesis is performed by the smooth muscle cells. This matrix deposition by "activated" smooth muscle cells is a key component of atherosclerotic lesion expansion, and is known as "positive remodeling".

ECM breakdown is catalyzed, in part, by matrix metalloproteinases, which are mainly secreted by foam cells (activated macrophages).

10. d Early in plaque formation, the plaque grows outward. This "positive remodeling" or "compensatory enlargement" involves the turnover of ECM molecules, and causes a minimal loss in the actual luminal diameter (where blood flow takes place). As such, coronary angiography (which only detects luminal loss) has low sensitivity in detecting early lesion formation and positive remodeling when compared with other modalities, such as intravascular ultrasound. Outward plaque growth eventually gives way to inward growth, leading to loss of lumen ("negative remodeling"). Luminal stenosis tends to occur only after the plaque burden exceeds about 40% of the cross-sectional area of the artery.

11. e Disruption of the atherosclerotic plaque is the underlying cause of acute thrombosis, leading to acute coronary syndromes. Two major processes are presumed to give rise to plaque disruption: superficial erosion of the intima (one third of cases) and fracture of the plaque's fibrous cap (two thirds of cases).

The mechanism of superficial erosion is not well understood; however, it appears to involve apoptosis of endothelial cells overlying the atherosclerotic plaque. The mechanism of plaque rupture is one of imbalance between forces that impinge on the plaque's cap and the mechanical strength of the fibrous cap. The processes involved are thought to be largely mediated by local inflammation within the vessel wall. For example, factors such as T-cell-derived interferon γ potently inhibit smooth muscle cell collagen synthesis, and macrophages in advanced human atheromas overexpress matrix metalloproteinases that can break down collagen and elastin in the arterial ECM. In contrast, collagen synthesis is increased by smooth muscle cells after transforming growth factor-β and PDGF stimulation through both paracrine and autocrine mechanisms. "Dropout" of smooth muscle cells from regions of local inflammation, influenced by both soluble and lymphocyte surface molecules within plaques, can also contribute to weakening of the fibrous cap.

The clinical relevance of these processes has been shown in recent studies, which have found the serum levels of a number of inflammatory factors and cytokines (IL-6, IL-7, IL-10, MCP-1, and CD-40) to be predictive of coronary events in patients with atherosclerosis.

In summary: the balance between proinflammatory cells and smooth muscle cells dictates the local milieu, determining the plaque lipid composition, matrix turnover, and fibrous cap stability, and leading to the formation of a stable versus unstable plaque. The cellular and molecular signals involved appear to form a complex system involving numerous factors (see **Figure 2**, overleaf). The identification of key steps in tipping the balance toward a stable plaque is an active area of research.

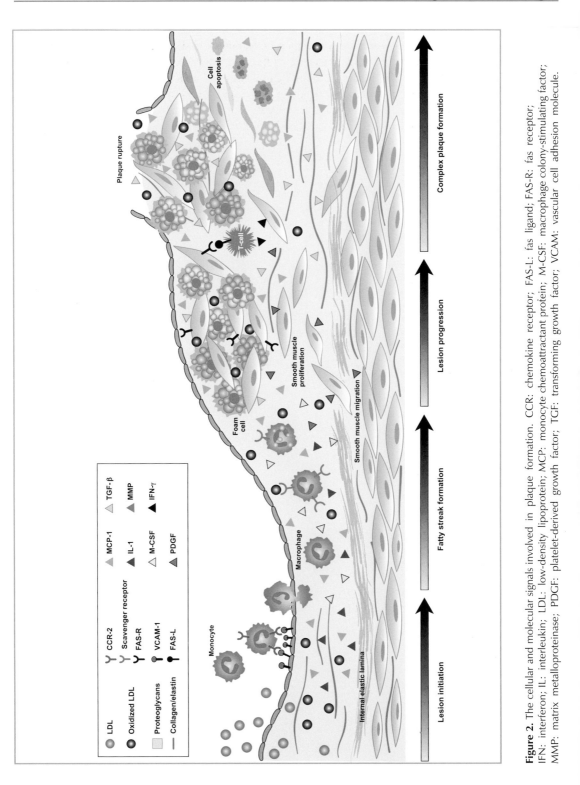

Figure 2. The cellular and molecular signals involved in plaque formation. CCR: chemokine receptor; FAS-L: fas ligand; FAS-R: fas receptor; IFN: interferon; IL: interleukin; LDL: low-density lipoprotein; MCP: monocyte chemoattractant protein; M-CSF: macrophage colony-stimulating factor; MMP: matrix metalloproteinase; PDGF: platelet-derived growth factor; TGF: transforming growth factor; VCAM: vascular cell adhesion molecule.

Youssef Al-Saghir and Ramtin Agah

29

Chapter 4

Evaluation of coronary lesions of intermediate severity

Khaled M Ziada and Deepak L Bhatt

Questions

1. A 59-year-old male patient presents with atypical chest pain 6 months after an anterior myocardial infarction (MI) and subsequent stenting of the mid left anterior descending (LAD) artery. Nuclear stress testing is equivocal regarding anterior ischemia. An angiogram of the LAD is shown in Figure 1. The stented segment (arrows) is not significantly obstructed, but there is a moderately severe stenosis proximal to the stent. Which of the following statements is *true*?

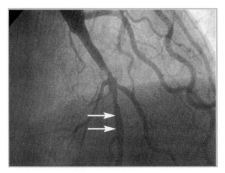

Figure 1. Angiogram of the LAD. The stented segment is not significantly obstructed (arrows).

   a. A fractional flow reserve (FFR) of 0.85 in the LAD implies that percutaneous coronary intervention (PCI) is indicated
   b. Coronary flow velocity reserve (CVR) is the best technique with which to assess the hemodynamic significance of the lesion
   c. Following FFR measurement, intravascular ultrasound (IVUS) imaging is required for anatomic evaluation of the LAD lesion proximal to the stent
   d. IVUS may reveal calcification or moderate stenosis at the lesion site

2. All of the following statements about CVR are false, *except*:
   a. A higher velocity at maximum hyperemia indicates a less severe stenosis
   b. Blood pressure has no significant impact on the measurement of CVR
   c. Relative (r)CVR is not as accurate as CVR in assessing lesion severity
   d. The Doppler wire measurement has excellent reproducibility

3. Compared with CVR, all of the following statements about rCVR are true, *except*:
   a. It is more time consuming
   b. It is a mechanism by which Doppler velocities can be used in diabetic patients
   c. It has significantly better correlation with FFR measurement
   d. It is the technology of choice for lesion evaluation in three-vessel disease

4. The angiogram in Figure 2a was obtained in a patient with an equivocal stress test because of repeated office visits and complaints of chest pain. The right coronary artery (RCA) and LAD were free of significant disease. Due to the moderate degree of stenosis in the left circumflex artery (LCX), FFR was measured. Based on the findings shown in Figures 2b and 2c, which of the following is *true*?

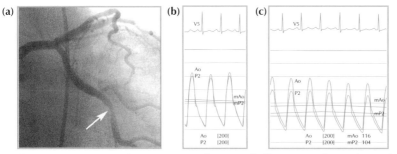

Figure 2. (a) Angiogram from a patient with repeated office visits and complaints of chest pain. Measurement of fractional flow reserve in the same patient (b) at baseline and (c) with hyperemia. mAo: mean aortic pressure; mP2: mean distal vessel pressure.

a. The patient will need a functional study with imaging to determine the significance of this lesion
b. Deferring PCI will not increase the risk of adverse events
c. Stenting may not reduce the risk of adverse events, but will reduce this patient's symptoms and use of antianginal medications
d. An additional imaging modality (eg, IVUS) is needed to decide whether stenting is required

5. A 74-year-old female presents with typical chest discomfort on heavy exertion. Work-up reveals previously undiagnosed hypertension and hyperlipidemia. She undergoes coronary angiography, which reveals a severe focal stenosis in the distal RCA. The lesion is amenable to PCI, but the angiogram of the left main trunk is suspicious. The worst view is shown in Figure 3. There is no "dampening" of the pressure tracing. What is the *most* appropriate next step?

Figure 3. Angiogram of the left main trunk.

a. End the diagnostic angiography procedure and consult with a cardiac surgeon
b. Obtain a ventriculogram to examine the location of the wall motion abnormality
c. Use IVUS imaging to better assess the left main trunk
d. Explain to the patient that a stress test will be needed before proceeding with PCI

6. A 70-year-old male presents with typical angina 3 years after coronary artery bypass grafting. His grafts to the LAD, posterior descending artery (PDA), and the first obtuse marginal (OM) branch of the LCX are patent. However, there is a tight lesion at the ostium of a second OM branch (see Figure 4a). The angiographic result following balloon angioplasty is shown in Figure 4b. The *most* evidence-based approach at this point would be:

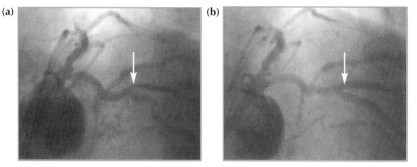

(a)　　　　　　　　　　　　(b)

Figure 4. Angiogram showing (a) a tight lesion at the ostium of an obtuse marginal branch (arrow). (b) The same lesion following balloon angioplasty.

a. End the procedure since the angiographic result appears excellent
b. Proceed with the interventional plan, ie, elective stenting
c. Perform quantitative coronary angiography (QCA) and measure CVR or FFR, then act accordingly
d. Perform QCA and use IVUS imaging, then act accordingly

7. A 59-year-old female presents with chest pain 8 months after undergoing stenting of the mid LAD in the setting of an acute anterolateral MI. Because her pain is "exactly the same" as her prior angina, she is taken directly to the catheterization laboratory. The RCA and LCX are free of significant obstructions. The LAD angiogram is shown in Figure 5, with arrows outlining the stented segment. What would be the *most* appropriate next step?

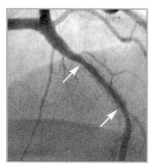

Figure 5. Angiogram of the LAD.
The arrows outline the stented segment.

   a. End the procedure, since the stenosis in the proximal stent is mild
   b. Cross the stented segment using a Doppler flow wire and measure CVR
   c. Image the stented segment using IVUS to rule out a more severe stenosis
   d. Cross the stented segment using a pressure wire and measure FFR
   e. Options "b" or "c"
   f. Options "c" or "d"

8. Which of the following is the *best* protocol by which to administer adenosine for the induction of coronary hyperemia during FFR measurement?
   a. Intracoronary injection of 24 µg
   b. Intravenous injection of 48 µg
   c. Continuous intravenous infusion of 24–48 µg/kg/min
   d. Continuous intravenous infusion of 140–180 µg/kg/min

9. Following stenting of a mid LAD lesion, an area of "haziness" is noted at the distal edge of the deployed stent (see Figure 6). The patient complains of chest discomfort, but there is TIMI 3 flow distally. Which of the following steps should be taken at this point?

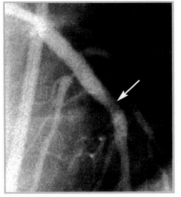

Figure 6. Angiogram following stenting of a mid LAD lesion. The arrow indicates "haziness" at the distal edge of the deployed stent.

a. End the procedure since there are no flow-limiting lesions
b. Measure the FFR and end the procedure if FFR equals 0.85
c. Image using IVUS to rule out an edge dissection
d. Measure the CVR and end the procedure if CVR equals 1.9

10. An elderly hypertensive patient presents 10 days after an episode of cough and breathlessness. Since that episode, he has been complaining of dyspnea on minimal effort. On admission, his serum troponin level was elevated, but creatine kinase MB (CKMB) was within the normal range. The right and circumflex coronary angiograms show mild disease. The LAD angiogram is shown in Figure 7. Which of the following statements is *true*?

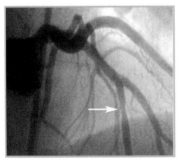

Figure 7. Angiogram of the LAD. The arrow indicates the lesion.

a. The lesion appears hazy, which means it is heavily calcified
b. IVUS imaging may alter the management of this patient
c. Measuring CVR will determine the significance of the lesion
d. If the diameter of the stenosis is at its worst in this projection, then there is no need for further evaluation

11. A patient is referred for coronary angiography prior to carotid endarterectomy. He complains of worsening shortness of breath, and his preoperative nuclear stress test reveals left ventricular cavity dilatation and multiple perfusion defects on stress. Figure 8 displays the observations made at the beginning of the procedure. What is the *most* appropriate next step?

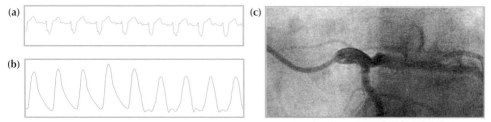

(a)

(b)

(c)

Figure 8. The first observations made at the beginning of coronary angiography prior to carotid endarterectomy. (**a**) Electrocardiogram; (**b**) central arterial pressure tracing; (**c**) left anterior oblique projection with cranial angulation.

a. Obtain the remaining standard angiographic views of the left coronary system
b. Obtain only one or two additional angiograms to define the distal vessels
c. Image the ambiguous lesion using IVUS
d. Measure FFR to define the hemodynamic significance of the lesion

12. A patient with multiple cardiovascular risk factors is referred for coronary angiography after a stress electrocardiogram test is interpreted as positive for ischemia. Figure 9a shows his LAD in an apical projection. His other vessels contain only mild irregularities. Because of the haziness in the mid vessel, an IVUS examination of the LAD is performed. Figures 9b and 9c show the IVUS findings. Which of the following conclusions is *true* regarding this patient?

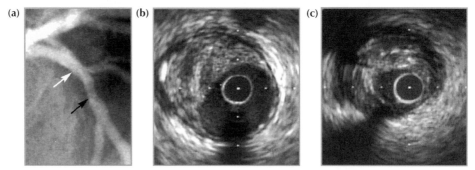

Figure 9. (a) The LAD in an apical projection. The black arrow indicates an area of "haziness" in the mid vessel. (b) Intravascular ultrasound (IVUS) at the site of the white arrow in Figure 9a; (c) IVUS at the site of the black arrow in Figure 9a. Reproduced with permission from Elsevier (Ziada KM, Kapadia SR, Tuzcu EM et al. The current status of intravascular ultrasound imaging. *Curr Probl Cardiol* 1999;24:541–66).

a. The IVUS images do not explain the discrepancy between the angiogram and the result of the stress test

b. The IVUS images explain why the stress test was a false positive

c. An FFR of 0.70 would confirm the findings of the IVUS images

d. In the distal LAD, a CVR value of 2.2 would be expected

13. A middle-aged, hypertensive female presents with chest pain precipitated by minimal effort. Coronary angiography reveals slight luminal irregularities in the left coronary system. Figure 10a is an angiogram of the RCA. On fluoroscopy, the RCA appears to be heavily calcified. Subsequently, a flow wire was advanced to the distal artery. The mean pressures of the aorta (mAo) and the distal vessel (mP2) were measured at baseline (see Figure 10b) and following injection of 36 µg adenosine (see Figure 10c). Based on these findings, which of the following statements is *true*?

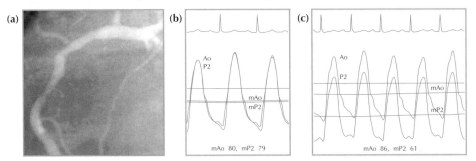

Figure 10. (a) Angiogram of the right coronary artery. The mean pressures of the aorta (mAo) and the distal vessel (mP2) (b) at baseline and (c) following injection of 36 µg adenosine.

    a. Deferring PCI on such lesions is not associated with an increased risk of events
    b. The lesion is not flow-limiting and is stable (since it is heavily calcified)
    c. Given the angiogram and the measured pressures, one would expect IVUS imaging of this lesion to reveal a minimum lumen area of 5 mm^2
    d. If mP2 increases to 72 mm Hg and mAo remains the same with adenosine injection after balloon angioplasty, the risk of restenosis will be high
    e. Measured FFR may be overestimated in calcified lesions

Answers

1. **d** Over the last decade, FFR measurement has become accepted as the most accurate technique for evaluating lesions of moderate severity. The technique is easy to perform and reproducible. The interpretation of the result is not affected by hemodynamic variables and the result is lesion specific – ie, it is not influenced by blood pressure, heart rate, or downstream microcirculation. An FFR value of <0.75 has been correlated with nuclear perfusion defects and the need for coronary revascularization [1,2].

   CVR is not a suitable modality when there is an infarct in the downstream distribution. The damaged microvasculature in the infarct distribution will result in decreased velocity, irrespective of the severity of the stenosis in the epicardial vessel. An alternate approach would be to compare the CVR in the LAD with the CVR in another normal or near-normal coronary artery – this is known as relative (r)CVR [3].

   IVUS imaging is helpful in quantifying lesion severity because it allows tomographic imaging and has a higher resolution than angiography. IVUS is also helpful in defining the lesion distribution, and in identifying calcification and possibly plaque rupture [4]. Several studies have compared IVUS findings with FFR cut-off values: a lumen area of <4 mm^2 correlates with an FFR of <0.75 [5,6]. Usually, only one modality is used in conjunction with angiography and the noninvasive tests that precede angiography to determine the functional severity of the lesion.

2. **a** See Answer 3 for a full explanation.

3. **d** CVR is a ratio of the maximum flow velocity that can be achieved in a coronary artery during maximum vasodilation and the flow velocity in the same artery at rest. An angioplasty wire with a Doppler transducer mounted at its tip can be used to measure flow velocity, which serves as a surrogate measurement for flow itself. Therefore, the Doppler wire is advanced beyond the lesion of interest and velocity is measured. A vasodilator agent is injected and the velocity is remeasured at maximum hyperemia. The ratio between the velocities expresses the degree to which the flow can increase. This is dependent on both the severity of the stenosis in the epicardial vessel and the adequacy of the microcirculation's response to the vasodilating agent. A coronary artery flow increase of <2-fold indicates an abnormal response and correlates with evidence of ischemia on nuclear scans [7]. This abnormality can be due to an epicardial flow-limiting lesion, but may also result from an impaired microcirculatory response. This is most commonly seen in patients with diabetes, severe ventricular hypertrophy, and/or in cases of MI with scarring of the downstream distribution. In addition, the measured velocities can vary significantly with changes in hemodynamic status (eg, blood pressure, heart rate, and low cardiac output states).

   Because most of the limitations of CVR are associated with diseases that affect the microcirculation, the concept of rCVR was devised. If the microcirculation is diffusely affected in multiple coronary beds (as is the case in left ventricular hypertrophy or diabetes), then CVR will be equally affected in all coronary arteries. In the presence of a moderate lesion, one can conclude that the difference in CVR between two coronary arteries in the same patient must be due to the epicardial component. For that to be established, CVR has to be measured at rest and with hyperemia in two coronaries: the one with the suspicious lesion and another with normal or near-normal angiographic appearance. Ideally, the ratio between CVR in two coronary beds should be 1.0, but >0.8 is accepted as normal. The lower that ratio, the more significant the epicardial coronary lesion under investigation. This ratio (rCVR) has a much better correlation with FFR than CVR measured in the suspected vessel alone [3]. Obviously, calculation of the rCVR requires twice the time needed to measure CVR. In addition, it cannot be used in a patient with moderate or severe three-vessel disease, since the denominator of the ratio always refers to the CVR in a normal or near-normal vessel.

4. **b** This is a typical case in which measurement of the FFR is of great clinical use. The patient's symptoms and noninvasive testing were not definitive. The angiogram revealed a moderately severe lesion in a large caliber vessel. At maximum hyperemia, the ratio of the mean aortic pressure (mAo) to the mean pressure distal to the lesion (mP2) is 0.89. The use of FFR in this scenario has been well validated. Studies by Pijls et al. have established that an FFR of >0.75 identifies epicardial lesions of less hemodynamic significance and a lower likelihood of causing symptoms [1,2]. In this case, deferring percutaneous revascularization would not be associated with any increase in adverse events. In addition, there has not been any increase in the frequency of chest pain or use of antianginal medications in patients with such lesions treated with medical therapy alone [8].

The addition of a functional study with imaging may have been an option before going to the catheterization laboratory as a method with which to overcome the limitations of the exercise electrocardiogram test. At this point, however, it will not add more information. Abnormal FFR measurements correlate well with evidence of perfusion defects on nuclear scanning, and *vice versa* [2]. Similarly, IVUS may reveal more anatomic information about the lesion severity and vessel size, but should not change the fact that the lesion was found to be hemodynamically insignificant, and that the risk of adverse events with medical therapy is quite low.

5. **c** Defining the severity of the ostial left main coronary artery, which is the problem in this patient, can be challenging. The difficulty arises from the lack of a proper "proximal" reference segment to which the stenosis can be compared. The filling of the aortic cusp is dependent on back flush of contrast around the catheter. An additional clue is the pressure waveform, which is "ventricularized" or shows "dampening" if the catheter is obstructing flow into the left main trunk. However, pressure dampening can be seen if the tip of the catheter is directed in such a way that it makes contact with the wall of the artery: a "false positive" pressure waveform.

IVUS imaging is very helpful in defining the severity of left main stenoses, as it accurately depicts the size of the lumen and the distribution of plaque *from within*. In such cases, the IVUS catheter is advanced over the guidewire and the guiding catheter is slightly withdrawn into the aortic cusp. This allows more accurate assessment of the true ostium.

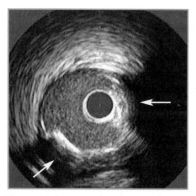

Figure 11. Intravascular ultrasound imaging of a patent left main trunk. Note the significant arc of calcification (arrows).

Although there have not been any rigorous studies to define the acceptable size of the left main trunk lumen, smaller patient series have suggested that a lumen area of <7 mm^2 is usually significant enough to warrant revascularization [4].

In this case, IVUS imaging revealed a patent left main trunk (see **Figure 11**) with a lumen area of >7 mm^2. There was a significant arc of calcification. The interventional cardiologist proceeded with angioplasty and stenting of the distal RCA.

It would have been premature to consult with a cardiac surgeon since the severity of the stenosis was not adequately assessed. A ventriculogram is not useful in defining the ischemic bed in the absence of documented old infarctions.

6. **c** Several potential effects may occur in the arterial segment subjected to balloon angioplasty. The vessel may stretch, with the atheroma remaining intact. This is the usual mechanism of immediate recoil, in which the lumen size appears to improve significantly, but recoils to its original size within minutes. If the recoil phenomenon is not recognized at the time of the procedure, this mechanism can explain at least some cases of early symptom recurrence following balloon angioplasty.

Another possibility is the creation of a small dissection or a "crack" in the fibrous cap of the atheroma, in addition to stretch and/or axial redistribution of atheroma material. This is the mechanism of a "successful" angioplasty, in which the balloon pressure is enough to irreversibly dilate the site of narrowing without causing deep arterial injury.

The third potential effect is the creation of more severe arterial wall damage in the form of longitudinal dissection and/or contained or open rupture. This is usually the result of balloon over-sizing or unusually high inflation pressures. Importantly, calcified lesions may be more susceptible to deep injury, even with reasonable balloon sizing and inflation. This is probably due to decreased compliance. Unrecognized deep arterial injuries can progress to acute arterial occlusions.

Angiographic evaluation of balloon angioplasty sites can be misleading. It is difficult to differentiate between the three possible scenarios following balloon angioplasty. In addition, mild stenoses are often underestimated when assessed visually. Although the angiogram may appear visually satisfactory, residual percentage stenosis is often found to be worse when assessed with QCA. Angiography alone is also not as accurate in determining the short- and long-term outcomes of balloon angioplasty. The combination of QCA and evaluation of flow reserve can be used to better evaluate the results of balloon angioplasty in the laboratory. The purpose is to identify angioplasty cases that will follow a "stent-like" course, ie, that are associated with fewer acute complications and lower restenosis. Using QCA in conjunction with the Doppler wire, restenosis was found to be significantly lower in patients in whom the residual diameter stenosis was <35% and the CVR >2.5 [9]. Similarly, a QCA residual stenosis of <35% and an FFR of >0.9 predict favorable outcomes [10].

In this case, assessment of the FFR may reveal an excellent hemodynamic result that predicts a restenosis rate similar to that of stenting. Although IVUS can provide an insight into the mechanism of action of balloon angioplasty and thus allow better decision making, there is no direct evidence linking IVUS findings to prognosis.

7. **f** Angiographic evaluation of in-stent restenotic segments can be difficult, especially with radio-opaque stents. In this case, the patient presented with typical angina and it would have been inappropriate to simply end the procedure because her stenosis appeared to be mild. An acceptable strategy would be to end the procedure and proceed with functional testing with an imaging modality to assess the evidence for peri-infarct ischemia in the anterior distribution. However, option "a" did not state that a stress test was planned. Since the patient had had an MI in the downstream distribution of the suspicious lesion, the use of CVR is not ideal. The finding of a lower CVR could be explained by the damaged microcirculatory bed in the infarction distribution, and would not necessarily be related to diminished epicardial flow. In this setting, FFR is a more lesion-specific test to determine the hemodynamic significance of the epicardial stenosis. An alternate approach would be to better define the anatomic severity of the stenosis using IVUS imaging. IVUS criteria that correlate with an abnormal FFR include a minimum lumen area of 3 mm^2 or percentage area stenosis >60% [5,6]. The presence of both criteria has a better correlation than each separately. IVUS imaging can usually provide more information about the precise length of the restenotic lesion as well as the mechanism of restenosis – ie, it can distinguish between adequate stent expansion with ensuing excessive intimal hyperplasia and suboptimal stent expansion [11].

In this case, the operator elected to use IVUS imaging. The image was obtained in the proximal segment of the stent (see **Figure 12a**). There is a significant discrepancy between the size of the vessel (see **Figure 12b**, white circle) and the size of the stent (marked by the echodense stent struts), indicating suboptimal expansion during the original procedure. In addition, there is evidence of significant intimal hyperplasia within the stent (gray area), leaving a very small lumen area around the IVUS catheter in the center.

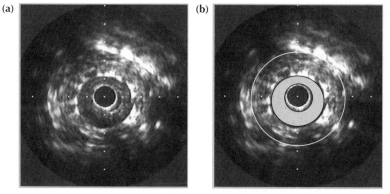

Figure 12. (a) Intravascular ultrasound imaging in the proximal segment of the stent. (b) The size of the vessel is indicated by the white circle, and the size of the stent is marked by the echodense stent struts. The gray area indicates significant intimal hyperplasia within the stent.

8. **d** When measuring the CVR and FFR, it is crucial to induce maximal hyperemia. Animal studies have demonstrated that the commonly used doses of intracoronary adenosine do not cause maximal hyperemia. This can result in an underestimation of CVR since flow velocity is directly proportional to the degree of hyperemia. Conversely, FFR will be overestimated since the gradient across the epicardial lesion is proportional to the degree of vasodilation. Ideally, adenosine should be administered intravenously via a continuous infusion and through a "central" vein. The standard dose for FFR measurement ranges between 140 and 180 µg/kg/min as a continuous infusion. This achieves both an adequate dose and a "steady-state" that results in a degree of hyperemia that cannot be reached by intracoronary injection, especially if a diffusely diseased vessel needs some time to interrogate. Obviously, inserting a central venous line involves more time, risk, and expense. In most cases, an intravenous infusion via an antecubital vein is adequate to achieve a steady-state of hyperemia. A commonly employed alternative is to use a larger dose of intracoronary adenosine, of up to 50 µg [12,13].

9. **c** It is not uncommon to identify a "peri-stent" hazy segment following stent deployment. Several explanations have been proposed: residual atheroma burden, calcified vessel wall, edge dissections, and/or thrombus formation. If flow limitation is associated with this angiographic finding, most operators would agree that it should not be left untreated. In most cases, an additional stent would be placed to cover that segment, making sure that the stent overlaps with the proximal stent and entirely covers the hazy segment.

    In the absence of flow limitation, such as in this case, it is not clear how significant these findings are. However, the risk of stent thrombosis and in-stent restenosis may be increased with the presence of edge tears and/or moderate lumen compromise. Therefore, in this case, it would be prudent to proceed with an additional evaluation technique to better define the etiology underlying the angiographic haziness and better understand the risk of possible complications.

    IVUS imaging would be the most appropriate technique with which to visualize the anatomical details of the vessel segment just distal to the edge of the stent [14], but FFR

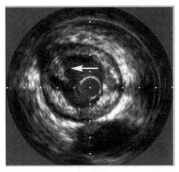

Figure 13. Intravascular ultrasound showing a linear dissection flap (arrow) on top of a moderate concentric stenosis.

and CVR have been used in this setting. An FFR of >0.95 and/or a CVR of >2.0 have been associated with a low risk of events. These values can provide the basis for deferring further intervention in such cases [15,16]. The values provided in statements "b" and "d" are not generally accepted as adequate after stenting. In fact, these values would prompt the operator to proceed with additional stenting. However, in the current era of drug-eluting stents, there is a tendency to use longer stents and cover all atheromata in the reference segments of the index lesions. This approach is buoyed by the fact that there is little penalty to pay in terms of restenosis with longer stents.

In this case, the operator elected to define the anatomical detail in the suspicious segment using IVUS (see **Figure 13**). A linear dissection flap was identified on top of a moderate concentric stenosis. Subsequently, an additional short stent was deployed, with resultant improvement of the angiographic appearance.

10. b The patient's history and elevated serum troponin level indicate that he has sustained an MI. Since his CKMB level was not elevated, this probably happened a few days previously, which is corroborated by the history starting 10 days earlier.

The angiographic haziness must be interpreted in that context. Haziness could be the result of calcification of the arterial wall, but that can usually be seen on fluoroscopy. In this setting, a more important differential diagnosis for haziness would be plaque rupture and/or overlying intraluminal thrombosis. Another possibility is an eccentric lesion that is more severe than is revealed by the angiogram in this projection. In this situation, IVUS can be very helpful in making a diagnosis.

IVUS of the mid LAD segment revealed a break in the fibrous cap of a relatively large atheroma, ie, plaque rupture (see **Figure 14**). The true lumen of the vessel (where the IVUS catheter is seen) is compromised (2×2 mm), explaining why the patient is continuing to have dyspnea on exertion, probably his anginal equivalent. There is no evidence of calcification. Flow reserve is not a suitable modality with which to assess the significance of a lesion in the setting of an MI. Accepting that the lesion is not angiographically significant is not appropriate, since the patient has both subjective and objective evidence of ischemia. Based on IVUS findings, the lesion was treated with direct stenting.

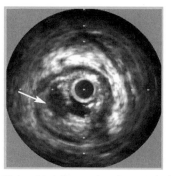

Figure 14. Intravascular ultrasound of the mid LAD segment. The arrow indicates plaque rupture.

11. b Adjunctive technologies such as IVUS imaging and pressure wires have had a significant impact on our ability to identify the significance of angiographically ambiguous stenoses. However, it is important to have a full understanding of the clinical context in which these technologies are utilized. In this case, it is clear from the clinical history (worsening shortness of breath, severe carotid disease) that there is a high likelihood of obstructive coronary disease. Moreover, the nuclear scan has two high-risk features (left ventricular dilatation with stress and perfusion defects in multiple segments). Therefore, one should anticipate the possibility of left main coronary stenosis and/or multivessel coronary disease, even prior to the beginning of cardiac catheterization. The aortic pressure tracing is "ventricularized" as soon as the catheter engages the left main trunk, suggesting an obstructive lesion at the tip of the catheter. Angiographically, there is no back flow of contrast into the aortic sinus, and there is ≈60% stenosis at the ostium. Based on the above indications, the diagnosis of obstructive ostial left main coronary disease can be made. Although FFR measurement and IVUS imaging do have a role in the diagnosis of indeterminate left main coronary stenoses [4,17], in this case there is enough clinical, angiographic, and hemodynamic evidence to make further testing unnecessary. In fact, in this situation, additional manipulation and unnecessary angiograms may increase the risk of complications. At this point, it is most prudent to refer the patient to cardiac surgery after defining the distal vessels using one or two angiograms at most.

12. c This is a case of a diffusely diseased LAD, in which the more severe segmental stenosis is superimposed on a moderate diffuse disease (see **Figures 9b** and **9c**). Coronary angiograms are traditionally interpreted in a segmental fashion, in which the least stenosed segment is assumed to be the "normal" reference to which other segments are compared. This logic falters if there is no true normal reference, as in this patient. Another problem with angiographic interpretation is the projection of usually complex lumen shapes within stenosed segments onto a flat surface. Angiographers compensate by obtaining orthogonal views, but the choice of those projections is still arbitrary. Therefore, it is conceivable that the projection perfectly perpendicular to the minimum lumen diameter may not be obtained [18].

In this case, IVUS imaging did demonstrate a severe stenosis in the mid vessel (see **Figure 9c**) with a minimum lumen diameter of <2 mm and an area of <3 mm^2. These measurements indicate hemodynamically significant stenosis, which is likely to cause ischemia on a stress test. This minimum lumen area by IVUS has a good correlation with an FFR of <0.75, which explains why the stress test was positive, despite the apparently "mild" narrowing on angiography [5,6]. It is unlikely that flow reserve would be normal in this patient, given the severity of the stenosis by IVUS and the positive result of the stress test.

13. d This patient has a moderately severe lesion in the mid RCA. Heavy calcification is probably responsible for the angiographic haziness of the diseased segment. The measured FFR across the mid vessel lesion was 0.71, indicating a hemodynamically significant (or flow-limiting) stenosis. The cut-off value of 0.75 usually identifies lesions with perfusion defects on nuclear scans [1,2]. There is no reason to believe that the FFR measured across calcified lesions is any less accurate than that measured in other settings. Comparative studies with IVUS imaging have shown that an FFR of <0.75 correlates with a minimum lumen area of 3 mm^2 and/or a percentage area stenosis of 60% [5,6]. Deferring revascularization for those lesions is associated with an increased risk of adverse events over the ensuing 1–2 years [8].

Following PCI, the risk of adverse events and/or restenosis is significantly reduced if the residual stenosis is minimal by QCA and the flow characteristics are optimal. Residual stenosis is considered acceptable if it is <35% after percutaneous transluminal coronary angioplasty (PTCA) or <11% after stenting [9,10,19]. Measures used to define optimal flow in the post-PTCA setting include an FFR of >0.9 or a CVR of >2.5 [9,10]. Following stenting, an FFR of >0.9 is acceptable, although a value >0.95 predicts an even lower risk of adverse events [15]. A CVR value of >2.0 after stenting is considered optimal [16]. An alternate indicator of a good stenting result is an rCVR of >0.88 [19]. In this case, if mAo remains unchanged at 86 mm Hg and mP2 increases to 72 mm Hg, the FFR will still be 0.83. This measurement should not be accepted as an optimal angioplasty result; in this particular case, the operator proceeded with stenting to improve the FFR to 0.96.

This approach of provisional stenting – ie, stenting only when the results of balloon angioplasty are suboptimal, as defined by QCA and flow characteristics (CVR) – has been tested in two large trials: DEBATE (Doppler Endpoints Balloon Angioplasty Trial Europe) and DESTINI (Doppler Endpoint Stenting International Investigation) [20,21]. The rationale behind this approach is to limit the use of stents to those cases in which the outcome of PTCA alone is expected to become significantly worse over time. Although there are some differences between the conclusions of the two trials, the provisional stenting strategy has never gained momentum in the interventional community. Optimal flow characteristics are not achieved by PTCA in a significant proportion of patients and, moreover, there is still some advantage of elective stenting over optimal PTCA in terms of restenosis.

References

1. Pijls NH, Van Gelder B, Van der Voort P et al. Fractional flow reserve. A useful index to evaluate the influence of an epicardial coronary stenosis on myocardial blood flow. *Circulation* 1995;92:3183–93.
2. Pijls NH, De Bruyne B, Peels K et al. Measurement of fractional flow reserve to assess the functional severity of coronary-artery stenoses. *N Engl J Med* 1996;334:1703–8.
3. Baumgart D, Haude M, Goerge G et al. Improved assessment of coronary stenosis severity using the relative flow velocity reserve. *Circulation* 1998;98:40–6.
4. Ziada KM, Kapadia SR, Tuzcu EM et al. The current status of intravascular ultrasound imaging. *Curr Probl Cardiol* 1999;24:541–66.
5. Takagi A, Tsurumi Y, Ishii Y et al. Clinical potential of intravascular ultrasound for physiological assessment of coronary stenosis: relationship between quantitative ultrasound tomography and pressure-derived fractional flow reserve. *Circulation* 1999;100:250–5.
6. Briguori C, Anzuini A, Airoldi F et al. Intravascular ultrasound criteria for the assessment of the functional significance of intermediate coronary artery stenoses and comparison with fractional flow reserve. *Am J Cardiol* 2001;87:136–41.
7. Miller DD, Donohue TJ, Younis LT et al. Correlation of pharmacological 99mTc-sestamibi myocardial perfusion imaging with poststenotic coronary flow reserve in patients with angiographically intermediate coronary artery stenoses. *Circulation* 1994;89:2150–60.
8. Bech GJ, De Bruyne B, Pijls NH et al. Fractional flow reserve to determine the appropriateness of angioplasty in moderate coronary stenosis: a randomized trial. *Circulation* 2001;103:2928–34.
9. Serruys PW, di Mario C, Piek J et al. Prognostic value of intracoronary flow velocity and diameter stenosis in assessing the short- and long-term outcomes of coronary balloon angioplasty: the DEBATE Study (Doppler Endpoints Balloon Angioplasty Trial Europe). *Circulation* 1997;96:3369–77.
10. Bech GJ, Pijls NH, De Bruyne B et al. Usefulness of fractional flow reserve to predict clinical outcome after balloon angioplasty. *Circulation* 1999;99:883–8.
11. Smith SC Jr, Dove JT, Jacobs AK et al. ACC/AHA guidelines for percutaneous coronary intervention (revision of the 1993 PTCA guidelines). *J Am Coll Cardiol* 2001;37:2215–39.
12. Kern MJ, Deligonul U, Tatineni S et al. Intravenous adenosine: continuous administration and low dose bolus administration for determination of coronary vasodilator reserve in patients with and without coronary artery disease. *J Am Coll Cardiol* 1991;18:718–29.
13. De Bruyne B, Pijls NH, Barbato E et al. Intracoronary and intravenous adenosine 5′-triphosphate, adenosine, papaverine and contrast medium to assess fractional flow reserve in humans. *Circulation* 2003;107:1877–83.
14. Ziada KM, Tuzcu EM, De Franco AC et al. Intravascular ultrasound assessment of the prevalence and causes of angiographic "haziness" following high-pressure coronary stenting. *Am J Cardiol* 1997;80:116–21.
15. Pijls NH, Klauss V, Siebert U et al.; Fractional Flow Reserve (FFR) Post-Stent Registry Investigators. Coronary pressure measurement after stenting predicts adverse events at follow-up: a multicenter registry. *Circulation* 2002;105:2950–4.
16. Nishida T, Di Mario C, Kern MJ et al. Impact of final coronary flow velocity reserve on late outcome following stent implantation. *Eur Heart J* 2002;23:331–40.
17. Bech GJ, Droste H, Pijls NH et al. Value of fractional flow reserve in making decisions about bypass surgery for equivocal left main coronary artery disease. *Heart* 2001;86:547–52.
18. Topol EJ, Nissen SE. Our preoccupation with coronary luminology. The dissociation between clinical and angiographic findings in ischemic heart disease. *Circulation* 1995;92:2333–42.
19. Haude M, Baumgart D, Verna E et al. Intracoronary Doppler- and quantitative coronary angiography-derived predictors of major adverse cardiac events after stent implantation. *Circulation* 2001;103:1212–7.
20. Serruys PW, de Bruyne B, Carlier S et al. Randomized comparison of primary stenting and provisional balloon angioplasty guided by flow velocity measurement. Doppler Endpoints Balloon Angioplasty Trial Europe (DEBATE) II Study Group. *Circulation* 2000;102:2930–7.
21. Di Mario C, Moses JW, Anderson TJ et al. Randomized comparison of elective stent implantation and coronary balloon angioplasty guided by online quantitative angiography and intracoronary Doppler. DESTINI Study Group (Doppler Endpoint STenting INternational Investigation). *Circulation* 2000;102:2938–44.

Chapter 5

Percutaneous revascularization versus bypass surgery versus medical therapy

Cameron Haery and Stephen G Ellis

Questions

1. A 56-year-old male with a history of coronary artery bypass graft (CABG) surgery and a left ventricular ejection fraction (LVEF) of 30% presents with persistent stable angina that is refractory to therapy with aspirin, beta-blockers, angiotensin-converting enzyme inhibitors, and oral nitrates. He has a resting heart rate of 56 bpm and a blood pressure of 116/72 mm Hg. Which of the following statements is *false* regarding this patient?
    a. Randomized controlled trials have demonstrated repeat CABG to be the preferred revascularization strategy for improved survival at 3 years
    b. The strongest predictor of survival in this patient is depressed LVEF
    c. Randomized controlled trials have not demonstrated repeat revascularization to be more likely with percutaneous coronary intervention (PCI) than with re-do CABG in this patient
    d. Significant atherosclerosis of a noncoronary artery or a nonbypass graft vessel is a potential cause of this patient's angina

2. A 55-year-old male with no significant medical history presents with stable angina of 2 weeks' duration. He is not on any medication. His physical exam and LV function are normal. Cardiac catheterization is performed (see Figure 1). Which of the following statements is *true* regarding this patient's treatment options and prognosis?

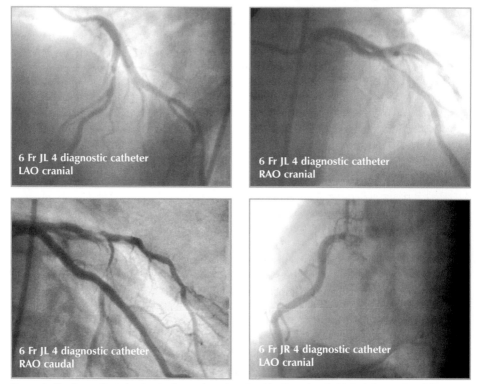

Figure 1. Cardiac catheterization in the patient described in Question 2. LAO: left anterior oblique; RAO: right anterior oblique.

a. Randomized controlled trials have demonstrated better 5-year survival with left internal mammary artery (LIMA) bypass graft surgery than with medical therapy alone

b. Freedom from myocardial infarction (MI) at 5 years is most likely with a strategy of LIMA bypass graft rather than either percutaneous transluminal coronary angioplasty (PTCA) or medical therapy alone

c. A sirolimus-eluting stent should not be considered at this time due to an increased risk of late thrombosis of the proximal left anterior descending (LAD) artery

d. Compared with PTCA without stenting, bare metal stenting offers increased event-free survival and reduced restenosis rates

e. If the patient refuses mechanical revascularization (PCI or CABG) and his symptoms are relieved after being placed on aspirin, nitrates, and beta-blockers, then no other evaluation or treatment is necessary at this time

3. Which of the following statements is *true* regarding sirolimus-eluting stents?

a. At 1 year, the RAVEL (Randomized Comparison of a Sirolimus-eluting Stent with a Standard Stent for Coronary Revascularization) trial demonstrated significantly reduced in-stent restenosis rates with sirolimus-eluting stents compared with bare metal stents. However, the reduction in cumulative event-free survival was not significant

b. Randomized controlled trials have demonstrated a significant reduction in target-vessel revascularization (TVR) at 1 year in *de novo* coronary lesions with a reference vessel diameter of <2.5 mm with sirolimus-eluting stents compared with bare metal stents

c. At 1 year, the SIRIUS (Sirolimus-eluting Stent in *De Novo* Native Coronary Lesions) trial demonstrated significant reductions in target-lesion revascularization for treatment of in-stent restenosis lesions with sirolimus-eluting stents compared with brachytherapy

d. In the RAVEL trial, there were no cases of stent thrombosis in the sirolimus-eluting stent arm at 1 year, despite aspirin and clopidogrel administration for only 2 months after the index procedure

e. The SIRIUS trial demonstrated significant reductions in target-lesion revascularization at 1 year in saphenous vein graft interventions with sirolimus-eluting stents compared with bare metal stents

4. Which of the following statements are *true* regarding PCI for CABG?

a. Distal anastomotic lesions of LIMA-to-LAD grafts are a common result of surgical technique and fibrosis. There is a high rate of procedural success and low restenosis rates when these are treated with balloon angioplasty alone, without stenting

b. In the SAVED (Saphenous Vein *De Novo*) trial, coronary stent deployment in saphenous vein bypass grafts resulted in a significantly reduced 6-month restenosis rate when compared with balloon angioplasty alone

c. In the SAVED trial, coronary stent deployment in saphenous vein bypass grafts resulted in a significantly lower adverse cardiac event rate at 6 months when compared with balloon angioplasty alone

d. In the SAFER (Saphenous Vein Graft Angioplasty Free of Emboli Randomized) trial, 30-day mortality was significantly reduced with stent deployment in saphenous vein grafts using the PercuSurge GuardWire compared with a conventional 0.014-inch guidewire

5. Which of the *following* treatment strategies has been proven in randomized controlled trials to prolong 3-year survival when added to a medical regimen of daily aspirin and statin?
    a. PTCA with adjunctive abciximab in a 50-year-old male with a type A mid-LAD stenosis, stable angina, and no other medical history
    b. CABG surgery (including use of a LIMA-to-LAD conduit to bypass a mid-LAD stenosis) in a 60-year-old male with two-vessel coronary artery disease (CAD) and a normal LVEF
    c. Clopidogrel and metoprolol in a 42-year-old normotensive female with chronic stable angina, isolated mid-LAD stenosis, and normal LV function who is not planning to undergo PCI
    d. None of the above

6. Which of the following statements is *true* regarding diabetes mellitus and coronary revascularization?
    a. In the BARI (Bypass Angioplasty Revascularization Investigation) trial, the survival advantage seen in diabetics undergoing CABG compared with PCI at 5 years after revascularization became nonsignificant at 7 years after randomization
    b. In the BARI trial, the survival advantage seen in diabetic patients undergoing CABG compared with PCI was independent of the type of graft conduits used
    c. The survival advantage seen in diabetics with multivessel CAD undergoing CABG when compared with PCI is largely due to greater limitation of infarct size and lower post-MI cardiac mortality after revascularization
    d. The survival advantage seen in diabetics with multivessel CAD undergoing CABG compared with PCI is largely due to a reduced rate of MI after revascularization
    e. The survival advantage seen in diabetics undergoing CABG compared with PCI is independent of the extent of disease (number of vessels involved) or whether complete revascularization can be achieved with PCI

7. Which of the following preoperative variables has the *least* power to predict in-hospital (short-term) mortality following CABG surgery?
    a. Age ≥65 years
    b. Prior CABG surgery
    c. Chronic obstructive pulmonary disease (COPD) (forced expiratory volume in 1 second / forced vital capacity <75% predicted)
    d. Child–Pugh class A liver cirrhosis
    e. Diabetes mellitus

8. A 62-year-old male presents with Canadian Cardiovascular Society (CCS) grade III angina. He is a smoker and hyperlipidemic, but has no other medical history of note. He is on the following medications: lopressor 50 mg PO bid, aspirin 325 mg PO qd, and simvastatin 40 mg PO qd. His physical examination is normal. Coronary angiography is performed and selected images are shown in Figure 2. LV function is normal. You feel that coronary stents can be successfully deployed in all three vessels. You consult with a cardiovascular surgeon, who believes complete revascularization can be achieved

using a LIMA-to-LAD graft and saphenous vein bypass grafts to the left circumflex and right coronary arteries. Which of the following statements is *true* regarding PCI with bare metal stent deployment versus CABG surgery in this patient?

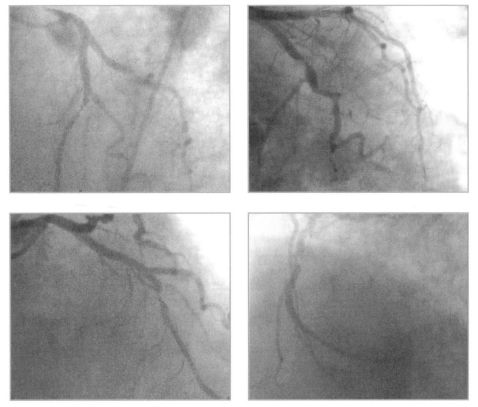

Figure 2. Coronary angiography from the patient is described in Question 8.

a. Survival over 1 year is better with CABG than with multivessel coronary stent deployment
b. Repeat revascularization within 3 years is less likely with CABG than with multivessel stenting
c. Randomized controlled trials have consistently demonstrated that cardiovascular event-free survival (freedom from death, MI, or angina requiring repeat revascularization) is more likely with CABG than with multivessel stenting
d. If this patient undergoes multivessel stenting, continued cigarette smoking is a major risk factor for in-stent restenosis

9. All of the following statements are true regarding the treatment of chronic stable angina due to single-vessel CAD, *except*:
   a. Initial treatment with aspirin, beta-blockers, and nitrates is an appropriate antianginal strategy before mechanical revascularization (PCI or CABG) is attempted
   b. PTCA with stenting is more likely to improve survival than medical therapy alone with aspirin and statin
   c. There is no significant difference in long-term mortality rates for patients undergoing PCI versus CABG
   d. PCI is likely to result in a longer exercise treadmill time and less angina than medical therapy alone with aspirin and statin

10. Which of the following statements are *true* regarding the management of chronic stable angina in the elderly (>75 years)?
    a. Randomized controlled trials have demonstrated improved 3-year survival in patients with multivessel disease who undergo bypass surgery compared with medical therapy alone
    b. For elderly patients with angina of CCS grade II or higher, PCI or surgical revascularization is more likely to reduce the incidence of major adverse cardiac events (death, nonfatal MI, hospitalization for acute coronary syndromes [ACS]) at 6 months than medical therapy alone
    c. For elderly patients with angina of CCS grade II or higher, PCI or surgical revascularization is more likely to improve quality of life and angina symptoms at 1 year than medical therapy alone
    d. The risk of periprocedural death after PCI is significantly higher in octogenarians than in younger patients

11. Which of the following statements are *true* regarding coronary artery bypass surgery?
    a. The risk of stroke after CABG increases incrementally with age
    b. Randomized controlled trials have demonstrated reduced rates of neurocognitive decline and stroke in patients undergoing off-pump coronary artery bypass (OPCAB) surgery compared with similarly matched patients undergoing cardiopulmonary bypass (CPB) surgery
    c. Early postoperative cognitive decline is a predictor of long-term neurocognitive impairment
    d. The risk of in-hospital mortality in patients who suffer a stroke after CABG is about 15%

Answers

1. **a** Randomized controlled trials designed to compare strategies of PCI versus CABG have largely excluded patients at high risk for bypass surgery, including those with prior bypass surgery, depressed LVEF, and medically refractory angina. The AWESOME (Angina with Extremely Serious Operative Mortality Evaluation) trial enrolled patients at high-risk of adverse outcomes with surgery, and with angina refractory to medical management, to a strategy of PCI or CABG [1–3]. The trial found 3-year survival to be comparable for PCI and CABG.

To date, randomized controlled trials have consistently demonstrated CABG to be more likely to reduce the need for repeat revascularization when compared with PCI alone. The era of drug-eluting stent implantation is likely to reduce PCI-related target-vessel revascularization (TVR), but head-to-head comparisons with CABG using TVR as a primary endpoint have not been performed. Whether contemporary interventional techniques or the introduction of antiproliferative drug-eluting stents will make PCI the more favorable approach for high-risk patients has yet to be demonstrated.

Any patient with a history of CABG and use of a left internal mammary artery (LIMA) or right internal mammary artery conduit and recurrent angina should be evaluated for the presence of subclavian artery stenosis. This may result in the so-called "subclavian coronary steal syndrome," which can usually be treated percutaneously with good angiographic and clinical results.

2. **d** The coronary angiogram reveals a discrete proximal LAD stenosis (proximal to the first diagonal and first major septal perforator). The reference vessel diameter is 3.2 mm (assessment aided by the 6F JL4 diagnostic catheter). The left circumflex and right coronary arteries are not significantly diseased.

To date, prospective randomized trials have not demonstrated a mortality benefit of mechanical revascularization (CABG or PCI) over medical therapy alone for patients with isolated proximal LAD stenosis and normal LV function. A meta-analysis of seven randomized trials comparing medical versus surgical management of patients with coronary artery disease (CAD) between 1972 and 1984 suggested that the subgroup of patients with isolated proximal LAD stenosis derives a 5-year survival benefit with CABG compared with medical management. However, this subgroup was relatively small in number, making definitive conclusions regarding survival benefit difficult. Also, this cohort probably did not receive universal antiplatelet and lipid-lowering therapies, which would be considered standard of care today.

More recently, MASS (Medicine, Angioplasty or Surgery Study) evaluated 214 patients with isolated proximal LAD disease and normal LV function who were randomized to CABG using a LIMA graft, PTCA (without stenting), or medical therapy [4]. At 5 years, CABG patients experienced a significant reduction in the primary composite endpoint of death, MI, and refractory angina requiring revascularization when compared with either medical therapy or PTCA. This was primarily due to a reduction in the need for subsequent revascularization – no significant difference was seen in mortality or infarction rates between groups. The cumulative survival rates at 5 years, however, were not significantly different amongst the three patient groups. Goy et al. reported on a similar group of 134 patients with isolated proximal LAD disease and normal LV function who were randomized to PTCA or bypass surgery using LIMA conduits [5]. The 5-year mortality rates were not significantly different.

The advantages of elective stenting over PTCA without stenting in patients with isolated proximal LAD disease have been demonstrated in subgroup analyses of the BENESTENT (Belgian Netherlands Stent Study), STRESS (Stent Restenosis Study) I, and BENESTENT II trials [6]. These advantages include increased event-free survival and reduced restenosis rates. The evaluation and management of patients with CAD must include risk factor modification, including dietary, exercise, and smoking cessation counseling (if relevant), as well as risk assessment, including fasting lipid profile and appropriate lipid-lowering therapy.

3.　d The RAVEL trial compared a sirolimus-eluting stent with a standard uncoated stent in single, primary lesions of native coronary arteries [7]. The overall rate of major cardiac events (death, MI, CABG, and target-lesion revascularization) was significantly reduced at 1 year with sirolimus-eluting stents (5.8% vs 28.8%, $P < 0.001$). This difference was entirely due to a reduction in target-lesion revascularization. There were no episodes of stent thrombosis in the sirolimus-eluting stent arm.

The results of drug-eluting stent trials are very encouraging. RAVEL primarily enrolled lower-risk patients, without complex angiographic anatomy. The SIRIUS trial, however, was designed to enroll a greater number of patients with higher-risk features for restenosis, including diabetes mellitus, small vessels, and long lesions [8]. The primary endpoint of target-vessel failure at 9 months was significantly reduced from 21% in the bare metal stent group to 8.6% in the sirolimus-eluting stent group ($P < 0.001$). Although subgroup analysis demonstrated a significant reduction in target-lesion revascularization rates in small vessels when sirolimus-eluting stents were compared with bare metal stents (7.3% vs 20.6%, $P < 0.001$), the study was not designed to demonstrate superiority in this subgroup, and definitive conclusions regarding the efficacy of sirolimus-eluting stents in small vessels should be reserved at this time.

The SIRIUS trial did not enroll patients with lesions in saphenous vein bypass grafts, nor did it compare the sirolimus-eluting stent with brachytherapy for treatment of in-stent restenosis. Whether the same impressive results will be borne out in higher-risk lesions – including small-caliber vessels, saphenous vein grafts, acute MI, and in-stent restenosis – is unconfirmed by prospective randomized trials to date.

4. a,c Although there are no randomized trials comparing restenosis rates of LIMA-to-LAD distal anastomosis lesions treated with PTCA alone versus stenting, numerous case series have reported high success rates and long-term patency when such lesions are treated with balloon angioplasty alone. The SAVED trial randomized 220 patients with *de novo* saphenous vein graft lesions to treatment with PTCA versus stenting [9]. Although the primary endpoint of restenosis at 6 months was not significantly reduced with stenting (37% vs 46% for PTCA, $P = 0.24$), cumulative event-free survival (death, MI, repeat bypass surgery, and target-lesion revascularization) was greater for patients receiving stents than for those who underwent PTCA (73% vs 58%, respectively, $P = 0.03$).

The SAFER trial randomized 801 patients undergoing stenting for saphenous vein bypass graft lesions to the PercuSurge GuardWire system versus a conventional 0.014-inch angioplasty guidewire. There was a significant 42% relative reduction in the primary endpoint of 30-day major adverse coronary events (composite of death, MI, emergent bypass surgery, or TVR) with use of the GuardWire system ($P = 0.004$). This improvement was driven by a reduction in MI and the "no-reflow" phenomenon. Although there was a trend towards reduced mortality at 30 days with the PercuSurge GuardWire, this trial was not powered to detect a mortality difference, and an independent survival benefit has not been shown.

5.   d One of the major limitations of the initial randomized controlled trials performed to compare strategies of medical therapy versus PCI or CABG is that many of them did not incorporate optimal contemporary treatment modalities. The major trials comparing PTCA and medical therapy include RITA (Randomized Intervention Treatment of Angina)-2, ACIP (Asymptomatic Cardiac Ischemia Pilot), ACME (Angioplasty Compared to Medicine), and AVERT (Atorvastatin Versus Revascularization Treatment). These trials have not demonstrated a survival advantage of PTCA over medical therapy for patients with stable angina and single-vessel disease. Particularly, results of the AVERT study indicate that revascularization does not provide a mortality benefit when compared with aggressive lipid-lowering therapy (atorvastatin 80 mg/day) in low-risk patients with CAD. Pooled data from the EPIC (Evaluation of c7E3 for Prevention of Ischemic Complications), EPILOG (Evaluation in PTCA to Improve Long-term Outcome with Abciximab GPIIb/IIIa Blockade), and EPISTENT (Evaluation of Platelet IIb/IIIa Inhibitor for Stenting Trial) studies have demonstrated a long-term mortality benefit of adjunctive abciximab use in patients undergoing PCI; however, these trials primarily enrolled clinically and angiographically high-risk patients [10,11]. A definitive long-term survival advantage for abciximab has not been demonstrated at 3 years in low-risk populations.

Early trials comparing medicine versus CABG did not universally treat with aspirin and statin in the medical therapy arms. Nonetheless, an independent survival advantage at 3 years has not been proven for patients with two-vessel CAD (without proximal LAD involvement) and normal LVEF when CABG has been added to aspirin and statin therapy.

Whether the combination of aspirin and clopidogrel confers additive benefits for patients without acute coronary syndromes (ACS), who are not undergoing PCI, yet are at risk for ischemic events, has not been adequately studied. Despite well-established benefits in specific patient populations, to date, no long-term mortality benefit has been proven in randomized controlled trials by adding clopidogrel or beta-blocker therapy to a regimen of aspirin and statin for stable patients *without* a history of MI, ACS, hypertension, or LV dysfunction.

6. c The survival advantage seen in diabetics undergoing CABG in the BARI study was significant beyond 7 years after revascularization (7-year survival: 55.7% for PTCA vs 76.4% for CABG, $P = 0.0011$) [12]. This benefit was primarily accounted for by patients assigned to CABG who received at least one internal mammary artery graft. In fact, the 7-year survival rate of diabetics undergoing multivessel PTCA was nearly identical to that of those undergoing CABG who received only saphenous vein bypass grafts. The survival advantage with bypass surgery seems to be associated with limited infarct size and improved post-MI survival rates, rather then a reduction in the frequency of MI.

    The judgment of the cardiologist and the cardiovascular surgeon is important in determining the appropriate revascularization strategy for diabetics with CAD. In the diabetic subgroup of the BARI registry selected to undergo PCI, 7-year survival was similar to that of CABG. Diabetic patients with less diffuse and less severe CAD may fare as well with contemporary PCI techniques as with CABG. Conclusions regarding optimal revascularization strategies in the era of contemporary PCI and CABG techniques should not be generalized.

    Pooled data from the EPIC, EPILOG, and EPISTENT trials show a reduction in 1-year mortality in diabetic patients undergoing PCI who are treated with abciximab versus placebo [13]. The risk of death in diabetic patients undergoing PCI and receiving abciximab was similar to that of nondiabetic patients receiving placebo.

    Based on existing data, PCI may be an effective alternative to CABG in diabetic patients with two- or three-vessel CAD where complete revascularization can be achieved. The future BARI 2 trial in diabetics (BARI 2D) is designed to address some of the questions regarding optimal medical and revascularization strategies for diabetics with CAD, in the modern era.

7. d To determine the most appropriate revascularization strategy for patients with CAD, evaluation of the short- and long-term risks of CABG surgery is critically important. Although numerous clinical and angiographic factors have been identified as influencing in-hospital mortality, Jones et al. reviewed seven large datasets of patients undergoing CABG between 1986 and 1994, and identified seven core variables with the greatest power to predict short-term mortality [14]. They are:

    *   patient age
    *   prior heart surgery
    *   female gender
    *   depressed LVEF
    *   urgency of operation (unstable angina with ischemia resolving with intra-aortic balloon pump or intravenous nitroglycerin, ongoing ischemia despite medical therapy, acute MI within 24 hours, or shock)
    *   number of significantly diseased vessels
    *   >50% stenosis of the left main coronary artery

Other variables were identified as having a modest influence on predicting short-term survival, including: recent MI; PTCA during the index admission; ventricular arrhythmia; congestive heart failure; mitral regurgitation; diabetes mellitus; peripheral vascular disease; cerebrovascular disease; COPD; and elevated creatinine.

Liver disease was not clearly shown to directly relate to short-term CABG mortality. There are limited prospective data regarding patients with liver disease undergoing bypass surgery; however, perioperative risk seems to correlate most with the severity of liver dysfunction, acuity of liver disease, and urgency of bypass surgery. Patients with Child–Pugh class A cirrhosis are generally well-compensated hemodynamically and CABG may be performed with acceptable risk as long as the patient is closely monitored during and after the procedure. Patients with more severe liver disease (Child–Pugh class B/C) are at higher risk and may be unable to tolerate the hemodynamic shifts associated with cardiopulmonary bypass. Other factors that influence morbidity in patients with hepatic dysfunction who are undergoing bypass surgery include coagulopathy, electrolyte imbalances, hepatic encephalopathy, variceal bleeding, pulmonary insufficiency, malnutrition, and special considerations for anesthesia.

8. b The figure reveals three-vessel CAD involving discrete lesions of the mid LAD, mid left circumflex, and mid right coronary artery. The best revascularization strategy for patients with multivessel CAD remains a topic of debate. To date, three major randomized trials have compared multivessel stenting with CABG. They are the ERACI (Argentine Randomized Trial of Percutaneous Transluminal Coronary Angioplasty versus Coronary Artery Bypass Surgery in Multivessel Disease)-II, ARTS (Arterial Revascularization Therapies Study), and SoS (Stent or Surgery) trials [15–17].

ERACI-II primarily enrolled patients with multivessel CAD and high-risk clinical features, including severe lifestyle-limiting angina, unstable angina, post-MI angina, or a large ischemic burden. Up to 3 years after enrollment, there was a significant survival advantage for multivessel stenting compared with CABG (96.9% vs 92.4%, respectively, $P = 0.017$). This survival advantage was demonstrated within the first 30 days; however, the Kaplan–Meier survival curves remained largely parallel beyond this point. The higher short-term mortality rate in the surgical arm has been attributed, in part, to the relatively high surgical risk population enrolled. This finding should be interpreted very cautiously when assessing the mortality advantage of PCI over CABG to lower-risk patients with multivessel CAD.

These results from ERACI are in contrast with those from the SoS trial. The SoS study demonstrated a survival advantage for CABG over PCI at a median of 2 years follow-up (mortality: 2% vs 5%, respectively, $P = 0.01$). These data should also be interpreted cautiously since survival was a secondary endpoint and the study was not designed or powered to assess mortality. In the ARTS trial, no significant survival difference was demonstrated at 1 year for either revascularization strategy.

Trials have consistently demonstrated a reduced need for repeat revascularization procedures with CABG compared with bare metal stenting for multivessel disease.

The composite primary endpoint of the ERACI-II trial was the occurrence of major adverse cardiac events, defined as death, Q-wave MI or stroke, and repeat revascularization. The authors reported a similar incidence of the composite endpoint of death, MI, or repeat revascularization at late clinical follow-up.

In the SoS trial, the primary endpoint was the rate of repeat revascularization. Although the rate of repeat revascularization was significantly less for patients undergoing CABG compared with multivessel stenting, the cumulative risk of death or MI was similar for the two groups. The composite endpoint of freedom from death, MI, or repeat revascularization was not reported in the SoS trial. In the ARTS trial, there was a significant reduction at 1 year in the composite endpoint of freedom from death, MI, cerebrovascular events, or repeat revascularization for patients who underwent CABG compared with multivessel stenting (74% vs 88%, respectively, $P < 0.001$).

Although the deleterious effects of smoking on cardiovascular health are well established, the specific impact of cigarette smoking on risk for in-stent restenosis has not been definitively demonstrated [18].

9. b PCI has not been shown to improve survival when compared with medical therapy in patients with single-vessel CAD and stable angina. The American College of Cardiology/American Heart Association 2002 guidelines for the management of patients with chronic stable angina and without high-risk features or large areas of myocardium at risk recommend an initial medical treatment strategy, before mechanical revascularization is attempted [19]. Aspirin, beta-blockers, and nitrates are all class I recommendations for the pharmacologic treatment of patients with chronic stable angina.

Early studies that compared medical therapy with PTCA did not demonstrate a mortality benefit of mechanical revascularization; however, reduced severity of angina and improved treadmill exercise times have been shown with PTCA compared with medical treatment alone. Aspirin and statin therapies have a proven mortality benefit in patients with coronary atherosclerosis. PCI has not been shown to provide any additive survival advantage to medical treatment with aspirin and statins for patients with single-vessel disease. Neither bypass surgery nor PCI has been shown to significantly impact long-term mortality in low-risk populations with single-vessel disease.

10.b,d As the elderly population continues to increase, so too does the prevalence of CAD in this population. Therefore, cardiovascular specialists will inevitably be faced with a growing number of patients with advanced cardiovascular disease and the attendant comorbid conditions that accompany this population.

Unfortunately, early randomized controlled trials comparing revascularization and medical treatment strategies excluded the very elderly, and outcomes data are limited. Although patients with three-vessel disease are more likely to derive a survival benefit with surgical revascularization than with medical therapy if they have "high-risk" characteristics (diabetes mellitus, LV dysfunction, a large territory of myocardium at risk, resting ST-segment depression), this advantage has been demonstrated predominantly in middle-aged cohorts, and not in the very elderly. The risk of perioperative morbidity and mortality for CABG is increased in the very elderly, and may mitigate survival advantages seen in earlier randomized trials of younger patients.

TIME (Trial of Invasive versus Medical Therapy in Elderly Patients with Chronic CAD) was a prospective, randomized, multicenter study comparing strategies of invasive revascularization (PCI or CABG) with optimal medical therapy in patients ≥75 years, with angina of CCS grade II or greater [20]. The primary endpoints were survival free of cardiac events, reduction in angina, and improvement in quality of life. At 6 months after randomization, the invasive management group had significantly fewer major adverse cardiac events than those undergoing medical treatment alone ($P < 0.0001$). Although the invasive revascularization group reported a greater improvement in quality of life scores than their medical treatment counterparts at 6 months, this outcome was not sustained at 1-year follow-up [21].

Despite advances in interventional cardiovascular techniques, tools, and adjunct pharmacotherapy, comorbidities in the very elderly represent a significant risk for periprocedural complications. A large-scale analysis comparing clinical characteristics and in-hospital outcomes of PCI in octogenarians (mean age: 83 years) versus patients younger than 80 years (mean age: 62 years) between 1994 and 1997 observed a 3-fold increased risk of periprocedural death in octogenarians (odds ratio: 3.6, 95% confidence interval: 3.2–4.1, $P < 0.001$) [22].

11. a,c The frequency of neurologic complications after CABG surgery increases with age. The incremental risk of stroke with age appears to be disproportionate to increases in other cardiac morbidities [23]. OPCAB may reduce the incidence of neurocognitive decline and stroke compared with CPB; however, randomized controlled trials have not consistently demonstrated this benefit. It appears that the greatest risk reduction may occur in patients at high risk for perioperative stroke. Stroke after CABG surgery can be devastating. The in-hospital mortality rate for patients who suffer a perioperative stroke is about 30%. Decline in cognitive function after coronary artery bypass surgery has been associated with an increased risk of long-term cognitive decline and a reduced level of overall cognitive functioning.

References

1. Morrison D, Sethi G, Sacks J et al. Percutaneous coronary intervention versus repeat bypass surgery for patients with medically refractory myocardial ischemia. AWESOME randomized trial and registry experience with post-CABG patients. J Am Coll Cardiol 2002;40:1951–4.
2. Morrison D, Sethi G, Sacks J et al. Percutaneous coronary intervention versus repeat bypass surgery for patients with medically refractory myocardial ischemia and risk factors for adverse outcomes with bypass: A multicenter randomized trial. J Am Coll Cardiol 2001;38:143–9.
3. Morrison D, Sethi G, Sacks J et al. Percutaneous coronary intervention versus repeat bypass surgery for patients with medically refractory myocardial ischemia and risk factors for adverse outcomes with bypass. The VA AWESOME multicenter registry: Comparison with the randomized trial. J Am Coll Cardiol 2002;39:266–73.
4. Hueb W, Bellotti G, Almeida et al. The medicine, angioplasty or surgery study (MASS): A prospective randomized trial of medical therapy, balloon angioplasty or bypass surgery for single proximal left anterior descending artery stenoses. J Am Coll Cardiol 1995;26:1600–5.

5.   Goy J, Eeckout E, Burnand B et al. Coronary angioplasty versus left internal mammary artery grafting for isolated proximal left anterior descending artery stenosis. *Lancet* 1994;343:1449–53.
6.   Serruys P, van Hout B, Bonnier H et al. for the BENESTENT study group. Randomized comparison of implantation of heparin coated stents with balloon angioplasty in selected patients with coronary artery disease (BENESTENT II). *Lancet* 1998;352:673–81.
7.   Morice M, Serruys P, Sousa E et al. A randomized comparison of a sirolimus-eluting stent with a standard stent for coronary revascularization. *N Engl J Med* 2002;346:1773–80.
8.   Moses JW. Transcatheter Cardiovascular Therapeutics (TCT) 2002 scientific session. Washington D.C. SIRIUS I: Clinical and Angiographic Outcomes. Available from URL: http://www.tctmd.com/expert-presentations/table-2.html?product_id=2922. Accessed on September 5, 2003.
9.   Savage M, Douglas J, Fischman D et al. Stent placement compared with balloon angioplasty for obstructed coronary bypass grafts. *N Engl J Med* 1997;337:740–7.
10. Topol E, Lincoff M, Kereiakes D et al. Multi-year follow-up of abciximab therapy in three randomized, placebo-controlled trials of percutaneous coronary revascularization. *Am J Med* 2002;113:1–6.
11. Kerieiakes D, Lincoff AM, Anderson K et al. Abciximab survival advantage following percutaneous coronary intervention is predicted by clinical risk profile. *Am J Cardiol* 2002;90:628–30.
12. The BARI Investigators. Seven-year outcome in the Bypass Angioplasty Revascularization Investigation (BARI) by treatment and diabetic status. *J Am Coll Cardiol* 2000;35:1122–9.
13. Bhatt DL, Marso SP, Lincoff AM et al. Abciximab reduces mortality in diabetics following percutaneous coronary intervention. *J Am Coll Cardiol* 2000;35:922–8.
14. Jones R, Hannan E, Hammermeister K et al. Identification of preoperative variables needed for risk adjustment of short-term mortality after coronary artery bypass graft surgery. *J Am Coll Cardiol* 1996;28:1478–87.
15. Rodriguez A, Bernardi V, Navia J et al. Argentine randomized study: Coronary angioplasty with stenting versus coronary bypass surgery in patients with multiple vessel disease (ERACI II): 30-day and one-year follow-up results. *J Am Coll Cardiol* 2001;37:51–8.
16. The Stent or Surgery (SoS) Trial Investigators. Coronary artery bypass surgery versus percutaneous coronary intervention with stent implantation in patients with multivessel coronary artery disease (the Stent or Surgery trial): a randomized controlled trial. *Lancet* 2002;360:965–70.
17. Abizaid A, Costa MA, Centemero M et al. Arterial Revascularization Therapy Study Group. Clinical and economic impact of diabetes mellitus on percutaneous and surgical treatment of multivessel coronary disease patients: insights from the Arterial Revascularization Therapy Study (ARTS) trial. *Circulation* 2001;104:533–8.
18. Hasdai D, Garratt K, Grill D et al. Effect of smoking status on the long-term outcome after successful percutaneous coronary revascularization. *N Engl J Med* 1997;336:755–61.
19. ACC/AHA Task Force Guidelines for Coronary Artery Bypass Graft Surgery. *J Am Coll Cardiol* 1999;34:1262–347.
20. Pfisterer M, Bertel O, Erne P et al. Trial of invasive versus medical therapy in elderly patients with chronic symptomatic coronary artery disease (TIME): a randomized trial. *Lancet* 2001;358:951–7.
21. Pfisterer M, Buser P, Osswald S et al. Outcome of elderly patients with chronic symptomatic coronary artery disease with an invasive vs. optimized medical treatment strategy. One-year results of the randomized TIME trial. *JAMA* 2003;289:1117–23.
22. Batchelor W, Anstrom K, Muhlbaier L et al. Contemporary outcome trends in the elderly undergoing percutaneous coronary interventions: results in 7,472 Octogenarians. *J Am Coll Cardiol* 2000;36:723–30.
23. Newman M, Kirchner J, Phillips-Bute B et al. Longitudinal assessment of neurocognitive function after coronary-artery bypass surgery. *N Engl J Med* 2001;344:395–402.

Chapter 6

Percutaneous coronary intervention in acute coronary syndromes

Jay S Varanasi, Shane Darrah, and Steven R Steinhubl

Questions

1. Early revascularization in patients with acute myocardial infarction (MI) and cardiogenic shock is associated with *which* of the following?
   a. Decreased 30-day mortality in patients aged >75 years
   b. No benefit; however, intra-aortic balloon bump (IABP) counterpulsation results in decreased 6-month mortality
   c. Decreased 6-month mortality
   d. Increased incidence of renal failure secondary to contrast nephropathy
   e. Higher mortality when the revascularization strategy is early coronary artery bypass graft (CABG) rather than percutaneous coronary intervention (PCI)

2. In the setting of acute MI, balloon angioplasty with stenting (rather than angioplasty alone) improves clinical outcomes through *which* of the following actions?
   a. Decreased mortality
   b. Decreased acute closure
   c. Decreased restenosis
   d. All of the above

3. In acute MI, administration of abciximab prior to catheterization and coronary stenting is associated with *which* of the following?
   a. Higher rates of preprocedural TIMI (Thrombolysis in Myocardial Infarction) grade 3 flow
   b. Higher rates of postprocedural TIMI grade 3 flow
   c. Increased left ventricular ejection fraction (LVEF) at 6 months
   d. A decreased need for urgent target-vessel revascularization (TVR) at 30 days
   e. All of the above

4. In patients hospitalized for suspected acute coronary syndromes (ACS), the *greatest* improvement in clinical outcomes after treatment with a glycoprotein (GP)IIb/IIIa inhibitor is seen in patients with which biological or clinical marker?
   a. Positive troponin levels
   b. Increased soluble CD40 ligand (>5.0 µg/L)
   c. Diabetes mellitus
   d. All of the above

5. In nearly 30,000 patients undergoing primary angioplasty for the treatment of ST-segment elevation MI (STEMI) in NRMI (National Registry of Myocardial Infarction)-2, a longer door-to-balloon time was associated with all of the following conditions/characteristics, *except*:
   a. Female gender
   b. Presence of diabetes
   c. Age >65 years
   d. Cardiogenic shock
   e. Contraindications to thrombolytics

6. Based on findings from the NRMI-2 dataset, patients presenting with an acute STEMI and treated with primary angioplasty did not experience a significant increase in in-hospital mortality until door-to-balloon times *exceeded*:
a. 60 minutes
b. 90 minutes
c. 120 minutes
d. 150 minutes
e. 180 minutes

7. According to the FRISC (Fast Revascularization during Instability in Coronary Artery Disease)-II study, the benefit of an early invasive strategy compared with a conservative approach in ACS patients *occurs*:
a. Immediately
b. Between 1 month and 1 year
c. After 1 year
d. From 1 month onwards

8. Angiographically, which of the following is the *strongest* predictor of long-term survival in patients treated with primary PCI in the setting of STEMI?
a. Preprocedure TIMI grade flow
b. Residual stenosis
c. Postprocedure TIMI grade flow
d. Preprocedure stenosis

9. Which of the following *best* describes the relationship between inflammatory changes involving the coronary vascular bed and ACS?
a. An ACS always involves the rupture of a solitary vulnerable plaque, which initiates a thrombotic cascade that can either occlude the vessel and/or shower emboli downstream
b. Coronary inflammation is present and specifically localized to the "infarct-related artery"
c. There is an inverse relationship between plasma levels of C-reactive protein (CRP) and complex lesions at angiography
d. All of the above
e. None of the above

10. What is the *most* effective method of evaluating the re-establishment of myocardial perfusion after PCI?
a. TIMI epicardial flow grade
b. Intravascular ultrasound
c. Myocardial perfusion grading
d. Coronary pressure wire
e. Both options "a" and "c"
f. Both options "a" and "d"

11. Which of the following statements is *most* accurate regarding the treatment of chronic coronary total occlusions (CTOs)?
   a. In randomized trials, PCI for CTOs has been shown to improve mortality rates at 1 year when compared with medical management
   b. Long-term follow-up of patients who underwent PCI for CTOs demonstrates a trend towards decreased major adverse coronary event (MACE) rates among patients treated with stents compared with percutaneous transluminal coronary angioplasty (PTCA)
   c. Final lumen diameter is the best predictor of reocclusion after PCI with stent placement in a CTO
   d. All of the above
   e. None of the above

12. In randomized trials comparing revascularization procedures (PCI with stent or CABG) in patients with multivessel coronary artery disease, which of the following would be *expected*?
   a. A higher percentage of complete revascularization in the CABG group
   b. No difference with regards to cost (including repeat revascularization) at 1 year
   c. No significant difference in the percentage of patients who reach a common endpoint of death, stroke, or nonfatal MI at 1 year
   d. All of the above

Answers

1. c The SHOCK (Should we Emergently Revascularize Occluded Coronaries in Cardiogenic
      Shock) investigators compared emergency revascularization with initial medical stabilization
      in 302 patients with evidence of ST-segment elevation MI (STEMI), left bundle branch
      block (LBBB) or Q-wave infarction, or cardiogenic shock secondary to left ventricular
      dysfunction [1]. Most patients had anterior infarctions, and the median time to onset of
      shock was 5.6 hours post-MI. IABP was used in 86% of patients from each of the early
      revascularization and medical treatment groups, thus allowing the effects of early
      revascularization to be determined.

      At 30 days post-MI, there was no statistically significant difference in mortality between
      patients treated with early revascularization and medical therapy (46.7% vs 56%,
      respectively, $P = 0.11$). At 6 months, however, mortality was significantly lower in the
      early revascularization group (50.3% vs 63.1%, $P = 0.027$). In the subgroup of patients
      aged <75 years, early revascularization significantly decreased mortality at both 30 days
      ($P = 0.02$) and 6 months ($P = 0.002$). Patients >75 years did not receive the same
      benefit from early intervention at these time intervals ($P = 0.16$, $P = 0.09$, respectively).

      In the early revascularization group, 64% of patients underwent angioplasty and 36%
      required bypass surgery. The mean time from randomization to angioplasty was 0.9 hours
      and the mean time to surgery was 2.7 hours. The mortality rate of patients undergoing
      angioplasty was 45.3%, while the mortality rate associated with CABG was 42.1%. Acute
      renal failure, defined by serum creatinine >3.0 mg/dL, occurred more frequently in
      patients assigned to intensive medical therapy than in those undergoing early
      revascularization (24% vs 13%, $P = 0.03$).

2. c Historical limitations of balloon angioplasty alone in the setting of STEMI are a reported
      10%–15% acute reocclusion rate and a 35%–40% restenosis rate. By tacking up intimal
      flaps or small dissections, and through the resultant larger luminal diameters, it was
      hoped that coronary stenting would address both of these limitations. Stent PAMI
      (Primary Angioplasty in Myocardial Infarction) was the largest trial to randomly assign
      patients having an acute MI to primary percutaneous transluminal coronary angioplasty
      (PTCA) alone or PTCA with stent implantation [2].

      Heparin-coated Palmaz-Schatz® (Cordis Corp., Miami Lakes, FL, USA) stents were used
      in 98% of patients undergoing stent implantation. The combined primary endpoint was
      the incidence of death, nonfatal MI, disabling stroke, or target-vessel revascularization (TVR)
      for ischemia during the 6 months following the intervention. Immediately after the
      procedure, stenting was associated with significantly larger lumens, significantly less
      residual stenosis, and significantly fewer coronary dissections. Surprisingly, there was
      no difference in the rate of subacute thrombosis: this occurred in 1.1% of patients
      in the angioplasty group and 0.9% of patients in the stent group ($P = 0.75$).

      At 6 months, routine angiography revealed significantly less restenosis (defined as a
      ≥50% stenosis) in the stent group (20.3% vs 33.5%, $P < 0.001$). The primary combined

endpoint occurred in 12.6% of stented patients and 20.1% of the PTCA-alone group
(P < 0.01). This difference was driven solely by a significantly lower incidence of TVR
for ischemia in the stent group (7.7% vs 17.0%, P < 0.001). There were no differences
in death, nonfatal MI, or disabling stroke between the two groups.

3. e The benefits of the upstream use of GPIIb/IIIa inhibitors are controversial. The only
data showing improved outcomes are from a trial by the ADMIRAL (Abciximab Before
Direct Angioplasty and Stenting in Myocardial Infarction regarding Acute and Long-term
Follow-up) investigators, which have not, as yet, been duplicated. The effects of
abciximab, administered prior to angiography and stent placement, were studied in
patients with STEMI [3]. A total of 300 patients were randomized to either abciximab
and stent or placebo and stent. The primary endpoint was comprised of death,
reinfarction, or urgent TVR at 30 days. Secondary endpoints included outcomes
at 6 months, TIMI grade flow rates, and LVEF at various intervals.

Patients who received abciximab instead of placebo had a lower incidence of the
primary endpoint (6.0% vs 14.6%, P = 0.01). At 6 months, the difference continued to
be significant (7.4% vs 15.9%, P = 0.02). This benefit was driven by decreased urgent
TVR in the abciximab-treated patients. Patients treated earlier with abciximab received
a greater benefit. In addition, there was a statistically insignificant mortality benefit seen
in abciximab-treated patients. The advantage these patients gained was thought to be
due to improved coronary blood flow. Preprocedural TIMI grade 3 flow rates were
more common in patients receiving abciximab (16.8% vs 5.4%, P = 0.01), as were
postprocedure TIMI grade 3 flow rates (95.1% vs 86.7%, P = 0.04). LVEF was also better
in the abciximab group at both 24 hours and 6 months after the procedure (P < 0.05,
P = 0.05, respectively).

A larger trial performed by the CADILLAC (Controlled Abciximab and Device Investigation
to Lower Late Angioplasty Complications) investigators randomized 2,082 patients into
four groups to compare angioplasty with stenting, with or without abciximab, in acute
MI [4]. This trial confirmed the results of the Stent PAMI trial, which found a decreased
incidence of restenosis associated with stents when compared with balloon angioplasty.
Stent PAMI also found higher postprocedure TIMI grade 3 flow rates after PTCA alone
compared with stenting, although this was contradicted by the larger CADILLAC trial
results, which showed similar postprocedure TIMI 3 grade flow rates for both procedures
(P = 0.16).

In the CADILLAC trial, abciximab reduced the rate of subacute thrombosis (P = 0.01),
but had no effect on TIMI flow rates, restenosis, or late cardiac events. In this study,
however, abciximab was given after catheterization had begun. It is possible that the
earlier administration of abciximab in the ADMIRAL trial allowed for earlier reperfusion
and thus better outcomes. The benefits of upstream abciximab, and other GPIIb/IIIa
inhibitors, will have to be studied in larger clinical trials before they become standard
practice in the treatment of acute MI. Such trials are ongoing.

4. **d** The CAPTURE (c7E3 Fab Antiplatelet Therapy in Unstable Refractory Angina) trial enrolled 1,265 patients with chest pain, electrocardiogram (ECG) changes, and an identifiable culprit lesion at initial diagnostic angiography [5]. Patients were randomized to receive either abciximab or placebo for 18–24 hours prior to intervention. Serum samples of 890 patients were available for troponin T analysis. In this subgroup, the measured outcomes comprising the reported (combined) event rate were death and nonfatal MI.

Troponin T was <0.1 ng/mL in 615 patients at randomization (307 patients receiving placebo and 308 receiving abciximab) [6]. In this population of troponin-negative patients, there was no significant difference in event rates between the abciximab and placebo groups before intervention (1.0% vs 0.7%, $P = 1.0$), 72 hours after randomization (4.5% vs 4.2%, $P = 1.0$), at 30 days (5.2% vs 4.9%, $P = 1.0$), or at 6 months (9.4% vs 7.5%, $P = 0.47$).

Troponin T levels were >0.1 ng/mL in 275 patients. In this cohort, there were significant differences in event rates between the abciximab and placebo groups before intervention (0.7% vs 6.6%, $P = 0.02$), at 72 hours after randomization (3.6% vs 17.4%, $P = 0.007$), at 30 days (5.8% vs 19.6%, $P = 0.001$), and at 6 months (9.5% vs 23.9%, $P = 0.002$). Therefore, a clear decrease in the combined endpoint was only shown in the troponin-positive patients. The benefit of GPIIb/IIIa antagonists in troponin-positive, but not troponin-negative, patients has been demonstrated in every placebo-controlled randomized trial of GPIIb/IIIa antagonists in ACS patients in which troponin levels were measured.

The CAPTURE trial also analyzed serum samples for soluble CD40 ligand, a proinflammatory marker of platelet activation that promotes coagulation [7]. Patients with a CD40 ligand level >5 μg/L had a significantly higher incidence of death or MI compared with those with lower levels prior to revascularization (4.1% vs 0.9%, $P = 0.02$), after 72 hours (13.1% vs 4.3%, $P < 0.001$), after 30 days (14.5% vs 5.3%, $P < 0.001$), and after 6 months (18.6% vs 7.1%, $P < 0.001$). The increased risk indicated by soluble CD40 ligand elevation persisted in troponin T-negative patients.

In patients with soluble CD40 ligand levels <5 μg/L, no significant difference was seen in patients treated with abciximab or placebo. Patients with higher levels of CD40 ligand who were treated with abciximab had a significantly lower rate of cardiac events prior to intervention (hazard ratio [HR] 0.12, $P = 0.01$), at 72 hours (HR 0.19, $P < 0.001$), and at 6 months (HR 0.37, $P = 0.001$) when compared with placebo. Thus, soluble CD40 ligand is another marker by which patients who will benefit from GPIIb/IIIa inhibitors can be identified, although there is currently no bedside test to measure this marker.

A meta-analysis of 6,458 diabetic patients enrolled in six different ACS trials (PRISM [Platelet Receptor Inhibition for Ischemic Syndrome Management], PRISM-PLUS [PRISM in Patients Limited by Unstable Signs and Symptoms], PARAGON [Platelet IIb/IIIa Antagonist for the Reduction of Acute Coronary Syndrome Events in a Global Organization Network] A and B, PURSUIT [Platelet Glycoprotein IIb/IIIa in Unstable Angina: Receptor Suppression using Integrilin Therapy], and GUSTO [Global Use of

Strategies to Open Coronary Arteries] IV) found significantly decreased 30-day mortality rates in diabetic patients treated with GPIIb/IIIa inhibitors compared with placebo (6.2% vs 4.6%, $P = 0.007$) [8].

The only individual trial to show a mortality benefit was GUSTO IV (odds ratio [OR] 0.62, $P = 0.022$), which used abciximab. Statistically significant decreases were seen in the composite endpoint of death and MI in the trials using tirofiban: PRISM (OR 0.50, $P = 0.038$) and PRISM-PLUS (OR 0.27, $P = 0.001$). These trials confirmed the benefits of GPIIb/IIIa inhibitors in diabetic patients with ACS.

5.   **d**   See Answer 6 for a full description of the NRMI-2 registry data.

6.   **c**   The NRMI-2 registry was an observational study that evaluated the relationship between in-hospital survival and time to treatment in patients undergoing angioplasty as their primary reperfusion therapy [9]. Overall, 27,080 patients with STEMI or LBBB on ECG were enrolled at 1,474 hospitals. In-hospital mortality was not significantly increased in patients whose door-to-balloon time ranged from 61 to 120 minutes when compared with the reference value of 0–60 minutes. Patients with a door-to-balloon time of 121–150 minutes had a multivariate-adjusted odds ratio in-hospital mortality of 1.41, which represents a significant increase in mortality ($P = 0.01$). Waiting times to angioplasty of >150 minutes were also associated with an increased risk of death. In this study, the median door-to-balloon time was 116 minutes.

A similar analysis was performed with data from GUSTO IIb, which compared primary angioplasty with thrombolytic therapy using tissue plasminogen activator in acute MI. Patients who underwent primary PTCA were divided into groups by trial enrollment-to-balloon times: <60 minutes, 61–75 minutes, 76–90 minutes, and >90 minutes [10]. A fifth group was comprised of patients who did not undergo PTCA, despite being randomized to that treatment arm. The highest 30-day mortality was seen in this fifth group of patients (14.1%). The lowest 30-day mortality was seen in those with an enrollment-to-balloon time <60 minutes (1.0%). For each subsequent 15-minute time interval, there was a 1.6-fold greater risk of death compared with the preceding time interval ($P = 0.008$). Symptom onset-to-balloon time did not correlate with mortality.

In the NRMI-2 study, the median time from symptom onset to presentation to hospital was 1.6 hours. However, adjusted mortality was not changed by symptom-to-balloon time; this may be because this value relies on patient recall, and is therefore not particularly accurate. Patients who died prior to presenting to the hospital were not included in the calculations, creating a survivor-cohort effect. The median symptom-to-balloon time was 3.9 hours.

Patients treated earlier were more likely to be younger (<61.5 years), male, have presented in cardiogenic shock, or had a history of previous angioplasty. Patients with a longer door-to-balloon time were more likely to be female, older (>61.5 years), diabetic, have a contraindication to thrombolytics, or have been transferred from another institution.

7. **d** The FRISC-II trial compared an early invasive strategy with a conservative strategy in patients with ischemic symptoms and ECG changes or elevated cardiac biomarkers [11]. Overall, there was a significant decrease in recurrent MI and death in the invasive group at 6 months (9.4% vs 12.1%, $P = 0.031$), as well as in the rates of angina and readmission to hospital. However, there was a higher combined rate of death and MI in the invasive group during the first 2 weeks after randomization. After 2 weeks, the event rate was higher in the noninvasive group, with the hazard curves crossing at about 4 weeks.

Two-year follow-up revealed a significantly lower mortality rate in patients treated with an early invasive strategy (3.7% vs 5.4%, $P = 0.038$). The reduced MI rate that was noted after 1 year was also sustained during the second year of follow-up [12].

The TACTICS (Treat Angina with Aggrastat and Determine Cost of Therapy with an Invasive or Conservative Strategy) TIMI 18 trial compared invasive and noninvasive strategies in patients with non-STEMI ($N = 2,220$) [13]. At 6 months, the primary endpoint of death, MI, and readmission occurred less frequently in the invasively treated group than in the conservatively managed patients (15.9% vs 19.4%, $P = 0.025$), confirming the FRISC-II results. However, in contrast to the data obtained from FRISC-II, no initial increased rate of cardiac events was seen in the invasive arm. This difference may stem from a shorter median time to PCI in TACTICS (22 hours vs 96 hours). The GPIIb/IIIa inhibitor tirofiban was used in all patients enrolled in TACTICS.

8. **a** An analysis of 2,507 STEMI patients enrolled in the four PAMI trials aimed to determine the impact of preprocedural TIMI 3 flow on outcomes [14]. At initial angiography, 375 patients (15.7%) were found to have TIMI 3 flow in the infarct-related artery because of spontaneous reperfusion. TIMI 2 flow was seen in 12.6%, and the remainder were occluded. In-hospital mortality was significantly lower in patients with preprocedural TIMI 3 flow compared with those with TIMI 2 and TIMI 0/1 flow (0.5%, 1.5%, and 2.6%, respectively, χ^2 P for trend $= 0.027$). Six-month mortality was also lower with a higher preprocedural TIMI flow (0.5%, 2.8%, and 4.4%, respectively, P for trend $= 0.009$).

Postprocedure TIMI flow was a predictor of 6-month survival. Mortality rates for patients with TIMI 3, TIMI 2, and TIMI 0/1 flow after PTCA were 2.2%, 6.1%, and 22.2%, respectively ($P < 0.0001$). Although preprocedure and postprocedure TIMI flow grades are predictive of survival, the risk ratio for the former is higher (1.9 vs 1.6, respectively) – ie, patients with a low preprocedure flow fare worse than those with a similarly low postprocedure flow.

PACT (Plasminogen-activator Angioplasty Compatibility Trial) examined low-dose fibrinolytic therapy prior to angioplasty in an attempt at earlier reperfusion compared with PTCA alone [15]. Patients ($N = 606$) were randomized to receive low-dose recombinant tissue-type plasminogen activator and PTCA, or placebo and PTCA. TIMI 3 flow was observed in 33% of patients receiving fibrinolytics versus 15% of those receiving placebo, a rate comparable with the spontaneous reperfusion rate seen in the PAMI analysis ($P < 0.001$). Echocardiograms performed 5–7 days postprocedure showed a significantly higher LVEF (62.4% vs 57.9%, $P = 0.04$) in patients who had achieved TIMI 3 flow at initial angiography.

Despite high success rates for primary angioplasty in attaining epicardial artery patency, its major disadvantage is the time delay that occurs before treatment. Fibrinolytics can be given quickly, but are less effective at achieving TIMI 3 flow. These studies confirm preprocedure TIMI 3 flow as a strong predictor of enhanced myocardial salvage and survival.

9. e A broad base of data now challenges the concept of a single plaque rupture as the cause of ACS. While option "a" describes a major component of ACS, and the primary etiology of the presenting signs and symptoms, we now understand that the process is not localized – instead, it involves a more systemic response.

A recent study examined the angiograms of 253 patients who presented with acute transmural MI [16]. Of these patients, 39% had multiple complex coronary plaques as opposed to a single complex plaque. During the year after MI, the patients with multiple complex plaques were more likely to present with recurrent ACS (19% vs 2.6%, $P < 0.001$).

In another study of 24 patients referred for PCI after a first ACS with troponin elevation, 72 vessels were evaluated with intravascular ultrasound [17]. Fifty plaque ruptures were diagnosed, based on finding a ruptured capsule with intraplaque activity. Plaque rupture on the culprit lesion was found in nine patients (37.5%). In 79% of patients, at least one plaque rupture was found at a location other than on the culprit lesion. These lesions were in an artery other than the culprit in 70.8% of cases and in both other arteries in 12.5%.

In a study of 228 patients who presented with primary unstable angina, plasma CRP levels were measured at admission [18]. Multivariate analysis showed plasma CRP to be independently associated with the presence of either multiple complex lesions (relative risk [RR] 1.8, 95% confidence interval [CI] 1.5–2.2, $P < 0.001$) or angiographically apparent thrombus-containing complex lesions (RR 1.4, 95% CI 1.2–1.7, $P = 0.03$).

A further trial, which confirmed the systemic nature of the inflammatory response associated with ACS, measured neutrophil myeloperoxidase content in the cardiac circulation of several groups of patients with ACS [19]. This marker is an index of inflammatory activation, with negative values representing depletion of the enzyme due to activation of neutrophils. The authors found that neutrophil activation was not confined to the vascular bed perfused by the vessel with the culprit coronary stenosis of patients presenting with unstable angina. Indeed, marker levels were similar in both the culprit and nonculprit artery beds. Marker levels were higher (suggesting less inflammation) in patients who did not present with ACS.

10. e Assessment of angiographic success after PCI should incorporate an evaluation of myocardial perfusion, as well as of epicardial flow.

TIMI epicardial flow grade has been the standard method for evaluating the establishment of reperfusion after thrombolysis and/or PCI in patients being treated for ACS. However, this grading system is not an accurate measure of myocardial perfusion at the microvascular level. Analysis of TIMI frame counts in patients who have had thrombolysis demonstrates that microvascular perfusion is compromised in many patients with TIMI 3 epicardial flow, as evidenced by longer TIMI frame counts [20].

The myocardial blush grade (MBG) is a simple and effective method of evaluating myocardial perfusion. MBG is an independent predictor of mortality in patients after primary angioplasty, irrespective of TIMI epicardial flow [21,22]. Additionally, infarct size and residual LVEF are associated with MBG [21,23].

Strategies are being studied to determine which interventions can improve myocardial perfusion after PCI. Currently, many investigators are evaluating distal protection devices and their impact on the prevention of distal embolization of vascular debris during PCI. Many of these devices are still under investigation, but a few have demonstrated safety and utility as an adjunct during PCI. Additionally, some devices have been associated with an improvement in markers of myocardial perfusion, including MBG and TIMI frame counts [24].

While intravascular ultrasound and coronary pressure wire are important tools with which to evaluate the epicardial artery, they are unable to evaluate the myocardial microvasculature.

11. b A 6-year follow-up of patients enrolled in the GISSOC (Gruppo Italiano di Studio sullo Stent nelle Occlusioni Coronariche) trial compared MACE events in patients treated with PTCA versus PCI and stent placement for CTO [25]. MACE events were defined as cardiac death, MI, TVR, and anginal status. At 6 years, freedom from MACE was found in 76.1% of the stent group compared with 60.4% of the PTCA group ($P = 0.0555$). While not statistically significant, this demonstrated a strong trend in favor of the stent group. The difference was driven mainly by TVR-free survival rates (85.1% vs 65.5% for the stent and PTCA groups, respectively, $P = 0.0165$). The SICCO (Stenting in Chronic Coronary Occlusion) trial also found a significant reduction in TVR with stenting of CTOs, which persisted through a mean of 33 months of follow-up.

The TOAST (Total Occlusion Angioplasty Study)-GISE (Italian Society for Invasive Cardiology) trial was a multicenter prospective observational study that followed 419 patients who underwent PCI of CTOs. In attempted PCI of a CTO, technical success (defined by restoration of TIMI flow grade 2 or 3 with residual stenosis <50% as assessed by quantitative coronary angiography) and procedural success (defined as technical success without in-hospital MACE) were obtained in 77.2% and 73.3% of lesions, respectively [26].

Another study evaluated 665 patients (716 coronary lesions) who had successful recanalization and stent implantation for CTO. A total of 375 patients underwent repeat angiography within 6 months of treatment. Restenosis was defined as ≥50% lumen narrowing and was observed in 151 (37.3%) lesions; 43 (10.6%) of those lesions had complete reocclusion. Multivariate analysis revealed that the only independent predictor of reocclusion was total stent length [27].

There are no good data comparing PCI of CTO with conservative management in a prospective fashion. However, ongoing trials are addressing this important question.

12. c ARTS (Arterial Revascularization Therapeutics Study) randomly assigned 1,205 patients to either PCI with stent placement or CABG. At 1 year, there was no difference among the two groups with regard to the primary endpoint of death, stroke, or MI; this was maintained at 2 years [28,29].

There was no significant difference between the groups with respect to completeness of revascularization. In the stenting group, a per patient mean of 2.8 ± 1.0 lesions with stenosis >50% of the luminal diameter was detected on diagnostic angiogram. A mean of 2.6 ± 1.1 lesions had stents placed (89%) or were treated with balloon angioplasty alone (11%). In the surgery group, a mean of 2.8 ± 1.0 lesions were found per patient, and a mean of 2.6 ± 1.0 anastomoses were performed with the use of 2.5 ± 0.7 conduits.

At 1 year, the net difference in cost favored angioplasty by $2,973 [28]. Of note, this trial did not involve the use of drug-eluting stents, which would impact the cost of the stent group and revascularization rates.

At 1 year, diabetic patients treated with stenting had a lower event-free survival rate (63.4%) than diabetic patients treated with CABG (84.4%, $P < 0.001$) and nondiabetic patients treated with stenting (76.2%, $P = 0.04$) [30]. This was mainly due to a higher incidence of revascularization in the diabetic patients who were treated with stents compared with the other groups. The 1-year event-free survival rates were similar among diabetic and nondiabetic patients treated with CABG (84.4% and 88.4%).

Unlike prior reports, ARTS found similar rates of death, stroke, or MI in patients with unstable versus stable angina who underwent multivessel percutaneous intervention (or multivessel CABG).

References

1. Hochman JS, Sleeper LA, Webb JG et al. Early revascularization in acute myocardial infarction complicated by cardiogenic shock. SHOCK Investigators. *N Engl J Med* 1999;341:625–34.
2. Grines CL, Cox DA, Stone GW et al. Coronary angioplasty with or without stent implantation for acute myocardial infarction. Stent Primary Angioplasty in Myocardial Infarction Study Group. *N Engl J Med* 1999;341:1949–56.
3. Montalescot G, Barragan P, Wittenberg O et al.; the ADMIRAL Investigators. Platelet glycoprotein IIb/IIIa inhibition with coronary stenting for acute myocardial infarction. *N Engl J Med* 2001;344:1895–903.
4. Stone GW, Grines CL, Cox DA et al.; Controlled Abciximab and Device Investigation to Lower Late Angioplasty Complications (CADILLAC) Investigators. Comparison of angioplasty with stenting, with or without abciximab, in acute myocardial infarction. *N Engl J Med* 2002;346:957–66.
5. The CAPTURE investigators. Randomized placebo-controlled trial of abciximab before and during coronary intervention in refractory unstable angina: the CAPTURE study. *Lancet* 1997;349:1429–35.

6. Hamm CW, Heeschen C, Goldmann B et al. Benefit of abciximab in patients with refractory unstable angina in relation to serum troponin T levels. c7E3 Fab Antiplatelet Therapy in Unstable Refractory Angina (CAPTURE) Study Investigators. *N Engl J Med* 1999;340:1623–9.

7. Heeschen C, Dimmeler S, Hamm CW et al.; CAPTURE Study Investigators. Soluble CD40 ligand in acute coronary syndromes. *N Engl J Med* 2003;348:1104–11.

8. Roffi M, Chew DP, Mukherjee D et al. Platelet glycoprotein IIb/IIIa inhibitors reduce mortality in diabetic patients with non-ST-segment-elevation acute coronary syndromes. *Circulation* 2001;104:2767–71.

9. Cannon CP, Gibson CM, Lambrew CT et al. Relationship of symptom-onset-to-balloon time and door-to-balloon time with mortality in patients undergoing angioplasty for acute myocardial infarction. *JAMA* 2000;283:2941–7.

10. Berger PB, Ellis SG, Holmes DR Jr et al. Relationship between delay in performing direct coronary angioplasty and early clinical outcome in patients with acute myocardial infarction: results from the Global Use of Strategies to Open Occluded Arteries in Acute Coronary Syndromes (GUSTO-IIb) trial. *Circulation* 1999;100:14–20.

11. FRagmin and Fast Revascularisation during InStability in Coronary artery disease Investigators. Invasive compared with non-invasive treatment in unstable coronary-artery disease: FRISC II prospective randomised multicentre study. *Lancet* 1999;354:708–15.

12. Lagerqvist B, Husted S, Kontny F et al.; Fast Revascularization during InStability in Coronary artery disease-II Investigators. A long-term perspective on the protective effects of an early invasive strategy in unstable coronary artery disease: two year follow-up of the FRISC II invasive study. *J Am Coll Cardiol* 2002;40:1902–14.

13. Cannon CP, Weintraub WS, Demopoulos LA et al.; TACTICS (Treat Angina with Aggrastat and Determine Cost of Therapy with an Invasive or Conservative Strategy)–Thrombolysis in Myocardial Infarction 18 Investigators. Comparison of early invasive and conservative strategies in patients with unstable coronary syndromes treated with the glycoprotein IIb/IIIa inhibitor tirofiban. *N Engl J Med* 2001;344:1879–87.

14. Stone GW, Cox D, Garcia E et al. Normal flow (TIMI-3) before mechanical reperfusion therapy is an independent determinant of survival in acute myocardial infarction: analysis from the primary angioplasty in myocardial infarction trials. *Circulation* 2001;104:636–41.

15. Ross AM, Coyne KS, Reiner JS et al. A randomized trial comparing primary angioplasty with a strategy of short-acting thrombolysis and immediate planned rescue angioplasty in acute myocardial infarction: the PACT trial. PACT investigators. Plasminogen-activator Angioplasty Compatibility Trial. *J Am Coll Cardiol* 1999;34:1954–62.

16. Goldstein JA, Demetriou D, Grines CL et al. Multiple complex coronary plaques in patients with acute myocardial infarction. *N Engl J Med* 2000;343:915–22.

17. Rioufol G, Finet G, Ginon I et al. Multiple atherosclerotic plaque rupture in acute coronary syndrome: a three-vessel intravascular ultrasound study. *Circulation* 2002;106:804–8.

18. Zairis MN, Papadaki OA, Manousakis SJ et al. C-reactive protein and multiple complex coronary artery plaques in patients with primary unstable angina. *Atherosclerosis* 2002;164:355–9.

19. Buffon A, Biasucci LM, Liuzzo G et al. Widespread coronary inflammation in unstable angina. *N Engl J Med* 2002;347:5–12.

20. Gibson CM, Cannon CP, Daley WL et al. TIMI frame count: a quantitative method of assessing coronary artery flow. *Circulation* 1996;93:879–88.

21. Henriques JP, Zijlstra F, van't Hof AW et al. Angiographic assessment of reperfusion in acute myocardial infarction by myocardial blush grade. *Circulation* 2003;107:2115–9.

22. Stone GW, Peterson MA, Lansky AJ et al. Impact of normalized myocardial perfusion after successful angioplasty in acute myocardial infarction. *J Am Coll Cardiol* 2002;39:591–7.

23. Poli A, Fetiveau R, Vandoni P et al. Integrated analysis of myocardial blush and ST-segment elevation recovery after successful primary angioplasty: real-time grading of microvascular reperfusion and prediction of early and late recovery of left ventricular function. *Circulation* 2002;106:313–8.

24. Limbruno U, Micheli A, De Carlo M et al. Mechanical prevention of distal embolization during primary angioplasty: safety, feasibility and impact on myocardial perfusion. *Circulation* 2003;108:171–6.

25. Rubartelli P, Verna E, Niccoli L et al.; Gruppo Italiano di Studio sullo Stent nelle Occlusioni Coronariche Investigators. Coronary stent implantation is superior to balloon angioplasty for chronic total occlusions: six-year clinical follow-up of the GISSOC trial. *J Am Coll Cardiol* 2003;41:1488–92.

26. Olivari Z, Rubartelli P, Piscione F et al.; TOAST-GISE Investigators. Immediate results and one-year clinical outcome after percutaneous coronary interventions in chronic total occlusions: data from a multicenter, prospective, observational study (TOAST-GISE). *J Am Coll Cardiol* 2003;41:1672–8.

27. Sallam M, Spanos V, Briguori C et al. Predictors of re-occlusion after successful recanalization of chronic total occlusion. *J Invasive Cardiol* 2001;13:511–5.

28. Serruys PW, Unger F, Sousa JE et al. Comparison of coronary-artery bypass surgery and stenting for the treatment of multivessel disease. *N Engl J Med* 2001;344:1118–24.

29. Unger F, Serruys PW, Yacoub MH et al. Revascularization in multivessel disease: comparison between two-year outcomes of coronary bypass surgery and stenting. *J Thorac Cardiovasc Surg* 2003;125;809–20.

30. Abizaid A, Costa MA, Centemero M et al. Clinical and economic impact of diabetes mellitus on percutaneous and surgical treatment of multivessel coronary artery disease patients: insights from the Arterial Revascularization Therapy Study (ARTS) trial. *Circulation* 2001;104:533–8.

Chapter 7
Equipment selection

Roberto A Corpus and Steven P Marso

Introduction

Advances in the field of interventional cardiology have been accompanied by a veritable explosion in intravascular technology. Technological breakthroughs – such as atherectomy, lasers, intracoronary brachytherapy, distal protection devices, and drug-eluting stents – have transformed the technique of coronary angioplasty from a much maligned and dangerous undertaking into a commonly performed, exceptionally effective, and exceedingly safe procedure. Despite such advancements, a fundamental understanding of basic angioplasty equipment is essential to successful percutaneous coronary intervention.

The basic angioplasty setup is composed of:

- a sheath for vascular access
- a guiding catheter to provide access and support for device delivery into the coronary artery
- a guidewire, which is navigated through the lesion(s) of interest and over which the device is delivered
- the angioplasty balloon, stent, or other device

The purpose of this chapter is to discuss the fundamental aspects and appropriate selection of intravascular sheaths, guiding catheters, and guidewires. Coronary interventional devices are discussed in Chapter 8.

Questions

1. Which of the following statements about sheaths is *false*?
   a. Small diameter sheath sizes minimize arterial trauma and bleeding risk
   b. When iliac tortuosity is encountered, longer sheaths are useful
   c. For radial artery cannulation, hydrophilic sheaths should be considered
   d. For radial artery sheaths, shorter sheaths are preferred over longer sheaths

2. Which of the following statements about guiding catheters is *false*?
   a. Diagnostic catheters are much stiffer than guiding catheters
   b. A guide catheter with side holes may be useful in patients with ostial disease
   c. Guide support tends to be more important for left circumflex artery intervention than for left anterior descending (LAD) artery intervention
   d. For right coronary arteries with an inferior orientation, a short-tip Judkins right (JR) catheter may be utilized for engagement
   e. During a left internal mammary artery (LIMA) intervention, smaller guide catheters may minimize trauma to the ostium

3. Which of the following statements about guidewires is *false*?
   a. Single core wires generally provide greater torque response than dual core wires
   b. Compared with nitinol wire cores, stainless steel cores tend to resist kinking
   c. Stiffer wires offer greater torque control than more flexible wires
   d. The radio-opaque tip of a wire is designed to enhance visibility

Answers

1. **d** The vascular sheath allows the operator to gain access to the central circulation in order to facilitate the passage of coronary catheters while maintaining hemostasis at the puncture site and continued perfusion to the distal extremity. The prototypical vascular sheath is a thin-walled structure with adequate stiffness to allow passage over a tapered dilator and guidewire into the vascular space. Sheaths are typically constructed of Teflon or polyurethane in order to minimize thrombogenicity. A one-way valve inside the sheath provides hemostasis during the procedure.

    For transfemoral coronary interventions, typical sheath sizes range from 6 to 10 Fr, with larger sheath sizes up to 20 Fr available for special applications such as percutaneous cardiopulmonary bypass. With the advent of newer low-profile devices, catheter sizes and subsequently sheath sizes have begun to decrease. The size of the sheath selected is dependent on the size of the guiding catheter to be used. Generally, in order to minimize the size of arterial puncture and attendant potential vascular complications, the smallest sheath size possible should be used [1].

    Sheath lengths range from 6 to 23 cm, with the majority of sheaths being 10 cm in length. In order to minimize the amount of material in the vascular space and maintain distal perfusion, routine use of shorter sheaths is recommended. However, in patients with tortuous iliac and femoral vessels, longer sheaths may facilitate improved guiding catheter support and manipulation [2].

    For interventions via the radial artery, hydrophilic sheaths, typically 5–6 Fr, are recommended. Longer sheaths (23 cm) are preferred over shorter sheaths (6–10 cm) to prevent radial artery spasm and facilitate unimpeded passage of the guiding catheter [3].

2. **a** Guiding catheter selection is of paramount importance in successful percutaneous coronary intervention. Important considerations in selecting the appropriate guiding catheter include:

    *   appropriate flexibility, torquability, and shape to facilitate catheter delivery for stable, coaxial, and atraumatic engagement of the target-vessel ostium
    *   sufficient internal diameter and lubricity to allow unimpeded passage of various devices
    *   ability to provide sufficient stiffness for back-up support during device delivery

    Compared with diagnostic catheters, guiding catheters are stiffer and have a larger internal diameter, reinforced construction, and a shorter, more angulated tip design [2]. Guiding catheters are constructed of a polyurethane or polyethylene outer layer to provide stiffness, a wire matrix for transmission and control of torque, and an inner layer composed of a thromboresistant Teflon coating to facilitate the passage of devices.

    The operator must choose a guiding catheter that will provide reliable pressure measurements and adequate contrast delivery. While larger diameter guides generally provide superior support, vessel opacification, and pressure monitoring compared with smaller diameter guides, this is at an increased risk of ostial trauma and vascular complications [4]. In cases involving small caliber vessels or vessels with ostial disease, damping or ventricularization of the pressure waveform may be observed. In such a situation, consideration should be given to using a guide catheter with side holes to allow for passive entry of blood into the vessel upon engagement of the catheter. Disadvantages to this approach include escape of contrast through the side holes, resulting in poor visualization and weakening of the catheter tip, which may cause kinking and loss of guide support [2].

Left coronary artery intervention

The Judkins left (JL) 4 guide catheter will usually fit coaxially into the left main and provide adequate guide support in the majority of cases. However, changes in aortic root geometry may affect the ability of the catheter to satisfactorily intubate the left main ostium. In situations in which the aortic root is dilated, or if the aortic arch is wide or elongated, the standard JL 4 catheter may be directed upward ("roofed") or not reach the ostium at all. In these cases, guides with a longer secondary curve may be used, such as a JL 4.5 or JL 5. Similarly, if the aortic root is small or the arch narrow, a JL 3.5 or shorter catheter is required. Superiorly directed left main coronary arteries may be successfully cannulated with an Amplatz left (AL), JL 3.5, or El Gamal (ELG) catheter. Using standard JL 4 catheters in short left main coronary arteries will usually result in preferential selection of the guide tip into the circumflex coronary artery. In such a situation, selection of a JL 3.5 or shorter catheter will usually allow direction into the LAD. Similarly, in a left main artery of normal length, selection of a shorter tip JL catheter will usually result in preferential selection of the LAD. In patients with ostial lesions of the LAD, a short-tip JL catheter will keep the guide tip more coaxial to the ostium of the LAD, thereby facilitating easier access to lesions in this location [1].

Due to the inherent tortuosity of the circumflex coronary artery and its marginal branches, guidewire placement and device delivery to lesions in these areas may be challenging. For selective intubation of the circumflex artery, Judkins-type catheters with longer secondary curves and shorter tips may be used. Other catheters that can be used to preferentially engage the circumflex artery include the Amplatz, hockey stick (HS), and multipurpose (MP) catheters [1].

When extra support is needed for left coronary artery interventions (ie, in cases of tortuous/angulated vessels, calcified lesions, total occlusions, or bulky devices), the AL, Voda left (VL), extra-backup (XB, EBU), and geometric left (GL) catheters are extremely useful. While the Judkins-type and Amplatz catheters rely on support from the contralateral sinus of Valsalva, the Voda, extra-backup, and GL catheters derive their added support from the contralateral aortic wall.

Right coronary artery intervention

For the majority of right coronary interventions, the JR 4 catheter will provide adequate ostial placement and guide support. For added support, Amplatz (AL or AR), HS, or Arani (DA 75/90) catheters may be employed. A fairly common variation of the right coronary artery origin is the so-called "shepherd's crook" orientation, in which the origin of the vessel is superiorly oriented. In such cases, use of an AL (0.5, 0.75, 1.0), shepherd's crook right (SCR), HS, ELG, or DA guide may be useful. For right coronary arteries with an inferior orientation, short-tip JR, HS, AR, right coronary bypass (RCB), or MP guide catheters can be used [1,2].

Bypass graft intervention

Saphenous vein bypass grafts (SVGs) subserving the right coronary artery territory typically arise anteriorly and several centimeters above the aortic root, while grafts to the left coronary artery system tend to arise superiorly and lateral to the right coronary artery grafts. JR, AL, AR, RCB, and MP guides are useful in cannulating right-sided grafts. Left-sided grafts are often successfully cannulated using a JR guide. However, owing to the variability in graft location and orientation, Amplatz, HS, left coronary bypass, ELG, and MP guides may be required [1,2]. Internal mammary artery (IMA) conduits may be successfully engaged with IMA catheters or a JR catheter. As the origin of the IMA may be friable and prone to dissection with excessive manipulation, smaller (6 Fr) guide catheters are recommended for interventions involving the IMA or its distal run-off vessels. **Table 1** depicts commonly used guiding catheter configurations and their intended applications.

Equipment selection

Table 1. Guiding catheter configurations and applications. DCA: directional coronary atherectomy; LAD: left anterior descending; LIMA: left internal mammary artery; RIMA: right internal mammary artery; SVG: saphenous vein graft. Adapted with permission from Physicians' Press (Safian RD, Freed MS, editors. *The Manual of Interventional Cardiology*).

Catheter (description)	Configuration	Sizes	Application
Left coronary			
JL (Judkins left)		3.0, 3.5, 4.0, 4.5, 5.0, 6.0	Standard left coronary artery interventions • shorter JL catheters (JL ≤3.5) to select LAD • longer JL catheters (JL ≥4.5) to select circumflex
SL (short left)		3.0, 3.5, 4.0, 4.5, 5.0, 6.0	Standard left coronary artery interventions Left main interventions Ostial LAD interventions (short tip keeps catheter coaxial to ostium of LAD)
FL (femoral left)		3.0, 3.5, 4.0, 4.5, 5.0, 6.0	Standard left coronary artery interventions
JCL (Judkins "C" left)		3.0, 3.5, 4.0, 4.5, 5.0, 6.0	Offers more gentle transition than standard JL catheters to enable passage of stiff and/or bulkier devices (large stents, DCA catheters, rotablator)
VL (Voda left)		3.0, 3.75, 4.0, 4.5, 5.0	Excellent backup support from contralateral aortic wall. Useful for tortuous or angulated vessels, total occlusions, and severely stenosed or calcified lesions in left coronary artery. Variant available for superiorly oriented left coronary: VHLT (Voda left high take-off)

82

Catheter (description)	Configuration	Sizes	Application
XB (extra-backup)		3.5, 3.75, 4.0, 4.5, 5.0	Excellent backup support from contralateral aortic wall. Useful for tortuous or angulated vessels, total occlusions, and severely stenosed or calcified lesions in left coronary artery
EBU (extra-backup)		3.5, 3.75, 4.0, 4.5, 5.0	Excellent backup support from contralateral aortic wall. Useful for tortuous or angulated vessels, total occlusions, and severely stenosed or calcified lesions in left coronary artery
GL (geometric left)		3.5, 3.75, 4.0, 4.5, 5.0	Excellent backup support from contralateral aortic wall. Useful for tortuous or angulated vessels, total occlusions, and severely stenosed or calcified lesions in left coronary artery
Right coronary			
JR (Judkins right)		3.0, 3.5, 4.0, 4.5, 5.0, 6.0	Standard right coronary artery interventions SVG interventions (particularly left coronary artery grafts)
FR (femoral right)		3.0, 3.5, 4.0, 4.5, 5.0, 6.0	Standard right coronary artery interventions SVG interventions (particularly left coronary artery grafts)

Catheter (description)	Configuration	Sizes	Application
SR (short right)		3.0, 3.5, 4.0, 4.5, 5.0, 6.0	Standard right coronary artery interventions SVG interventions
JCR (Judkins "C" right)		3.5, 4.0	Offers more gentle transition than standard JR catheters to enable passage of stiff and/or bulkier devices (large stents, DCA catheters, rotablator) into right coronary artery or left-sided SVGs with horizontal or inferiorly directed origins
NTR (no torque right)			Right coronary artery interventions. Does not require significant catheter manipulation
VR (Voda right)		3.5, 3.75, 4.0, 4.5, 5.0	Excellent backup support from contralateral aortic wall. Useful for tortuous or angulated vessels, total occlusions, and severely stenosed or calcified lesions in right coronary artery Up-sloping (shepherd's crook) right coronary artery Left-sided SVGs with superiorly oriented take-off
DA 75 (double-loop Arani [75°])			Backup support from contralateral aortic wall. Useful for tortuous or angulated vessels, total occlusions, and severely stenosed or calcified lesions in right coronary artery Up-sloping (shepherd's crook) right coronary artery

Catheter (description)	Configuration	Sizes	Application
DA 90 (double loop Arani [90°])			Backup support from contralateral aortic wall. Useful for tortuous or angulated vessels, total occlusions, and severely stenosed or calcified lesions in right coronary artery Up-sloping (shepherd's crook) right coronary artery
SCR (shepherd's crook right)		3.5, 4.0, 5.0	Up-sloping (shepherd's crook) right coronary artery Left-sided SVGs with upward take-off
Multipurpose			
HS (hockey stick)		I, II, III	Right coronary arteries with horizontal or slightly vertical origin Left-sided SVGs
MP (multipurpose)		B1, B2	Right coronary arteries with downward take-off Right-sided SVGs with inferior origin Left-sided SVGs with slightly inferior or horizontal origin
AL (Amplatz left)		0.75, 1.0, 1.5, 2.0, 2.5, 3.0, 4.0	Shepherd's crook or high-anteriorly directed right coronary artery Excellent backup support for tortuous or angulated vessels, total occlusions, and severely stenosed or calcified lesions in left coronary artery All SVGs

Catheter (description)	Configuration	Sizes	Application
AR (Amplatz right)		1.0, 2.0	Right coronary artery interventions Right-sided SVGs with inferior origins
ALR 1–2 (Amplatz left–right)			Modified Amplatz. Longer reach than AR and shorter reach than AL
ELG (El Gamal)			Horizontal or slightly superior right coronary artery origin Left-sided SVGs
Bypass grafts			
IMA (internal mammary artery)			Target lesions in LIMA, RIMA, or distal run-off vessels Right coronary artery with superior origin SVGs with superior origin
LCB (left coronary bypass)			Left-sided SVGs with horizontal to slightly superior origin
RCB (right coronary bypass)			Right-sided SVGs with slightly superior origin Left-sided SVGs with horizontal to slightly superior origin

3. b During percutaneous coronary intervention, a guidewire is passed into the target vessel, negotiated through the lesion(s) of interest, and ultimately placed in a distal position. Subsequently, various devices are delivered to the target area by being passed over the wire. To facilitate this process, coronary guidewires have been designed to maximize trackability, radiographic visibility, torque control, and device support in numerous situations.

Guidewires are composed of three components:

- a central core (usually stainless steel or nitinol) ground to a progressive distal taper
- a flexible, spring-like coil (usually stainless steel followed by radio-opaque material at the distal-most tip) welded to the tapered core either directly or via a shaping ribbon to allow the operator to shape the distal wire
- a hydrophilic coating (silicone, PTFE, or other material) to facilitate passage of the wire through tortuous vessels or severe stenoses [2,5]

The core of the guidewire can be classified as either single core (composed of a single material) or dual core (composed of two different materials). Wires of single core construction generally provide greater torque response and are less likely to prolapse in vessels with angulated or tortuous characteristics [2]. The tapering characteristics of the core influence the degree of torque control. A longer, more gradual taper leads to increased tip control. Furthermore, wires with a thicker inner core offer more support to enable passage of bulky and/or stiff devices. Compared with cores composed of stainless steel, nitinol cores tend to retain their shape and resist kinking.

In general, while flexible wires may inherently negotiate obstructions favorably, they tend to be less deliverable in terms of steerability and precise torque control. Stiffer wires offer greater torque control, the ability to straighten out curves in vessels, and more support for the advancement of various devices; they may, however, suffer due to their inability to negotiate through angulated or tortuous vessels. Wires come in either standard (180–190 cm) or exchange (300 cm) lengths. Several extension devices (DOC™, Linx™, Cinch®) are offered. All guidewires have either a short (2–3 cm) or long (25–40 cm) radio-opaque tip for enhanced radiographic visibility. Furthermore, several guidewires have marker bands along the distal tip to facilitate more precise measurement of lesion length.

When choosing a wire, the operator must be cognizant of the selected wire's properties and match them to the particular target vessel's characteristics. Wire specifications, characteristics, and applications are listed in **Table 2**.

Table 2. Coronary guidewires and applications. PCI: percutaneous coronary intervention; PTFE: polytetrafluoroethylene. Adapted with permission from Physicians' Press (Safian RD, Freed MS, editors. *The Manual of Interventional Cardiology*).

Name	Description	Length (cm)	Wire support	Tip flexibility	Tip shape	Radio-opaque segment (cm)	Application
Boston Scientific							
Trooper™ Floppy	Full spring coil, stainless steel, unibody core	185, 300	Light	Soft	Straight J tip	2	General use, simple to complex PCI
Trooper™ Extra Support	Full spring coil, stainless steel, unibody core	185, 300	Moderate	Soft	Straight J tip	2	More support for added pushability and device delivery
Trooper™ Intermediate	Full spring coil, stainless steel, unibody core	185, 300	Moderate–stiff	Soft	Straight J tip	2	More support for added pushability and device delivery
Trooper™ Standard	Full spring coil, stainless steel, unibody core	185, 300	Stiff	Moderate	Straight J tip	2	Total occlusions
Forte™ Floppy	Full spring coil, stainless steel, enhanced tip radio-opacity	185, 300	Light	Soft	Straight J tip	2	General use, simple to complex PCI
Forte™ Moderate Support	Full spring coil, stainless steel, enhanced tip radio-opacity, added support	185, 300	Moderate	Soft	Straight J tip	2	More support for added pushability and device delivery
Forte™ Floppy with Markers	Full spring coil, stainless steel, enhanced tip radio-opacity, marker bands	185, 300	Light	Soft	Straight J tip	2	Simple to complex PCI, lesion length assessment
Forte™ Moderate Support with Markers	Full spring coil, stainless steel, enhanced tip radio-opacity, marker bands	185, 300	Moderate	Soft	Straight J tip	2	More support for added pushability and device delivery, lesion length assessment
Patriot™	Full spring coil, unibody core	185, 300	Moderate	Soft	Straight J tip	2	Stent delivery
Luge™	Spring tip, hydrophilic-coated polymer sleeve, unibody core, moderate support	182, 300	Moderate	Soft	Straight J tip	3	General use, more support for added pushability and device delivery

Name	Description	Length (cm)	Wire support	Tip flexibility	Tip shape	Radio-opaque segment (cm)	Application
Boston Scientific							
CholCE® Floppy	Spring tip, hydrophilic-coated polymer sleeve, unibody core	182, 300	Light	Soft	Straight J tip	3	Simple to complex PCI
CholCE® Extra Support	Spring tip, hydrophilic-coated polymer sleeve, unibody core	182, 300	Moderate	Soft	Straight J tip	3	Extra rail support for stent delivery
CholCE® Intermediate	Spring tip, hydrophilic-coated polymer sleeve, unibody core	182, 300	Moderate–stiff	Soft	Straight J tip	3	More support for added pushability and device delivery
CholCE® Standard	Spring tip, hydrophilic-coated polymer sleeve, unibody core	182, 300	Stiff	Moderate	Straight J tip	3	Total occlusions
CholCE® PT	Polymer tip, hydrophilic-coated polymer sleeve	182, 300	Moderate	Soft	Straight J tip	35	Simple to complex PCI, crossing tortuous vessels and tight/complex lesions
PT Graphix™	Polymer tip, hydrophilic-coated polymer sleeve, enhanced support	182, 300	Moderate	Soft	Straight J tip	35	More support for device delivery, crossing tortuous vessels and tight/complex lesions
PT2™ Light Support	Polymer tip, hydrophilic-coated polymer sleeve, stainless steel shaping ribbon, nitinol distal core	185, 300	Light	Soft	Straight J tip	2	Enhanced tip shaping/retention, crossing tortuous vessels and tight/complex lesions
PT2™ Moderate Support	Polymer tip, hydrophilic-coated polymer sleeve, stainless steel shaping ribbon, nitinol distal core	185, 300	Moderate	Soft	Straight J tip	2	Enhanced tip shaping/retention, added support for crossing tortuous vessels and tight/complex lesions

Name	Description	Length (cm)	Wire support	Tip flexibility	Tip shape	Radio-opaque segment (cm)	Application
Boston Scientific							
Platinum Plus™	Full spring coil, highly supportive	185, 300	Stiff	Firm	Straight	5	Device delivery in tortuous vessels, vessel straightening
Mailman™	Spring tip, hydrophilic-coated polymer sleeve, unibody core, super support	182, 300	Stiff	Soft	Straight J tip	3	Device delivery in tortuous vessels, vessel straightening
Terumo/Boston Scientific							
Crosswire™	Platinum coil, hydrophilic-coated sleeve, nickel/titanium core	180	Stiff	Firm	Straight	5	Total occlusions
Glidewire® Gold	Gold tip, hydrophilic-coated polymer sleeve, nitinol core	180	Stiff	Moderate	45°, 75°	5, 8	Total occlusions, difficult to steer
Cordis/Johnson & Johnson							
ATW™ All Track Wire	Single corewire, broad distal transition, PTFE sleeve	195, 300	Moderate	Supersoft (floppy)	Straight J tip	3	Simple to complex PCI, stent delivery
ATW™ Marker Wire	ATW™ wire with markers	195, 300	Light	Straight	Straight J tip	3 (four markers, 10 mm apart, marker width 1 mm)	Simple to complex PCI, lesion length assessment
Wizdom® Supersoft	Very steerable and flexible single corewire, PTFE sleeve	180, 300	Light	Supersoft (floppy)	Straight J tip	3	Stent delivery, vessel straightening
Wizdom® Soft	Very steerable and flexible single corewire, PTFE sleeve, stiffer tip	180, 300	Light	Supersoft (floppy)	Straight J tip	3	Added pushability with stiffer tip

Name	Description	Length (cm)	Wire support	Tip flexibility	Tip shape	Radio-opaque segment (cm)	Application
Cordis/Johnson & Johnson							
Wizdom® ST	Single piece corewire, short distal transition, PTFE sleeve	180, 300	Light	Supersoft (floppy)	Straight J tip	3	Light support, simple PCI, stent delivery
Stabilizer® Balanced Performance Supersoft	Single piece corewire, broad distal transition, PTFE sleeve	180, 300	Moderate	Supersoft	Straight J tip	3	Simple to complex PCI, stent delivery
Stabilizer® Balanced Performance Soft	Single piece corewire, broad distal transition, PTFE sleeve, stiffer tip	180, 300	Moderate	Soft	Straight J tip	3	Simple to complex PCI, added pushability with stiffer tip
Stabilizer® Marker Wire	Stabilizer® Balanced Performance wire with markers	180, 300	Moderate	Supersoft, soft	Straight J tip	3 (six markers, 15 mm apart)	Simple and complex PCI, stent delivery, lesion length assessment
Stabilizer® Plus	Single piece corewire, PTFE sleeve, added support	180, 300	Stiff	Supersoft	Straight J tip	3 (six markers, 15 mm apart)	Added support for device delivery, vessel straightening
Stabilizer® XS	Single piece corewire, PTFE sleeve, extra support	180, 300	Stiff	Supersoft	Straight J tip	3 (six markers, 15 mm apart)	Extra support for device delivery, vessel straightening
Reflex® (0.012/0.014)	Silicone-coated, radio-opaque wire with 0.012" distal taper for flexibility	175, 300	Moderate	Supersoft	Straight J tip	25	Simple to complex PCI, tight lesions
Reflex® Supersoft (0.014)	Silicone-coated, radio-opaque wire	175, 300	Moderate	Supersoft	Straight J tip	25	Simple to complex PCI
Reflex® Soft (0.014)	Silicone-coated, radio-opaque wire with stiffer tip	175, 300	Moderate	Soft	Straight J tip	25	Simple to complex PCI, added pushability with stiffer tip

Name	Description	Length (cm)	Wire support	Tip flexibility	Tip shape	Radio-opaque segment (cm)	Application
Cordis/Johnson & Johnson							
Shinobi®	Single piece broad transition corewire, PTFE sleeve	180, 300	Stiff	Moderate	Straight	3	Uncrossable lesions
Shinobi® Plus	Single piece broad transition corewire, PTFE sleeve, extra support	180, 300	Stiff+	Moderate	Straight	3	Uncrossable lesions, extra support
Guidant							
Hi-Torque Balance Middleweight Universal™	Stainless steel shaft, nitinol distal core, hydrocoat hydrophilic tip coating	190, 300	Moderate	Soft	Straight J tip	3	Frontline use, simple to complex PCI
Hi-Torque Balance™	Long tapering nitinol core, hydrocoat hydrophilic or microglide tip coating	190, 300	Light	Soft	Straight J tip	3	Broad tip transition allows for enhanced trackability through angulated take-offs and tortuous vessels
Hi-Torque Balance Middleweight™	Stainless steel shaft, nitinol distal core, hydrocoat hydrophilic or microglide tip coating	190, 300	Moderate	Soft	Straight J tip	3	Frontline use, simple to complex PCI
Hi-Torque Balance Trek™	Stainless steel shaft, nitinol distal core, hydrocoat hydrophilic tip coating	190, 300	Moderate	Soft	Straight J tip	3	Frontline use, simple to complex PCI, slimmer profile allows easier distal vessel access
Hi-Torque Floppy II™	Stainless steel core, microglide-coated intermediate coils	190, 300	Light	Soft	Straight J tip	2, 30	Frontline use, simple to complex PCI
Hi-Torque Floppy II Extra Support™	Stainless steel core, microglide-coated intermediate coils	190, 300	Moderate	Soft	Straight	2, 30	Simple to complex PCI, added support

Name	Description	Length (cm)	Wire support	Tip flexibility	Tip shape	Radio-opaque segment (cm)	Application
Guidant							
Hi-Torque Traverse™	Long tapering stainless steel core, hydrocoat hydrophilic or microglide tip coating	190, 300	Light	Soft	Straight J tip	3, 30	Broad tip transition allows for enhanced trackability through angulated take-offs and tortuous vessels
Hi-Torque Whisper™	Transitionless profile, Durasteel™ core, hydrocoat hydrophilic tip coating	190, 300	Light	Moderate	Straight J tip	3	Transitionless tip profile allows for device delivery around acute angles
Hi-Torque Balance Heavyweight™	Nitinol core, hydrocoat hydrophilic tip coating	190, 300	Moderate–stiff	Moderate	Straight J tip	4.5	Added rail support for vessel straightening
Hi-Torque Extra Support™	Stainless steel core, microglide-coated intermediate section, flattened distal core	190, 300	Stiff	Moderate	Straight J tip	3	Added rail support for vessel straightening, high support
Hi-Torque All Star™	Stainless steel core, microglide tip coating, soft tip	190, 300	Stiff	Soft	Straight J tip	3	Added rail support for vessel straightening, tip transition enhances steerability, high support
Hi-Torque Iron Man™	0.014″ stainless steel core, microglide tip coating, flattened distal core	190, 300	Stiff+	Soft	Straight J tip	3	Maximum rail support for vessel straightening, highest support
Hi-Torque Cross-It (100, 200, 300) XT™ series	Increasing levels of torque transmission and tip support, hydrocoat hydrophilic tip coating	190, 300	Moderate	Moderate/firm to firm	Straight J tip	3	Tapered distal tip and gradual core taper allow access through tortuous vessels and severe stenoses
Hi-Torque Wiggle™	Stainless steel core, microglide tip coating	190, 300	Moderate	Moderate	Wiggle design	2, 30	Unique "wiggle" design of distal tip allows change in orientation of device tip to move in and around obstructions

Name	Description	Length (cm)	Wire support	Tip flexibility	Tip shape	Radio-opaque segment (cm)	Application
Medtronic AVE							
Fusion	Stainless steel hypotube, nitinol core	180, 300	Moderate	Soft	Straight J tip	30	General use, simple to complex PCI
Floppy	Stainless steel, double-barrel support zone, floppy tip allows for tip prolapse	180, 300	Light	Supersoft	Straight J tip	30	Simple PCI not requiring significant device support
Direct	Stainless steel, double-barrel support zone for added support	180, 300	Moderate	Soft	Straight J tip	30	Simple to complex PCI, added device support
Light Support	Stainless steel, modified core design for more support, atraumatic tip	180, 300	Moderate	Soft	Straight J tip	30	General use, simple to complex PCI
Support	Stainless steel, modified core design for more support, atraumatic tip	180, 300	Moderate	Moderate	Straight J tip	30	Extra support for device delivery
Standard	Stainless steel, unibody core, core wire extends to tip	180	Stiff	Firm	J tip	25	Maximum torquability and pushability, chronic total occlusions
Hi-per Flex	Double-barrel core design modified for maneuverability	180, 300	Moderate–stiff	Soft	Straight J tip	30	Enhanced maneuverability for tortuous vessels, more support for device delivery
Silk	Continuously tapering single core extends to tip	180, 300	Light	Supersoft	Straight J tip	2	Excellent maneuverability for tortuous vessels, PCI not requiring significant device support
Lumisilk	Transitionless core, dual round steering wires	180	Light	Soft	Straight J tip	2	Excellent maneuverability for tortuous vessels, PCI not requiring significant device support

References

1.   Tenaglia AN, Tcheng JE, Phillips HR. Coronary Angioplasty: Femoral Approach. In: Stack RS, Roubin GS, and O'Neill WW, editors. *Interventional Cardiovascular Medicine: Principles and Practice*. Philadelphia, PA: Churchill Livingstone, 2002:477–90.

2.   Safian RD, Freed MS. Coronary Intervention: Preparation, Equipment & Technique. In Safian RD, Freed MS, editors. *The Manual of Interventional Cardiology*. Royal Oak, MI: Physicians' Press, 2001:1–32.

3.   Almany SL, Yakubov SJ, George BS. Brachial and Radial Approach to Coronary Intervention. In Safian RD, Freed MS, editors. *The Manual of Interventional Cardiology*. Royal Oak, MI: Physicians' Press, 2001:33–46.

4.   Safian RD, Freed MS. Coronary Intervention: Preparation, Equipment & Technique. In Freed MD, Grines CL, Safian RD, editors. *The New Manual of Interventional Cardiology*. Royal Oak, MI: Physicians' Press, 2001:1–64.

5.   Baim DS. Coronary Angioplasty. In Baim DS, Grossman W, editors. *Grossman's Cardiac Catheterization, Angiography, and Intervention*. Philadelphia, PA: Lippincott Williams & Wilkins, 2000:211–56.

Chapter 8
Interventional devices

Aamer H Jamali and Rajendra R Makkar

Questions

1. A 38-year-old male with hypercholesterolemia, hypertension, and a history of
exertional chest pain undergoes coronary angiography. The right coronary artery (RCA)
and circumflex are normal. The left anterior descending (LAD) artery shows a lesion
(see Figure 1).

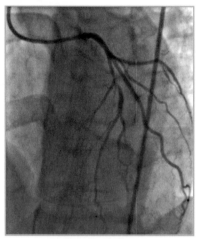

Figure 1. Angiogram showing a lesion in the LAD.

  i. *How* would you approach this lesion?
    a. Primary stenting
    b. Provisional stenting
    c. Rotational atherectomy
    d. Directional coronary atherectomy (DCA)

  ii. Would a drug-eluting stent be *preferable* in this situation?
    a. No, not advisable
    b. No, never studied
    c. Yes

  iii. If you decided to use a drug-eluting stent, are there any *other*
procedural considerations?
    a. Predilation would be recommended
    b. A shorter stent size is preferred
    c. A longer stent size is preferred
    d. Both options "a" and "b"
    e. Both options "a" and "c"

2. A 67-year-old male with a history of coronary artery bypass grafting (CABG) presents with a non-ST-elevation myocardial infarction (MI) and undergoes urgent coronary angiography. All grafts, except as noted below, appear to be open and the distal vessels do not have significant stenoses. There is a lesion on the saphenous vein graft (SVG) to the obtuse marginal (see Figure 2).

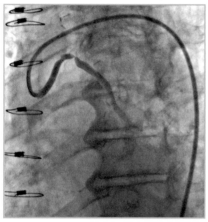

Figure 2. Angiogram showing a lesion on the saphenous vein graft to the obtuse marginal.

i. *How* would you approach this lesion?
   a. Plain balloon angioplasty
   b. Stenting with a bare metal stent
   c. Stenting with a drug-eluting stent
   d. Stenting with a covered stent

ii. What other devices are considered to *decrease* the rate of clinically significant distal embolization?
   a. A distal protection device
   b. Rheolytic thrombectomy
   c. Pharmacologic therapy with vasodilators
   d. All of the above

3. A 58-year-old diabetic patient with a history of CABG had a stent placed in the SVG to the RCA 2 years ago for symptoms of unstable angina. She now presents with exertional chest pain. Thallium scintigraphy post-adenosine infusion reveals a reversible defect in the inferior wall. Coronary angiography reveals a lesion within the area of the previous stent (see Figure 3).

Figure 3. Angiogram showing a lesion within the area of a previous stent.

i. What would be your *first* intervention on this lesion?
   a. Plain balloon angioplasty
   b. Stenting with a bare metal stent
   c. Cutting balloon angioplasty
   d. Atherectomy

ii. Would you perform any *other* interventions?
   a. No
   b. Stenting with a bare metal stent
   c. Brachytherapy
   d. Atherectomy

4. A 57-year-old female with a history of stable angina pectoris presents with new unstable angina. After ruling out MI, she is taken to the cardiac catheterization laboratory. Angiography of the left system shows mild nonobstructive disease in the LAD and the first obtuse marginal. The RCA is shown in Figure 4a. Immediately after direct stenting with a 3×15-mm stent, the patient develops transient ventricular fibrillation and pulseless electrical activity. Pericardiocentesis reveals 1,200 cc of red blood. After resuscitation with inotropes and intra-aortic balloon counterpulsation, repeat angiography is performed (see Figure 4b). What is your *next* step?

(a)   (b)

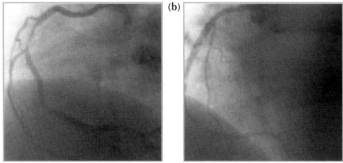

Figure 4. (a) Angiogram of the RCA. (b) Repeat angiography after resuscitation with inotropes and intra-aortic balloon counterpulsation.

   a. Cardiothoracic surgical consultation
   b. Perfusion balloon tamponade
   c. Stenting with a bare metal stent
   d. Stenting with a covered stent

5. A 52-year-old male experiences progressive dyspnea on exertion. He undergoes exercise and rest thallium scintigraphy, which shows a reversible defect in the distribution of the LAD. Angiography reveals a normal coronary RCA. Figure 5 shows the angiogram of the left coronary system. Note the highly eccentric obstructive lesion in the LAD. Which of the following would represent the *most* reasonable approach to this lesion?

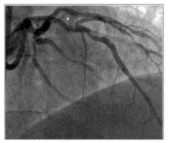

Figure 5. Angiogram of the left coronary system.

   a. DCA
   b. Rotational atherectomy
   c. Laser atherectomy
   d. Surgical consultation

6. A 78-year-old diabetic female presents with symptoms of stable angina. Angiography reveals a 30%–40% lesion in the mid LAD, mild plaquing in the mid left circumflex artery, and a long, heavily calcified lesion in the RCA (see Figure 6).

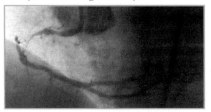
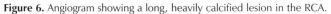

Figure 6. Angiogram showing a long, heavily calcified lesion in the RCA.

i. *How* would you approach this lesion?
   a. Rotational atherectomy
   b. DCA
   c. Direct stenting
   d. Percutaneous transluminal coronary angioplasty (PTCA)

ii. What *other* options are appropriate?
   a. Laser atherectomy
   b. Cutting balloon angioplasty
   c. CABG
   d. All of the above

7. Diagnostic coronary angiography in a 54-year-old male with 4 years of stable angina reveals a chronic total occlusion (CTO) of the proximal RCA, no significant left coronary system disease, and grade 2 collaterals from the LAD to distal RCA (see Figure 7). A previous attempt at recanalization with multiple guidewires was unsuccessful. Which device should be considered in a fresh attempt to *recanalize* the RCA?

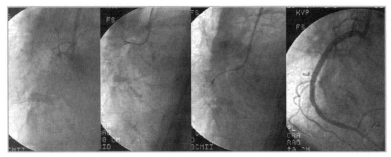

Figure 7. Angiogram showing grade 2 collaterals from the LAD to the distal RCA.
   a. Rotablator
   b. Laser wire
   c. LuMend Frontrunner™ device
   d. Flexi-Cut® DCA system

Answers

1i. a This lesion is ideal for primary stenting. In *de novo* coronary lesions, stenting results in lower rates of restenosis than plain balloon angioplasty alone. Primary stenting may involve balloon predilation, a strategy that is useful if the lesion is very tight, calcified, or if sizing of the stent is problematic. The vessel distal to a significant stenosis may appear to be underfilled, potentially misleading the operator into using a smaller stent than is necessary and hence increasing the restenosis rate. Primary stenting may alternatively involve direct stenting, ie, stent deployment without balloon predilation.

Provisional stenting refers to a strategy by which stenting is reserved for use in the case of a suboptimal angioplasty result. In this young patient, stent usage is preferred because stenting is associated with better long-term outcomes.

Atherectomy techniques, such as rotational atherectomy and DCA, should be reserved for more complex lesion subsets. Rotational atherectomy, in particular, is useful for treating highly calcified lesions, as well as for chronic total occlusions (CTOs). DCA can play a major role in bifurcation lesions.

1ii. c Drug-eluting stents have been hailed as a major breakthrough in percutaneous intervention. While many styles of stents with different medications will soon become available, they all (to date) elute antiproliferative agents into the vessel wall. These medications prevent in-stent restenosis. The SIRIUS (Sirolimus-eluting Stent in *De Novo* Native Coronary Lesions) trial is the largest study to date utilizing the only commercially available drug-eluting stent (at the time of publication): the sirolimus (rapamycin)-eluting CYPHER™ stent (Cordis Corp., FL, USA) [1]. This trial included patients with *de novo* coronary lesions. Compared with a bare metal stent, the sirolimus-eluting stent decreased target-vessel failure at 9 months (21% vs 8.6%, $P < 0.001$). The frequency of major adverse cardiac events (MACE) was also reduced at this time point (18.9% vs 7.1%, $P < 0.001$).

Similar results have been found in trials with other stent designs and pharmacologic agents. An example is the TAXUS-IV trial, in which the paclitaxel-eluting Express™ stent (Boston Scientific, MA, USA) reduced target-vessel revascularization when compared with an identical bare metal stent (12% vs 4.7%, $P < 0.0001$) at 9 months.

Patients undergoing stenting for single-vessel disease in a *de novo* lesion should be considered as suitable candidates for a drug-eluting stent. This is especially true for patients at high risk for restenosis, such as diabetics. Drug-eluting stents can also be considered for more complex subsets, such as patients with CTOs (see **Figure 8**) and bifurcation lesions (see **Figure 9**) or multivessel interventions, pending clinical trials.

1iii. e As drug-eluting stents work by releasing medications into the vessel wall, proper apposition with the vessel wall is critical to avoid uneven drug delivery. Therefore, lesion predilation, prior to stent deployment, is currently recommended by stent manufacturers. This recommendation is at least partially based on the fact that many studies with drug-eluting stents have included predilation as part of the protocol.

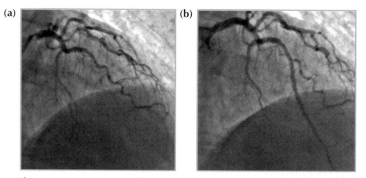

Figure 8. (**a**) Angiogram of the left coronary system demonstrating a chronic total occlusion of the LAD. (**b**) After recanalization. A significant lesion at the area of previous occlusion has been stented with a drug-eluting stent.

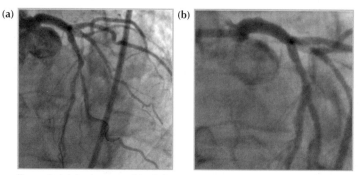

Figure 9. (**a**) Angiogram of the left coronary system in a patient with unstable angina. Note the lesion in the diagonal branch, which extends back to compromise the LAD. (**b**) A magnified view of the bifurcation after bifurcation stenting with drug-eluting stents using the "crush" (simultaneous positioning, sequential deployment) technique.

Many operators err on the side of shorter bare metal stents to decrease restenosis rates. However, as drug-eluting stents have much lower rates of restenosis than traditional bare metal stents, one can afford to use relatively longer stents to assure complete lesion coverage. The stent chosen should be long enough to confidently include the entire lesion, while minimizing injury to the surrounding normal tissue. The patients in the SIRIUS trial had a stent-to-lesion ratio of 1.7, whereas those in the smaller RAVEL (Randomized Comparison of a Sirolimus-eluting Stent with a Standard Stent for Coronary Revascularization) trial had a stent-to-lesion ratio of 1.9 [2]. Both of these are significantly larger ratios than traditionally used. The final result for this patient is shown in **Figure 10**.

2i. d SVG atherosclerosis is fundamentally different from native coronary atherosclerosis. Lesions are cholesterol-rich and tend to be very soft and friable. Instrumentation of these lesions often leads to significant distal embolization, resulting in the no-reflow phenomenon or, occasionally, clinically significant infarction. A covered stent, which is a traditional stent covered by a Dacron or pericardial membrane, can compensate for some of the friability; preliminary data confirm that the use of such stents may decrease the risk of embolization and restenosis.

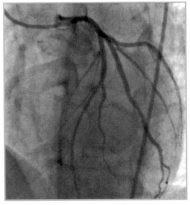

Figure 10. The final angiographic result for the patient described in Question 1.

Covered stents can also be useful in more complex lesions, such as excluding aneurysmal dilatation of coronary arteries, repairing coronary artery rupture, or sealing fistulous connections. Covered stents (eg, Jomed stent, CardioVasc stent, Symbiot) are available in both self-expanding and balloon-expanding forms (see **Figure 11**). Self-expanding stents are frequently preferred in vein-graft lesions as they often conform better to the tortuous vessel wall.

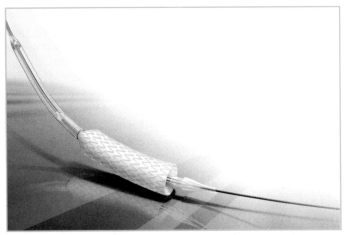

Figure 11. The Symbiot stent graft (figures courtesy of Boston Scientific).

(a)   (b)

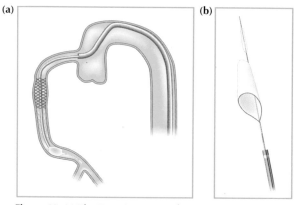

Figure 12. (a) The PercuSurge GuardWire system (Medtronic).
(b) The filter from the FilterWire system (Boston Scientific).

With regards to the case presented, restenosis rates are highest with plain balloon
angioplasty. Until recently, traditional stent placement has achieved satisfactory results,
but covered stents may offer a better alternative. Drug-eluting stents have not yet been
thoroughly studied in vein-graft lesions, but early experience appears promising.

2ii. b Distal protection devices are relatively new – to date, only two are approved for use in
the coronary system: the PercuSurge™ device (Medtronic, MN, USA) and FilterWire EX™
(Boston Scientific) (see **Figure 12**). Their size and method of use have, for the most part,
limited their application at this point to SVGs. When using the PercuSurge device, a
special guidewire, which includes a distal occlusion balloon at its tip, is advanced to a
distal portion of the vein graft. The balloon is then inflated, temporarily occluding flow.
Once the intervention of choice (eg, PTCA, stent) has been performed, a small catheter
is advanced to the area of the balloon and debris is suctioned out. The balloon is then
deflated. Care must be taken to minimize ischemic time.

Most trials of this system have been quite favorable. The SAFER (Saphenous Vein Graft
Angioplasty Free of Emboli Randomized) trial, which studied patients with vein-graft
lesions, found an approximately 50% reduction in the frequency of MACE with
PercuSurge usage as compared with a conventional 0.014-inch angioplasty guidewire
(9.6% vs 16.5%, $P = 0.004$) [3]. Filter devices carry the advantage of not occluding
blood flow completely, but there are theoretical concerns that they may miss smaller
embolic particles, and that they cannot remove already-present clots in occluded vein
grafts. The FIRE (FilterWire During Transluminal Intervention of Saphenous Vein Grafts)
trial compared FilterWire EX with the PercuSurge system; there was no significant
difference in MACE at 30 days [4].

The rheolytic thrombectomy system (AngioJet®, Possis Medical Inc., MS, USA) consists of
a stainless steel catheter tip connected to a hypotube (see **Figure 13**). Saline is injected
under high pressure from the tip back into the catheter and, by the Venturi–Bernoulli

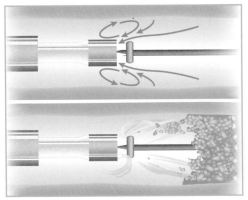

Figure 13. The rheolytic thrombectomy system (AngioJet, Possis Medical Inc.).

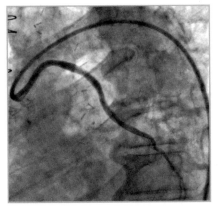

Figure 14. Angiogram of a patient after covered stent implantation with distal protection.

principle, an area of near-perfect vacuum is obtained. This causes surrounding blood and thrombus to be pulled back into the tip. There, the thrombotic particles are broken up by the saline jets, which propel them back into the catheter and out of the body. VEGAS (Vein Graft AngioJet Studies) I and II confirmed a high success rate in decreasing thrombus burden with this technique compared with thrombolytic infusion [5]. Thus, the AngioJet system should be considered in vein-graft lesions with significant acute thrombotic burdens.

Vasodilator therapy will not decrease the rate of distal embolization, but can decrease the clinical sequelae by dilating small vessels and increasing flow. As such, it can be used in treatment or prophylaxis for the no-reflow phenomenon caused by distal embolization. This patient underwent covered stent implantation with distal protection. The result is shown in **Figure 14**.

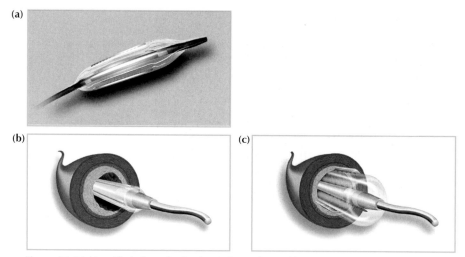

Figure 15. (**a**) Magnified view of a Cutting Balloon. (**b**) Predilatation; (**c**) postdilatation (figures courtesy of Boston Scientific).

3i. c The patient is suffering from in-stent restenosis. Unlike *de novo* lesions, restenotic lesions represent a fibrous, neointimal proliferation within the stent. Balloon angioplasty alone can slip out of the lesion (called "watermelon seeding"). This can increase the area of injury and paradoxically increase future risk for restenosis. Another problem with plain balloon angioplasty is that it can fail to generate the compressive force needed to enlarge the lumen, leading to significant recoil.

A Cutting Balloon (Boston Scientific) has three blades placed lengthwise around the balloon, which score the fibrotic wall and help to prevent balloon slippage (see **Figure 15**). In addition, this scoring can ease compaction of the fibrotic material and thus decrease recoil. These factors make cutting balloon angioplasty ideal for restenosis lesions. For the same reasons, cutting balloon angioplasty can be considered for other lesions that are anticipated to be less responsive to angioplasty, such as heavily calcified lesions or bifurcation lesions.

3ii. c Intravascular brachytherapy is a method of delivering a radiation source to the lesion in question while minimizing exposure to the surrounding tissues. It promotes a milieu that decreases the risk of significant neointimal proliferation. The use of brachytherapy has been correlated with a 35%–70% relative risk reduction in target-vessel revascularization at 6 months. Either beta- or gamma-emitting systems can be used. Many systems use a centering balloon to make sure the radiation source is centered within the vessel lumen, and hence that radiation is delivered circumferentially in similar doses (see **Figure 16a**). A radiation-emitting source is then passed through the lumen, and its dwell time within the lesion is calculated based on vessel size and the radioactivity of the source (see **Figures 16b** and **16c**).

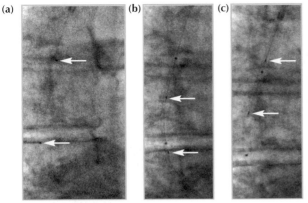

Figure 16. Intravascular brachytherapy. (**a**) Markers show the proximal and distal edges of the centering balloon (arrows). (**b**) Markers represent the proximal and distal edges of the radiation source (arrows). (**c**) Given the length of the injured area, a second more proximal dwell of the radiation source was performed (arrows).

Figure 17. Magnified view of the GALILEO intravascular radiation catheter showing the short wire lumen and the centering balloon, as well as the gold distal markers on the proximal and distal ends (figure courtesy of Guidant Corp.).

Medical therapy is often insufficient for symptom control. Atherectomy is unlikely to be successful given the tough, fibrous nature of the lesions. Reimplantation of a bare metal stent is likely to worsen the problem in the long term. The introduction of drug-eluting stents may provide more options for the treatment of restenosis. Indeed, the TAXUS-III registry data show that use of these stents is safe and feasible for the treatment of focal in-stent restenosis, and that recurrence rates are low [6]. These stents elute an inhibitor of neointimal proliferation, thus promising to arrest the problem of in-stent restenosis. This option should be considered in patients with refractory restenosis.

This patient underwent cutting balloon angioplasty, followed by intravascular brachytherapy with the GALILEO® beta-radiation system (Guidant Corp., IN, USA) (see **Figure 17**). The angiographic result is as shown in **Figure 18** (overleaf).

4. **b** This patient has experienced the rare complication of coronary artery rupture or laceration during percutaneous coronary intervention. The operator must quickly deal with this complication in order to avoid serious, potentially rapidly fatal consequences, such as massive MI, tamponade, or death. The first step is to stop the extravasation. While the extravasation can be stopped temporarily by any available balloon, a perfusion balloon is preferred if readily available. This balloon has an extra port, allowing passage of blood from a point proximal to the balloon to a point distal to the balloon. Thus, it allows continued coronary flow while still applying pressure to the area of vessel injury.

Figure 18. Angiogram after cutting balloon angioplasty and brachytherapy.

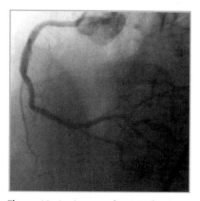

Figure 19. Angiogram after initial tamponade with a perfusion balloon, followed by stenting with a Jomed balloon-expanding covered stent.

The vessel should be periodically checked for evidence of healing. Cardiothoracic surgical consultation is appropriate, but only once the bleeding has been stopped, at least temporarily. Once the area of injury has stopped bleeding acutely, a membrane-covered stent should be considered to shield the area in question, and thus definitively seal the site of rupture. There is no role for a bare metal stent in this situation. This patient underwent initial tamponade with a perfusion balloon, followed by stenting with a Jomed balloon-expanding covered stent. The angiographic results are shown in **Figure 19**. She remained hemodynamically stable after stenting and did well.

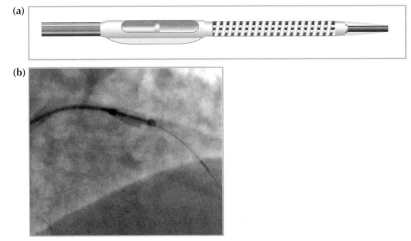

Figure 20. (a) Schematic of the Flexi-Cut directional coronary atherectomy (DCA) system (Guidant Corp.). **(b)** During DCA, a cup-shaped cutter is advanced to the lesion and plaque is removed into a nosecone collection chamber.

5. **a** The angiogram depicts a highly eccentric obstructive stenosis of the LAD. DCA is ideally suited for such eccentric lesions – as well as ostial and bifurcation lesions – in which conventional angioplasty techniques can fail due to uneven stent deployment or improper stent placement. In this case, primary stenting of the lesion without atherectomy would carry a high risk of uneven stent apposition, and increase the risk for stent thrombosis or restenosis. In the case of ostial lesions or bifurcation lesions, improper stent placement can lead to residual ostial stenosis, or protrusion of the stent into the main vessel. During DCA, a cup-shaped cutter is advanced to the lesion and plaque is removed into a nosecone collection chamber (see **Figure 20**). CAVEAT (Coronary Angioplasty Versus Excisional Atherectomy Trial) was the initial large series to look at this technique [7]. The trial showed that DCA resulted in better initial lumen diameters, as well as similar rates of restenosis and major complications as traditional coronary angioplasty. BOAT (Balloon versus Optimal Atherectomy Trial) suggested that initial success was higher with DCA versus balloon angioplasty and that 1-year restenosis rates were lower [8].

Although rotational atherectomy is a reasonable approach, it is ideally suited for lesions with significant superficial calcification, where DCA is unlikely to be as effective. Laser atherectomy has mostly fallen out of favor, largely due to unimpressive results and high rates of restenosis. Surgical revascularization could also be considered. Minimally invasive surgery utilizing the left internal mammary artery is becoming more popular, but surgical techniques are still being perfected. This patient underwent DCA of the proximal LAD lesion with the Flexi-Cut (Guidant Corp.) DCA system. The repeat angiogram is shown in **Figure 21**.

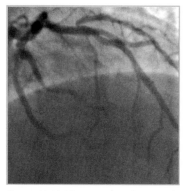

Figure 21. Angiogram after directional coronary atherectomy of the proximal left anterior descending lesion with the Flexi-Cut directional atherectomy system.

6i. **a** The lesion appears ideally suited for rotational atherectomy. In rotational atherectomy, a cutting tool called a "burr" is advanced through the lesion in question with high-velocity rotation (see **Figure 22**). It can thus ablate the atheroma, even if it is significantly calcified, into microscopic debris and restore flow. Distal embolization is a potential problem; however, the microscopic nature of the debris often renders this clinically irrelevant. There is a significant risk for conduction system abnormalities due to either atheroemboli in the nodal arteries or vagally mediated conduction block. Placement of a transvenous pacing wire into the right ventricle prior to the start of the procedure is recommended; alternatively, an aminophylline drip can be used.

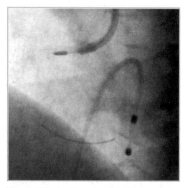

Figure 22. Angiogram showing rotational atherectomy.

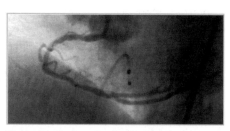

Figure 23. Angiogram after rotational atherectomy with a 1.25-mm burr, followed by stenting with a 3×18-mm CYPHER stent.

DCA can also be effective in the treatment of calcific lesions when the calcification is deeper within the plaque, but is less useful for concentric stenoses that do not involve bifurcations. Direct stenting should not be attempted on such heavily calcified lesions. Failure to crush the atheroma completely may lead to strut malapposition and increase the risk for stent thrombosis and restenosis. Balloon angioplasty without stenting may lead to adequate initial outcomes with high pressure inflations; however, the long-term rate of restenosis would be unacceptable. This patient underwent rotational atherectomy with a 1.25-mm burr, followed by stenting with a 3×18-mm CYPHER stent. The angiographic result is shown in **Figure 23**.

6ii. d In Excimer laser angioplasty systems, a fiberoptic laser system is guided to the area of stenosis. A laser in the ultraviolet spectrum is then applied to the lesion with a concomitant saline flush to avoid bubble formation. While the initial results of this system were quite promising, the cost of the equipment and a lack of convincing data showing superiority over other angioplasty or atherectomy techniques have limited its clinical use. Nonetheless, it can still be considered for use in the treatment of long lesions as well as in refractory restenosis.

Cutting balloon angioplasty may facilitate stenting of this lesion. As previously described, the cutting balloon can predilate the lesion, as well as cracking the lesion's calcific shell. This can lead to easier stent placement and better strut apposition. Given the patient's diabetic status, the complexity of this stenosis, and the moderate lesion in the LAD, surgical revascularization is also a reasonable option.

7. c The recently approved Frontrunner catheter (LuMend, Inc., CA, USA) is indicated for the revascularization of CTOs. The device is a 6-Fr guide-compatible catheter that employs blunt microdissection to navigate an intraluminal pathway through a CTO. Actuation of hinged jaws (opening diameter 3–4 mm) of the distal assembly in different orientations allows separation and/or fracture of the plaque. The newer version of this catheter is 0.039 inches thick and is used in conjunction with a microguide, which lends more support. Under fluoroscopic guidance and occasionally with visualization by contralateral artery injection, the Frontrunner catheter is advanced past the proximal cap of the CTO. The length of the CTO is traversed by blunt dissection, following which the device is removed and replaced by hydrophilic guidewires. More definitive therapy, such as stenting, is subsequently performed over the guidewire after balloon dilatation. This patient underwent PTCA and stenting after the lesion was crossed by the Frontrunner device. The final angiographic result is shown in **Figure 24**.

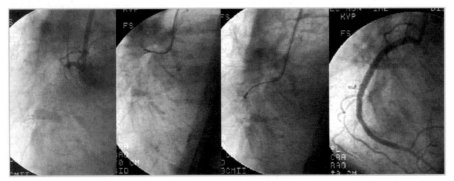

Figure 24. Repeat angiogram after percutaneous transluminal coronary angioplasty and stenting after the lesion was crossed by the Frontrunner device.

References

1. SIRIUS: Clinical and Angiographic Outcomes. Presented by Moses JW, Leon MB at Transcatheter Cardiovascular Therapeutics (TCT), September 24, 2002.

2. Morice MC, Serruys PW, Sousa JE et al. A randomized comparison of a sirolimus-eluting stent with a standard stent for coronary revascularization. *N Engl J Med* 2002;346:1773–80.

3. Baim DS, Waher, D, George B et al. Randomized trial of a distal embolic protection device during percutaneous intervention of saphenous vein aorto-coronary bypass grafts. *Circulation* 2002;105:1285–90.

4. Stone GW, Rogers C, Hermiller J et al. Randomized comparison of distal protection with a filter-based catheter and a balloon occlusion and aspiration system during percutaneous intervention of diseased saphenous vein aorto-coronary bypass grafts. *Circulation* 2003;108:548–53.

5. Kuntz RE, Baim DS, Cohen DJ et al. A trial comparing rheolytic thrombectomy with intracoronary urokinase for coronary and vein graft thrombus (the Vein Graft AngioJet Study [VeGAS 2]). *Am J Cardiol* 2002;89:326–30.

6. Tanabe K, Serruys PW, Grube E et al. TAXUS-III Trial: in-stent restenosis treated with stent-based delivery of paclitaxel incorporated in a slow-release polymer formulation. *Circulation* 2003;107:559–64.

7. Topol EJ, Leya F, Pinkerton CA et al. A comparison of directional atherectomy with coronary angioplasty in patients with coronary artery disease. The CAVEAT Study Group. *N Engl J Med* 1993;329:221–7.

8. Baim DS, Cutlip DE, Sharma SK et al. Final results of the balloon vs optimal atherectomy trial (BOAT). *Circulation* 1998;97:322–31.

Chapter 9
Adjunctive pharmacology

Rupal Dumasia and Debabrata Mukherjee

Questions

1. A 66-year-old male with diabetes and hypertension presents to the emergency room with waxing and waning substernal chest discomfort, at rest, for 4 hours. His blood pressure is 145/85 mm Hg and his heart rate is 85 bpm. An electrocardiogram (ECG) shows 2-mm horizontal ST depression in the right precordial leads. Cardiac enzymes are negative. A therapeutic regimen of aspirin, unfractionated heparin (UFH), intravenous (IV) nitroglycerin, and IV β-blocker is commenced. He is made pain free, and coronary angiography is performed the following morning. Which of the following is *false* regarding the use of clopidogrel in this patient?
    a. Clopidogrel is not indicated unless percutaneous coronary intervention (PCI) is performed
    b. Clopidogrel should be initiated in the emergency room and continued for 1 year after PCI
    c. Clopidogrel should be initiated in the emergency room and discontinued after angiography if coronary artery bypass graft (CABG) surgery is planned within the next 5 days
    d. Clopidogrel given more than 6 hours prior to PCI may decrease the risk of death, myocardial infarction (MI), or urgent target-vessel revascularization at 28 days
    e. Clopidogrel may cause increased risk of hemorrhagic complications after CABG

2. Which of the following is *true* regarding the use of oral glycoprotein (GP)IIb/IIIa inhibitors as an adjunct to PCI?
    a. Xemilofiban decreases mortality when used in conjunction with PCI
    b. Orbofiban decreases mortality from ischemic events when used in the setting of acute coronary syndromes (ACS)
    c. A pooled analysis of all five of the major oral GPIIb/IIIa inhibitors has shown an increase in mortality associated with the use of these drugs
    d. Sibrafiban is the only oral GPIIb/IIIa inhibitor to be associated with an increase in mortality when used in patients with ACS
    e. In general, oral GPIIb/IIIa inhibitors are comparable with IV GPIIb/IIIa inhibitors when used during PCI

3. Which of the following is *true* regarding percutaneous interventions for aorto-coronary vein grafts?
    a. Morbidity following vein-graft interventions is about the same as that with native coronary arteries
    b. Adjunctive GPIIb/IIIa inhibitors may not be beneficial in vein-graft interventions in the absence of mechanical emboli protection
    c. There is definitive evidence that GPIIb/IIIa inhibitors plus mechanical emboli protection decrease death, MI, and revascularization at 1 month compared with mechanical emboli protection alone
    d. GPIIb/IIIa inhibition decreases adverse cardiac outcomes to a greater extent than mechanical emboli protection
    e. If mechanical emboli protection is not used, GPIIb/IIIa inhibitors should be used to prevent adverse clinical outcomes

4. In patients with acute ST-segment MI in whom primary PCI is planned, which of the following is the *most* appropriate role for abciximab?
   a. There is no role for abciximab in this setting
   b. Abciximab (bolus and infusion) is appropriate when given in conjunction with standard therapy, including aspirin and heparin, while awaiting PCI
   c. Abciximab alone should be given in the emergency department while awaiting primary PCI
   d. Abciximab alone should be given in the catheterization laboratory once it has been determined that PCI will be undertaken
   e. Abciximab should only be given in the catheterization laboratory if the lesion is complex

5. A 65-year-old male presents with chest discomfort of 6 hours duration. An ECG in the emergency department reveals 3-mm ST elevation in the right precordial leads. He is given a regimen of aspirin, heparin, abciximab, and reduced-dose reteplase. Which of the following is *true* with regard to cardiac catheterization in this patient?
   a. There is no need for cardiac catheterization
   b. Early PCI is associated with an excess bleeding risk compared with no PCI
   c. TIMI (Thrombolysis in Myocardial Infarction)-3 flow at 90 minutes is less likely to occur in this patient than in a similar patient who receives full-dose reteplase without abciximab
   d. Cardiac catheterization is contraindicated
   e. Early PCI may be associated with less death, reinfarction, or urgent revascularization at 30 days compared with no PCI

6. A 55-year-old male with hypertension and diabetes presents to the emergency room with 8 hours of intermittent chest discomfort at rest. His blood pressure is 150/90 mm Hg and his heart rate is 75 bpm. Clinically, he is not in congestive heart failure. His ECG reveals dynamic 2-mm ST depression in the anterior leads. His cardiac biomarkers are positive. He is given aspirin, IV nitroglycerin, β-blockers, and subcutaneous (SC) enoxaparin. The discomfort is relieved, and the patient is admitted with plans for cardiac catheterization the following day. Which of the following is *not* accurate with regard to management of anticoagulation in the catheterization laboratory?
   a. Since enoxaparin cannot be monitored with the activated clotting time (ACT), PCI cannot be performed safely
   b. If PCI is performed within 8 hours of the last dose of SC enoxaparin, no additional anticoagulation is required for PCI, regardless of concomitant GPIIb/IIIa inhibitor use
   c. If PCI is performed 8–12 hours after the last SC enoxaparin dose, an additional dose of 0.3 mg/kg IV bolus enoxaparin should be administered, regardless of concomitant GPIIb/IIIa inhibitor use
   d. If PCI is performed 12 hours after the last SC enoxaparin dose, a bolus of UFH may be given
   e. If PCI is performed 12 hours after the last SC enoxaparin dose, and a GPIIb/IIIa inhibitor is used, the ACT target for UFH in the catheterization laboratory should be lower than if no GPIIb/IIIa inhibitor is used

7. If enoxaparin is used as a procedural anticoagulant, which of the following is *true* regarding the management of vascular sheaths after PCI?
   a. Sheaths must remain in place for 12 hours after the procedure
   b. Enoxaparin can be resumed immediately after sheath removal
   c. Femoral closure devices have been extensively studied in this population
   d. Sheaths can be safely removed 4–6 hours after the procedure
   e. Protamine should be administered prior to sheath removal

8. A 67-year-old female with hypertension and a long-standing history of smoking presents for elective PCI for a high-grade left anterior descending (LAD) artery stenosis. Which of the following is the *most* appropriate role for bivalirudin?
   a. Bivalirudin alone, rather than heparin, can be used as a procedural anticoagulant, with GPIIb/IIIa inhibitor to be used on a provisional basis
   b. Bivalirudin, rather than heparin, can be used with planned GPIIb/IIIa inhibitor
   c. Bivalirudin should only be used if the patient has a history of heparin-induced thrombocytopenia and thrombosis, because it confers a much higher bleeding risk
   d. Bivalirudin should be used rather than heparin, because it confers lower 30-day mortality
   e. Pretreatment with a thienopyridine is contraindicated if bivalirudin is used as a procedural anticoagulant

9. A 65-year-old male presents to the emergency department with 4 hours of chest discomfort at rest. An ECG reveals 2-mm ST elevation in the anterior leads. The patient is taken to the catheterization laboratory for primary PCI. After successful treatment of the occluded LAD with percutaneous transluminal coronary angiography (PTCA) and stent, the patient complains of more chest discomfort. The ST-segment elevation has failed to resolve. Repeat angiography reveals slow flow down the LAD. Which of the following is *least* accurate regarding the phenomenon present in this patient?
   a. Prolonged ischemia leads to damage to the microcirculation, including swelling of capillary endothelium, intravascular plugging with platelets and fibrin, and leukocyte plugging
   b. Reperfusion itself may lead to this phenomenon
   c. Downstream embolization of atherosclerotic debris and platelet plugs may contribute to this phenomenon
   d. This phenomenon is seen only in the setting of percutaneous revascularization, and not in CABG
   e. The presence of this phenomenon has been linked to malignant arrhythmias, a lower ejection fraction, and more cardiac death.

10. In the patient described in Question 9, which of the following would *not* be of benefit in either preventing or treating the condition?
    a. Early PCI
    b. GPIIb/IIIa inhibitors
    c. Intracoronary verapamil
    d. Intracoronary adenosine
    e. IV thrombolytics

11. A 65-year-old patient with known critical aortic stenosis by echocardiography presents to the cardiac catheterization laboratory for coronary angiography prior to aortic valve replacement. He is nervous about the procedure and requires 2 mg IV midazolam and 75 µg IV fentanyl for adequate sedation. After the first coronary injection, the patient suddenly becomes hypotensive and tachycardic, and 2-mm ST-segment depression is seen in the anterior and lateral leads. Which of the following is *least* likely to be the cause of this patient's acute decompensation?
    a. Excessive vasodilation from sedatives and narcotic agents
    b. Anaphylactoid reaction from the contrast injection
    c. Ischemia resulting from catheter engagement of a severely diseased left main coronary artery
    d. Vagally mediated hypotension
    e. Retroperitoneal bleeding

12. A 45-year-old female with no major coronary risk factors presents for cardiac catheterization after multiple episodes of severe substernal chest discomfort. They typically occur in the early morning and awake the patient from sleep. She has had a negative exercise ECG test. Her coronary arteries are free from angiographically significant disease. Which of the following tests in the catheterization laboratory would be *least* helpful in making a diagnosis?
    a. IV ergonovine maleate
    b. Intracoronary sodium nitroprusside
    c. Intracoronary acetylcholine
    d. Hyperventilation test
    e. Histamine receptor stimulation

13. A 69-year-old female presents to the emergency department with ST-segment elevation MI. She is treated with aspirin, β-blockade, IV heparin, and IV nitroglycerin infusion. She undergoes angiography and is noted to have a totally occluded right coronary artery with thrombus. She is given a bolus and infusion of abciximab, and undergoes PCI. Within a minute of balloon inflation, she becomes bradycardic and hypotensive. All of the following are reasonable treatment options, *except*?
    a. IV normal saline bolus
    b. IV atropine
    c. IV dobutamine
    d. IV epinephrine
    e. IV norepinephrine

14. Which of the following is *not* an acceptable sedation regimen for cardiac catheterization?
    a. Diazepam 2 mg IV and morphine 2 mg IV
    b. Etomidate 0.3 mg/kg IV
    c. Midazolam 1–2 mg IV and fentanyl 25 µg IV
    d. Lorazepam 2 mg IV and fentanyl 25 µg IV
    e. Lorazepam 2 mg IV

Answers

1. a This patient is clearly having an acute coronary syndrome (ACS). The role of clopidogrel in this setting and with regard to PCI has been reported in several clinical trials. In the CURE (Clopidogrel in Unstable Angina to Prevent Recurrent Ischemic Events) trial, 12,562 patients with non-ST-segment elevation ACS were randomized to aspirin plus clopidogrel or aspirin plus placebo, in addition to standard therapy [1]. The follow-up period was 3–12 months (average 9 months). The primary endpoint was the composite of cardiovascular death, MI, or stroke – this was reached in 11.47% of the placebo group and 9.28% of the clopidogrel group (relative risk [RR] 0.80, 95% confidence interval 0.72–0.89, $P < 0.001$). Of note, the Kaplan–Meier curves for this endpoint began to diverge 2 hours after the onset of therapy, and the benefit was maintained throughout the entire follow-up period. The benefit was present regardless of cardiac enzyme status or ECG findings. Among those who did not undergo revascularization, there was a 21% RR reduction (10.1% vs 8.1%) in the endpoint. Thus, early and continued use of clopidogrel is indicated for this patient.

The issue of when to start clopidogrel in the setting of PCI and how long to continue with this therapy has been studied in two trials. The PCI-CURE study assessed whether pretreatment with clopidogrel followed by long-term treatment after PCI would reduce adverse cardiac events [2]. Patients were randomized to either clopidogrel plus aspirin or placebo plus aspirin. The mean time from randomization to PCI was 6 days. After PCI, all patients received a thienopyridine plus aspirin for up to 4 weeks. The clopidogrel arm continued with this therapy for an average of 8 additional months; the aspirin-alone arm generally stopped clopidogrel at 4 weeks. At 30 days, the primary endpoint of cardiovascular death, MI, or urgent target-vessel revascularization was reached in 4.5% of the clopidogrel group and 6.4% of the aspirin-alone group ($P = 0.03$). Thus, clopidogrel plus aspirin was found to be better than aspirin alone if given in the days prior to PCI for ACS. At 1 year, there was a 31% reduction of cardiovascular death or MI in the clopidogrel arm compared with the aspirin-alone arm. Therefore, clopidogrel should be started prior to PCI and continued for at least 1 year afterwards.

The CREDO (Clopidogrel for the Reduction of Events During Observation) trial sought to evaluate the benefits of pretreatment with clopidogrel 3–24 hours prior to PCI and continued clopidogrel use for 1 year [3]. All patients received clopidogrel from PCI through to day 29. At 28 days, there was no difference in death, MI, or stroke in the pretreated group compared with placebo; however, in the prespecified subgroup of those pretreated more than 6 hours prior to PCI, there was a 38.6% RR reduction in the event rate. At 1 year, there was a 26.9% RR reduction of the composite endpoint in the clopidogrel arm. Thus, one can conclude that clopidogrel pretreatment more than 6 hours prior to PCI reduces the adverse event rate at 1 month, and continued treatment decreases the adverse event rate at 1 year.

The issue of hemorrhagic complications after CABG in those pretreated with clopidogrel was addressed in the CURE trial. Those who received clopidogrel and then underwent CABG within 5 days had more major bleeding events than those who had not received clopidogrel (9.6% vs 6.3%, $P = 0.06$). Conversely, those who underwent CABG more than 5 days after clopidogrel was stopped had no increase in major bleeding. In light of these data, a reasonable strategy is to treat with clopidogrel more than 6 hours prior to angiography in anticipation of PCI. If bypass surgery is the preferred revascularization option, clopidogrel can be discontinued and CABG can be performed after 5 days, in elective cases.

2. **c** There have been five major oral GPIIb/IIIa inhibitor trials and a pooled analysis of the five trials. The results have been quite discouraging. The EXCITE (Evaluation of Oral Xemilofiban in Controlling Thrombotic Events) trial showed a nonstatistically significant increase in mortality with the use of oral xemilofiban in patients undergoing elective PCI [4]. The OPUS-TIMI (Oral Glycoprotein IIb/IIIa Inhibition with Orbofiban in Patients with Unstable Coronary Syndromes-Thrombosis in Myocardial Infarction)-16, SYMPHONY (Sibrafiban versus Aspirin to Yield Maximum Protection from Ischemic Heart Events Postacute Coronary Syndromes), and SYMPHONY II trials studied patients with ACS [5,6]; BRAVO (Blockade of the GPIIb/IIIa Receptor to Avoid Vascular Occlusion) included patients with a variety of coronary, peripheral, or cerebrovascular diseases [7]. Of these, OPUS, SYMPHONY II, and BRAVO showed a statistically significant increase in mortality with the use of an oral GPIIb/IIIa inhibitor. A pooled analysis revealed a 35% increase in RR (0.7% increase in absolute risk) for death in the 45,523 patients studied.

The untoward effect of these agents may be explained by chronic low-level (<80%) platelet inhibition. This inhibition may lead to a paradoxical increase in platelet activity, which itself may lead to inflammatory cytokine release. For example, release of CD40 ligand by activated platelets leads to activation of monocytes and macrophages. Also, low levels of inhibition lead to platelet P-selectin expression, which mediates binding to leukocytes. In summary, oral GPIIb/IIIa inhibitors have been shown to be detrimental in both PCI and ACS.

3. **b** Vein-graft PCI has worse outcomes compared with PCI on native coronary arteries. Acute complications include distal embolization, no-reflow, and postprocedural MI. Patients also remain at risk for repeat revascularization. A pooled analysis of five major trials investigating GPIIb/IIIa inhibitors in PCI (though not all patients received stents) showed that the incidence of death, MI, or urgent revascularization at 30 days was 15.2% for graft PCI and 10% for native-vessel PCI ($P < 0.001$). At 6 months, the event rates were 37.1% for graft PCI and 25.4% for native-vessel PCI ($P < 0.001$). Many of the difficulties with vein-graft PCI are related to thrombosis, intimal hyperplasia, and diffuse atherosclerosis. The soft, friable nature of the plaque leads to distal embolization during PCI.

The same study found that when the outcomes of vein-graft PCI (without mechanical emboli protection) were separated according to use of GPIIb/IIIa inhibitor or placebo, the 30-day event rates were 16.5% and 12.6%, respectively ($P = 0.18$). At 6 months, the event rates were 39.4% and 32.7%, respectively ($P = 0.07$). On the basis of these data, one may conclude that GPIIb/IIIa inhibitors, in the absence of mechanical emboli protection, do not decrease rates of the composite endpoint at 30 days or 6 months.

Because of the friability of plaque in vein grafts, mechanical emboli protection is often used. Trial evidence has shown that mechanical emboli protection leads to a significant reduction in adverse events at 30 days. This is probably because the device traps particles of debris. There have been no trials comparing the strategy of both GPIIb/IIIa inhibitors and mechanical emboli protection versus either alone in vein-graft PCI. However, the rationale for using both seems reasonable – the device may trap the larger particles and the GP inhibitor may prevent filter thrombosis or protect the microvasculature.

4. **b** The role of abciximab in acute MI is well established. In patients in whom primary PCI is planned, the role of abciximab has been elucidated in several major clinical trials. In the CADILLAC (Controlled Abciximab and Device Investigation to Lower Late Angioplasty Complications) trial, patients with acute ST-segment elevation MI were randomized to PTCA (with or without abciximab) or primary stenting (with or without abciximab) and followed for 1 year [8]. Abciximab did not decrease mortality, but did decrease recurrent ischemia and target-lesion revascularization in both the PTCA and stent groups.

In the ADMIRAL (Abciximab Before Direct Angioplasty and Stenting in Myocardial Infarction Regarding Acute and Long-term Follow-up) trial, patients with acute ST-segment elevation MI in whom primary PCI was planned received standard therapy, including heparin, β-blockers, and aspirin [9]. The treatment arm received abciximab bolus and infusion, and the control arm received placebo. The abciximab arm had higher TIMI-3 flow prior to PCI (16.8% vs 5.4%, $P = 0.01$), at 24 hours (95.1% vs 86.7%, $P = 0.04$), and at 6 months (94.3% vs 82.8%, $P = 0.04$). In addition, there were fewer reocclusions in the abciximab group. At 30 days, there was no difference in mortality, but the composite endpoint of death, reinfarction, and urgent target-vessel revascularization was significantly lower in the abciximab group (6.0% vs 14.6%, $P = 0.01$). This benefit was maintained at 6 months. There was no difference in major bleeding, but there were more minor bleeding events and groin hematomas in the abciximab group. Of note, the subgroup who received abciximab early (in the ambulance or emergency room) had a greater benefit. The diabetic subgroup showed a mortality benefit at 6 months. In summary, in patients undergoing primary PCI for acute ST-elevation MI, there is adequate evidence that early use of abciximab improves angiographic and clinical outcomes.

5. **e** The best treatment strategy for acute ST-segment elevation MI is still evolving. The regimen of reduced-dose fibrinolytics in conjunction with GPIIb/IIIa inhibitors has recently been studied. In the GUSTO (Global Use of Strategies to Open Coronary Arteries) V and ASSENT (Assessment of the Safety and Efficacy of a New Thrombolytic) III trials, there was no difference in mortality associated with this strategy compared with full-dose fibrinolytics, but a decreased reinfarction and increased bleeding risk (intracranial hemorrhage) was noted [10,11].

In a retrospective analysis of the SPEED (Strategies for Patency Enhancement in the Emergency Department) trial (the GUSTO V pilot), those patients who received abciximab and reduced-dose reteplase and then underwent angiography/PCI were compared with those who received the same regimen, but did not undergo early

PCI [12]. Those who underwent early PCI had less death, reinfarction, or urgent revascularization at 30 days (5.6% vs 18.0%, $P = 0.001$). In addition, transfusion was less common in the early PCI group (16.0% vs 9.0%). In SPEED, the incidence of TIMI-3 flow at 60–90 minutes was 54% in the abciximab plus reduced-dose reteplase arm and 47% in the full-dose reteplase arm ($P = 0.32$). From these preliminary data, facilitated PCI appears to be a reasonable strategy. Several ongoing trials, including FINESSE (Facilitated Intervention with Enhanced Speed to Stop Events) [13] and ADVANCE-MI (Addressing the Value of Facilitated Angioplasty after Combination Therapy or Eptifibatide Monotherapy in Acute Myocardial Infarction), are testing this hypothesis.

6. a There is ample evidence that PCI is safe in patients who have received low molecular weight heparin (LMWH). In both the ESSENCE (Efficacy and Safety of Subcutaneous Enoxaparin in Non-Q-Wave Coronary Events) [14] and TIMI-11B trials, those patients who received enoxaparin and then underwent PCI had similar rates of major bleeding as those who received UFH. In these studies, enoxaparin was discontinued and UFH was used as the procedural anticoagulant. In another study, PCI was performed within 8 hours of the last LMWH dose, without additional anticoagulation during PCI. Immediately prior to catheterization, antifactor Xa activity was found to be above the arbitrary limit set for venous thromboembolism (0.6 IU/mL). There were no in-hospital abrupt closures or urgent revascularizations in the PCI patients. The rate of death/MI at 30 days was 3% in the PCI cohort. This is comparable with historical controls from the EPILOG (Evaluation in PTCA to Improve Long-term Outcome with Abciximab GPIIb/IIIa Blockade) trial [15].

The issue of management of anticoagulation if PCI is performed 8–12 hours after the last SC dose of LMWH was addressed in the PEPCI (Pharmacokinetic Study of Enoxaparin in Patients Undergoing Percutaneous Coronary Intervention) trial. In this pharmacokinetic study, supplemental enoxaparin (0.3 mg/kg IV) was administered in the catheterization laboratory and 96% of patients had antifactor Xa levels within the therapeutic range. The NICE (National Investigators Collaborating on Enoxaparin) registry studies provided further evidence that enoxaparin is a safe procedural anticoagulant. In NICE 3, 0.3 mg/kg IV enoxaparin was administered if more than 8 hours had elapsed after the last SC enoxaparin dose. Of these patients, those who had received GPIIb/IIIa antagonists had a 2% risk of non-CABG major bleeding. This was identical to historical controls who received UFH and GPIIb/IIIa inhibitors.

Another reasonable alternative, based on the strategy used in TIMI-11B and ESSENCE, is that patients receiving enoxaparin upstream for ACS can be switched to UFH in the catheterization laboratory after withholding the morning dose of enoxaparin. A lower ACT (>200 seconds) should be targeted if concomitant GPIIb/IIIa inhibitors are used.

7. d The issues of sheath removal and the management of enoxaparin after sheath removal were addressed in the NICE 1 and 4 registries. In these studies, sheaths were safely removed within 4 hours of IV enoxaparin or 6 hours of SC enoxaparin, with or without concomitant GPIIb/IIIa use. According to the package insert, enoxaparin should not be reinstituted until 6–8 hours after sheath removal.

Femoral closure devices have not been extensively studied in patients who have received enoxaparin, but there did not appear to be excess bleeding in patients receiving closure devices who had been randomized to enoxaparin compared with UFH during PCI in the CRUISE (Coronary Revascularization Utilizing Integrilin and Single-bolus Enoxaparin) trial [16].

8. a The role of bivalirudin, a direct thrombin inhibitor, has recently been elucidated in a large randomized controlled trial: REPLACE (Randomized Evaluation in PCI Linking Angiomax to Reduced Clinical Events)-2 [17]. In this trial, patients who underwent elective or urgent PCI for unstable angina, stable angina, or a positive stress test were randomized to one of two strategies for procedural anticoagulation. One group received a bivalirudin bolus and infusion to keep the ACT ≥ 225 seconds. In this group, a GPIIb/IIIa inhibitor could only be used on a provisional or "bailout" basis. The indications included, but were not restricted to, abrupt or side-branch closure, obstructive dissection, new thrombus, slow coronary flow, distal embolization, persistent residual stenosis, unplanned stent placement, or prolonged ischemia. The other group received a heparin bolus and infusion to keep the ACT ≥ 225 seconds. This group received a planned GPIIb/IIIa inhibitor. The choice of GPIIb/IIIa inhibitor (eptifibatide or abciximab) was left to the discretion of the operator, but had to be specified prior to randomization. Pretreatment with clopidogrel was strongly encouraged (85.4% in the heparin arm and 86.7% in the bivalirudin arm). All patients received aspirin; 96.5% of the heparin group and 7.2% of the bivalirudin group received a GPIIb/IIIa inhibitor.

The primary endpoint was the 30-day incidence of death, MI, urgent revascularization, or in-hospital major bleeding. In this, there was no significant difference between the two arms. In-hospital major bleeding was significantly reduced in the bivalirudin group (2.4% vs 4.1%, $P < 0.001$). However, when bleeding was classified according to the TIMI major bleeding criteria, there was no difference between the groups. There was a decrease in TIMI minor bleeding in the bivalirudin group. Overall, bivalirudin with provisional GPIIb/IIIa inhibitor appeared as effective as UFH with planned GPIIb/IIIa inhibitor as an antithrombotic during PCI, but was associated with less bleeding.

9. d The slow-reflow or no-reflow phenomenon is seen in patients with acute MI. Despite excellent reperfusion at the epicardial level, ischemia remains a problem at the tissue level. This is manifested by continued chest discomfort, failure of ST-segment elevation to resolve, and even hemodynamic compromise.

Prolonged ischemia leads to injury to the microcirculation. This is seen on a cellular level as swelling of the capillary endothelium and abnormal protrusions into the capillary lumen. There may also be edema of the capillary endothelium and occlusion of the capillary lumen. In addition, no-reflow can be exacerbated by reperfusion and downstream embolization of debris. This phenomenon has also been noted after CABG. With regard to outcomes, patients with no-reflow have more arrhythmias, a lower ejection fraction, and a higher risk of cardiac death.

10. e The no-reflow phenomenon is best prevented, rather than treated. One way to do this is to achieve early reperfusion. In addition, GPIIb/IIIa inhibition leads to less angiographic no-reflow and improved clinical outcomes. Once the no-reflow state is established, intracoronary verapamil has been shown to improve coronary flow and myocardial wall motion compared with placebo. Intracoronary adenosine may also be beneficial on the basis of microvascular vasodilation and preservation of endothelial integrity. Retrospective analysis has shown that those undergoing PCI for acute MI have less no-reflow if adenosine is given prophylactically. Intracoronary nitroprusside may also help to resolve no-reflow. IV nicorandil (an ATP potassium-channel opener) was found to improve microvascular flow when administered in acute MI. IV thrombolytics are useful in achieving reperfusion, but do not have an effect on the no-reflow phenomenon once it is established, though a low dose of intracoronary thrombolytic can help to dissolve a distal thrombus.

11. d Hypotension in the catheterization laboratory can occur for any number of reasons, including – but not limited to – hypovolemia, reduced cardiac output, contrast reaction, vagal response, and tamponade. Patients with critical aortic stenosis are at increased risk for hypotension and acute decompensation, and therefore require careful attention. Hypotension may develop when patients with critical aortic stenosis are excessively sedated. As a result of the hypotension, ischemia may occur. This is quite significant, because these patients often have hypertrophied myocardium and are prone to subendocardial ischemia. As a result of ischemia, further hypotension may occur, and a downward spiral can ensue. This cycle may be initiated by any cause of hypotension. In this case, vagally mediated hypotension is least likely, because this is typically accompanied by bradycardia and not tachycardia.

12. b This case illustrates a classic example of Prinzmetal's angina. This phenomenon is characterized by transient spasm of a proximal coronary artery leading to ischemia. There may be a fixed stenosis at a single site or the coronary arteries may be normal. The right coronary artery is the most frequent site, followed by the LAD. Spasm may occur at a single site, simultaneously at different sites, or sequentially at different sites.

The diagnosis can be made in the catheterization laboratory by provocative testing using IV injections of methylergonovine or ergonovine maleate. There is an inverse relationship between the dose of ergonovine required to provoke an attack and the frequency of spontaneous episodes. The effects can be quickly reversed with intracoronary nitroglycerin. Hyperventilation testing has been shown to be useful in making a diagnosis of vasospastic angina. Intracoronary acetylcholine infusions in escalating doses can also be used to induce severe vasospasm. A positive test is indicated by a focal spasm, rather than mild, diffuse vasoconstriction. Histamine, dopamine, and serotonin can also cause vasospasm. The mainstay of treatment of this disorder is calcium-channel blockade. Long-acting nitrates may also be used, and prazosin and nicorandil have been shown to be of benefit. In the absence of significant underlying coronary disease, the prognosis is excellent.

13. c This case illustrates a commonly observed phenomenon, known as the Bezold–Jarisch reflex. Within seconds of successful reperfusion, particularly in inferior infarctions, there is some degree of sinus bradycardia and hypotension. This is usually a transient effect and is quickly reversed with atropine or epinephrine. Temporary right ventricular apical pacing may be required.

Hypotension should be treated with an IV bolus of normal saline. If hypotension does not quickly respond to IV fluids, vasopressor therapy should be started promptly. Norepinephrine or epinephrine are reasonable choices. Dobutamine should not be used in this situation, because it may lower the blood pressure even further through β_2-agonist activity. Dopamine can also transiently lower blood pressure when first initiated.

Differential diagnoses include dissection or abrupt closure of the coronary artery, right ventricular infarction, or coronary perforation and tamponade. These should be suspected if bradycardia and hypotension are not quickly reversed with the above measures. Dissection leading to abrupt closure will usually be apparent on repeat coronary injections. Right ventricular infarction is a consideration if the lesion is proximal to the acute marginal branches of the right coronary artery, or if side-branch occlusion of the acute marginal branches occurs. The treatment consists of IV fluids. Coronary perforation leading to tamponade is a rare, but often catastrophic, complication. This may be noted on repeat coronary injections as contrast extravasation into the pericardium. It may be confirmed by echocardiography. It should be treated by stopping anticoagulants, performing immediate pericardiocentesis, and taking measures to seal the perforation, including immediate placement of a balloon across the perforation and, subsequently, a covered stent.

14. b The goal of medicating prior to cardiac catheterization is to achieve conscious sedation. An adequate level of sedation should relieve the patient's anxiety and prevent excessive movement. The patient should be able to protect his/her own airway and follow commands during the procedure. Vital signs, including pulse oximetry, should be continuously monitored. Blood pressure may decrease as a result of sedation, causing vasodilation and negative inotropy. In case the patient becomes overly sedated, drugs to reverse sedation, an airway kit, and personnel skilled in intubation must be readily available.

Any of the listed choices are acceptable except etomidate. Etomidate is most commonly used to induce general anesthesia. It may result in a deeper than desired level of sedation for catheterization. Some operators may prefer not to use conscious sedation prior to right heart catheterization because of its hemodynamic effects.

References

1. Yusuf S, Zhao F, Mehta SR et al.; Clopidogrel in Unstable Angina to Prevent Recurrent Events Trial Investigators. Effects of clopidogrel in addition to aspirin in patients with acute coronary syndromes without ST-segment elevation. *N Engl J Med* 2001;345:494–502.
2. Mehta SR, Yusuf S, Peters RJ et al.; Clopidogrel in Unstable Angina to Prevent Recurrent Events Trial (CURE) Investigators. Effects of pretreatment with clopidogrel and aspirin followed by long-term therapy in patients undergoing percutaneous coronary intervention: the PCI-CURE study. *Lancet* 2001;358:527–33.

3. Steinhubl SR, Berger PB, Mann JT 3rd et al.; CREDO Investigators. Clopidogrel for the Reduction of Events During Observation. Early and sustained dual oral antiplatelet therapy following percutaneous coronary intervention: a randomized controlled trial. *JAMA* 2002;288:2411–20.

4. O'Neill WW, Serruys P, Knudtson M et al. Long-term treatment with a platelet glycoprotein-receptor antagonist after percutaneous coronary revascularization. EXCITE Trial Investigators. Evaluation of Oral Xemilofiban in Controlling Thrombotic Events. *N Engl J Med* 2000;342:1316–24.

5. Cannon CP, McCabe CH, Wilcox RG et al. Oral glycoprotein IIb/IIIa inhibition with orbofiban in patients with unstable coronary syndromes (OPUS-TIMI 16) trial. *Circulation* 2000;102:149–56.

6. The SYMPHONY Investigators. Comparison of sibrafiban with aspirin for prevention of cardiovascular events after acute coronary syndromes: a randomised trial. Sibrafiban versus Aspirin to Yield Maximum Protection from Ischemic Heart Events Post-acute Coronary Syndromes. *Lancet* 2000;355:337–45.

7. Topol EJ, Easton JD, Amarenco P et al. Design of the blockade of the glycoprotein IIb/IIIa receptor to avoid vascular occlusion (BRAVO) trial. *Am Heart J* 2000;139:927–33.

8. Stone GW, Grines CL, Cox DA et al.; Controlled Abciximab and Device Investigation to Lower Late Angioplasty Complications (CADILLAC) Investigators. Comparison of angioplasty with stenting, with or without abciximab, in acute myocardial infarction. *N Engl J Med* 2002;346:957–66.

9. Montalescot G, Barragan P, Wittenberg O et al.; ADMIRAL Investigators. Abciximab before Direct Angioplasty and Stenting in Myocardial Infarction Regarding Acute and Long-Term Follow-up. Platelet glycoprotein IIb/IIIa inhibition with coronary stenting for acute myocardial infarction. *N Engl J Med* 2001;344:1895–903.

10. Simoons ML; GUSTO IV-ACS Investigators. Effect of glycoprotein IIb/IIIa receptor blocker abciximab on outcome in patients with acute coronary syndromes without early coronary revascularisation: the GUSTO IV-ACS randomised trial. *Lancet* 2001;357:1915–24.

11. Assessment of the Safety and Efficacy of a New Thrombolytic Regimen (ASSENT)-3 Investigators. Efficacy and safety of tenecteplase in combination with enoxaparin, abciximab, or unfractionated heparin: the ASSENT-3 randomised trial in acute myocardial infarction. *Lancet* 2001;358:605–13.

12. Herrmann HC, Moliterno DJ, Ohman EM et al. Facilitation of early percutaneous coronary intervention after reteplase with or without abciximab in acute myocardial infarction: results from the SPEED (GUSTO-4 Pilot) Trial. *J Am Coll Cardiol* 2000;36:1489–96.

13. Herrmann HC, Kelley MP, Ellis SG. Facilitated PCI: rationale and design of the FINESSE trial. *J Invasive Cardiol* 2001;13(Suppl A):10A–15A.

14. Goodman SG, Cohen M, Bigonzi F et al. Randomized trial of low molecular weight heparin (enoxaparin) versus unfractionated heparin for unstable coronary artery disease: one-year results of the ESSENCE Study. Efficacy and Safety of Subcutaneous Enoxaparin in Non-Q Wave Coronary Events. *J Am Coll Cardiol* 2000;36:693–8.

15. The EPILOG Investigators. Platelet glycoprotein IIb/IIIa receptor blockade and low-dose heparin during percutaneous coronary revascularization. *N Engl J Med* 1997;336:1689–96.

16. Bhatt DL, Lee BI, Casterella PJ et al.; Coronary Revascularization Using Integrilin and Single bolus Enoxaparin Study. Safety of concomitant therapy with eptifibatide and enoxaparin in patients undergoing percutaneous coronary intervention: results of the Coronary Revascularization Using Integrilin and Single bolus Enoxaparin Study. *J Am Coll Cardiol* 2003;41:20–5.

17. Lincoff AM, Bittl JA, Harrington RA et al.; REPLACE-2 Investigators. Bivalirudin and provisional glycoprotein IIb/IIIa blockade compared with heparin and planned glycoprotein IIb/IIIa blockade during percutaneous coronary intervention: REPLACE-2 randomised trial. *JAMA* 2003;289:853–63.

Chapter 10

Endovascular therapy for peripheral and cerebrovascular diseases

Ravish Sachar and Jay S Yadav

Questions

1. All of the following statements about carotid artery disease are true, *except*:
   a. Approximately 10% of patients with severe coronary artery disease (CAD) have high-grade carotid stenosis
   b. The presence of a carotid bruit prior to coronary artery bypass grafting (CABG) increases the risk of stroke after surgery
   c. Asymptomatic patients with ≥ 80% internal carotid artery (ICA) stenosis have a risk of stroke of approximately 5% per year
   d. In patients with both severe three-vessel CAD and high-grade ICA stenosis, the risk of stroke is lower if carotid revascularization is conducted first, rather than CABG
   e. The ECST (European Carotid Stenosis Trial) criteria tend to underestimate the degree of ICA stenosis as compared with the NASCET (North American Symptomatic Carotid Endarterectomy Trial) criteria

2. Which of the following statements is *false* regarding treatment for ICA disease?
   a. Based on data from randomized controlled trials, a patient without neurological symptoms and 60%–79% left ICA stenosis on Doppler ultrasound should be considered for carotid endarterectomy (CEA)
   b. A 63-year-old Caucasian male presents to your clinic because of a transient episode of right-sided weakness involving his upper and lower extremities. This lasted for less than 5 minutes and spontaneously resolved the night prior to presentation. He had one other such episode 2 weeks ago, which also spontaneously resolved within a few minutes. He has no symptoms at this time. His medical history is unremarkable, and he does not take any medications. His physical examination is within normal limits, except for a blood pressure of 158/92 mm Hg. There is no carotid bruit audible. Bilateral carotid Doppler studies show 50%–69% right ICA stenosis and 40%–59% left ICA stenosis. The optimal management is to start this patient on dual antiplatelet therapy with aspirin and clopidogrel and refer him for CEA
   c. A neurologically asymptomatic patient with CAD is found to have 80%–99% left ICA stenosis on Doppler ultrasound. He should be considered for CEA
   d. Based on data from ACAS (Asymptomatic Carotid Atherosclerosis Study), patients with asymptomatic high-grade right ICA stenosis should be considered for CEA only if the operative stroke/mortality rate in the surgical center is ≤ 5%
   e. You are consulted on a 73-year-old African American female who has had a recent stroke with left-sided hemiparesis. Bilateral carotid Doppler studies show 40%–59% right ICA stenosis and 20%–39% left ICA stenosis. An extensive noninvasive evaluation, including a transesophageal echocardiogram, reveals no embolus source and a hypercoaguable work-up is negative. She should be referred for carotid angiography

3. Characteristics that make a patient high-risk for CEA, and thus eligible for carotid stenting, include all of the following, *except*:
   a. ICA lesions above C2 or common carotid artery lesions below the clavicle
   b. Diabetes mellitus
   c. Left ventricular (LV) dysfunction (LV ejection fraction [EF] <30%)
   d. Age >80 years
   e. All of the above

4. Which of the following statements is *true* regarding carotid artery stenting?
   a. Carotid artery stenting is technically more challenging in patients with a type I aortic arch than in patients with a type III aortic arch
   b. The restenosis rate after carotid artery stenting is approximately the same as that after coronary artery stenting
   c. In order to avoid vessel trauma, distal emboli protection devices should only be used when carotid duplex studies have shown a radiolucent area in the carotid plaque, suggesting a lipid-rich core prone to distal embolization
   d. Bradycardia and hypotension secondary to pressure from balloon inflation on the carotid sinus can be well managed with atropine, fluids, and vasopressors as needed; deployment of a temporary pacemaker is rarely necessary
   e. In randomized trials, glycoprotein (GP)IIb/IIIa inhibitors (as with coronary interventions) have been shown to reduce the 30-day incidence of stroke and myocardial infarction (MI) after carotid stenting, probably by preventing distal embolization

5. Accepted indications for renal artery angiography in a patient undergoing cardiac catheterization include all of the following scenarios, *except*:
   a. New-onset hypertension in a 62-year-old, previously healthy male patient
   b. Increase in serum creatinine from 1.1 to 2.4 mg/dL on follow-up blood work 1 week after starting ramipril 5 mg bid
   c. A 54-year-old African American male with a long history of medically controlled hypertension on hydrochlorothiazide 25 mg qd, lopressor 50 mg bid, and amlodipine 10 mg qd presents with exertional angina. His blood pressure is 186/94 mm Hg during the physical examination. At cardiac catheterization 2 weeks later, his blood pressure is 193/91 mm Hg
   d. A 47-year-old female patient with abdominal bruit
   e. All of the above

6. A 70-year-old Caucasian male with hypertension refractory to hydrochlorothiazide 25 mg qd, long-acting metoprolol 200 mg qd, and clonidine 0.3 mg bid undergoes renal angiography (the results of which are shown in Figure 1). His serum creatinine has increased from 1.5 to 2.4 mg/dL over the past year. All of the following are true, *except*:

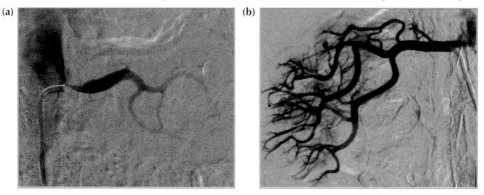

Figure 1. Angiogram of (**a**) left renal artery; (**b**) right renal artery.

    a. A history of recurrent episodes of acute pulmonary edema would justify renal artery intervention
    b. Stenting for this lesion is preferred over angioplasty alone
    c. His hypertension is entirely due to increased renin production by the left kidney
    d. In patients with bilateral renal artery stenosis (RAS) of ≥ 70% and deteriorating renal function, there is a high probability that renal artery revascularization will result in long-term stabilization or improvement of renal function
    e. With optimal results after percutaneous intervention, there is an approximately 70% chance that his hypertension will be better controlled and he may be able to reduce his antihypertensive medications

7. A 65-year-old Caucasian male is referred to you with a 2-year history of intermittent lower extremity claudication, which has recently worsened. He has a medical history of hypertension and hyperlipidemia, and currently smokes 10 cigarettes/day. His medications include enteric-coated aspirin 81 mg qd, simvastatin 20 mg qd, pentoxifylline (Trental) 400 mg tid, amlodipine 10 mg qd, hydrochlorothiazide 25 mg qd, and folic acid 1 mg qd. Lower extremity resting vascular studies reveal a right ankle brachial index (ABI) of 0.65 and a left ABI of 0.53. All of the following statements are true, *except*:
    a. Patients who continue to smoke after the onset of intermittent claudication have approximately 20% risk of limb loss
    b. In patients over 40 years, most cases of claudication are due to aorto-iliac disease
    c. When compared with placebo, randomized clinical trials have shown that cilostazol (Pletal) is more effective than pentoxifylline for improving symptoms of intermittent claudication
    d. An ABI of <0.9 is considered abnormal, and an ABI of <0.4 is compatible with rest pain
    e. 50% of lower extremity amputations resulting from peripheral vascular disease (PVD) occur in diabetics

8. All of the following statements are true regarding the treatment of lower extremity vascular disease, *except*:
   a. Treatment goals should be an improvement in pain-free walking distance, prevention of lower extremity ischemia, prevention of limb loss, and prophylaxis against future cardiac events
   b. In patients with intermittent claudication, medical therapy should be the primary method of treatment unless the patient progresses to rest pain, ulcers, or the development of gangrene
   c. In the first 5 years after the development of intermittent claudication, most patients will remain stable or improve with medical therapy and risk factor modification, despite angiographic disease progression
   d. Most patients with angiographic evidence of lower extremity atherosclerotic obstructive lesions do not have symptoms of intermittent claudication

9. Which of the following is *false* regarding percutaneous intervention for lower extremity atherosclerotic disease?

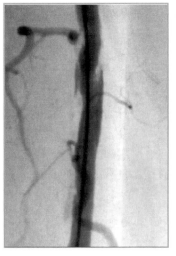 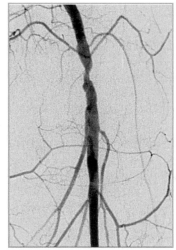

Figure 2. Angiogram showing a distal superficial femoral artery lesion after a 2-minute balloon inflation.

Figure 3. Angiogram of a mid superficial femoral artery lesion.

   a. A 73-year-old female presents with severe lifestyle-limiting intermittent claudication and is referred for percutaneous treatment of a severe distal superficial femoral artery (SFA) lesion 18 mm in length. The angiographic result after a 2-minute balloon inflation is shown in **Figure 2**. The next step should be deployment of a nitinol self-expanding stent
   b. A 65-year-old non-diabetic male presents with intermittent claudication. He is the manager at a local grocery store and feels that his symptoms interfere with his work. Angiography reveals a focal, noncalcified, mid-SFA stenosis of 8 mm in length with excellent distal run-off (see **Figure 3**). The immediate success rate and the 3-year patency of this lesion after balloon angioplasty alone are expected to be approximately 85% and 60%, respectively

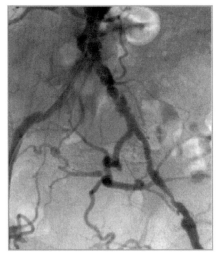

Figure 4. Angiogram of the left iliac artery.

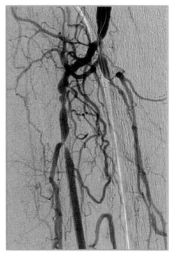

Figure 5. Angiogram of right-sided infrapopliteal disease.

c. With the femoropopliteal circulation, self-expandable stents are preferable; there is a risk of external compression and stent deformation with balloon-expandable stents

d. A 58-year-old female smoker presents with intermittent claudication and is referred for angiography (see **Figure 4**). Randomized trials have shown that an initial strategy of direct stent placement is equivalent to an initial strategy of balloon angioplasty, with stent placement only in the case of suboptimal results

e. A 68-year-old diabetic male is referred to you for recent-onset right calf pain with exertion. Angiography reveals infrapopliteal disease (see **Figure 5**). The primary modeof treatment should be percutaneous transluminal angioplasty (PTA) with stent placement if needed for suboptimal results.

10. A 53-year-old female secretary undergoes angiography (see Figure 6). Which of the following is *not* an indication for percutaneous revascularization in this patient?

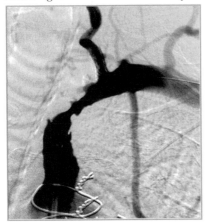

Figure 6. Angiogram of the left subclavian artery.

a. Left hand claudication while typing

b. Improvement of flow to the left internal mammary artery (LIMA) prior to CABG or in a patient after CABG with a LIMA graft to the coronary vasculature

c. History of thromboembolism to the left radial artery

d. Improvement of flow to a dialysis arteriovenous fistula in the left arm

e. ≥ 30 mm Hg gradient across the stenosis, with no other symptoms

Answers

1. **e** Atherosclerosis is a systemic disease process that often affects more than one vascular bed. For example, 25%–60% of patients with peripheral vascular disease (PVD) have carotid disease, and approximately 50%–60% of patients with carotid disease have severe CAD [1–3]. However, only 10% of patients with CAD have high-grade carotid disease. Among asymptomatic patients, the risk of stroke is proportional to the degree of carotid stenosis. Patients with ≥60% stenosis have a stroke risk of approximately 2% per year [4,5], while those with ≥80% stenosis have a risk of approximately 5% per year [6]. In one study, risk factors for stroke following CABG were identified as [7]:

    • the presence of a carotid bruit
    • prior stroke or transient ischemic attack (TIA)
    • severe carotid stenosis or occlusion

    In the majority of these patients, however, the location of the neurological deficit was not in the same vascular territory as the carotid stenosis. Similarly, in the Framingham study, the presence of a carotid bruit in asymptomatic patients doubled the risk of stroke, but the majority of these strokes occurred in vascular beds different from that of the carotid bruit [8]. The presence of a bruit may be a general marker for patients at higher risk of cerebrovascular events.

    It is not uncommon for patients who present with severe carotid disease to have concomitant symptoms secondary to coronary disease, and any surgery in this population is a high-risk procedure. It is unclear which vascular bed should be treated first, and each case must be evaluated on an individual basis. Options include simultaneous carotid endarterectomy (CEA) and CABG, staged CEA followed by CABG, or staged CABG followed by CEA. In staged procedures, conducting CEA first results in a high risk of perioperative myocardial infarction (MI), while conducting CABG first carries a high risk of perioperative stroke [9,10]. Some authors recommend a simultaneous procedure [11,12], but patient selection is likely to be an important factor in achieving a satisfactory outcome [13]. A meta-analysis of 16 studies with a total of 844 combined and 920 staged procedures reported a higher risk of death, stroke, or MI among patients undergoing the combined procedure [14]. Percutaneous intervention may be a better choice for this patient population, and can usually be safely performed in one setting or as staged procedures.

    There are two accepted criteria for determining the degree of ICA stenosis, and the difference between the two criteria lies in which anatomical segment is used as the reference diameter (see **Figure 7**). The ECST criteria use the largest diameter of the carotid bulb as the normal reference diameter. If the carotid bulb is not angiographically visible, an estimate is used. The NASCET criteria use the widest diameter of the ICA distal to the lesion as reference. Since the diameter of the carotid bulb will always be more than the maximal diameter of the ICA distal to the lesion, the ECST criteria will tend to overestimate the degree of stenosis as compared with the NASCET criteria.

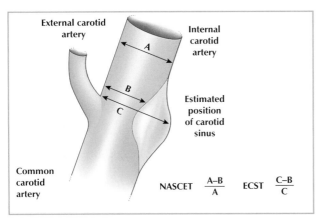

Figure 7. The NASCET (North American Symptomatic Carotid Endarterectomy Trial) and ECST (European Carotid Stenosis Trial) criteria for determining carotid stenosis. Due to a larger reference diameter, the ECST criteria tend to overestimate stenosis as compared with the NASCET criteria.

2. **d** At this time, CEA is the standard of care for the reduction of stroke/TIA in patients with symptomatic or high-grade asymptomatic carotid stenosis. Several trials have firmly established the effectiveness of CEA in preventing stroke in the presence of severe carotid stenosis as compared with medical therapy. It is important to note, however, that high-risk patients were not enrolled in these trials. Since the risk of surgery is likely to be increased among high-risk patients, extrapolation of these trial data to high-risk patients must be done with caution. Furthermore, the SAPPHIRE (Stenting and Angioplasty with Protection in Patients at High-risk for Endarterectomy) trial has already shown the superiority of carotid stenting over CEA among high-risk patients. Future trials may expand the patient population for which carotid stenting is preferred over, or is equivalent to, CEA.

Randomized trials of symptomatic patients
NASCET is the largest trial to have compared CEA with medical therapy. Symptomatic patients ($N = 659$) with severe carotid stenosis were randomized to medical therapy or CEA. The 30-day risk of stroke or death was 3.3% in the medical group and 5.8% in the surgical group. The 2-year risk of any stroke was 26% and 9% in the medical and surgical arms, respectively ($P < 0.001$), and the risk of major stroke or death was 18.1% and 8%, respectively ($P < 0.001$) [15]. Patients with 50%–69% stenosis had a 5-year stroke rate of 22.2% in the medical group and 15.7% in the surgical group ($P = 0.045$); those with 30%–49% stenosis had a 5-year stroke rate of 18.7% in the medical group and 14.9% in the surgical group ($P = 0.16$) [16].

In the Veterans' Affairs Cooperative Study, 189 men with $\geq 50\%$ ICA stenosis and ipsilateral symptoms were randomized to CEA or medical therapy. At a mean follow-up of 11.9 months, the incidence of stroke or TIA was 19.4% in the medical arm and 7.7% in the surgical arm ($P = 0.011$). Similarly, ECST randomized 2,518 patients with carotid stenosis to CEA or medical therapy with aspirin. At 3 years' follow-up, the stroke rate was 16.8% in the medical arm and 2.8% in the surgical arm.

Randomized trials of asymptomatic patients

ACAS randomized 1,659 patients with ≥60% asymptomatic carotid stenosis to CEA or medical therapy. At 2.7 years follow-up, the incidence of stroke or death was 11% in the medical treatment arm and 5.1% in the surgical arm ($P = 0.004$). The annualized risk of stroke among patients in the medical arm was 2.2%. The rate of perioperative stroke or death was very low at 2.3%. The Veterans' Affairs Asymptomatic Carotid Stenosis Study randomized 444 patients with ≥50% asymptomatic carotid stenosis to CEA versus medical therapy with aspirin. There was a nonsignificant trend towards better outcomes with surgery versus medical therapy (stroke or death: 4.7% vs 9.7%) [17].

American Heart Association recommendations [7]

Based on the above trials, CEA has proven benefit and is recommended for the prevention of stroke in symptomatic patients with ≥70% ipsilateral carotid artery stenosis. It is an acceptable recommendation in symptomatic patients with ipsilateral carotid artery stenosis of 50%–69%. CEA is an inappropriate recommendation for the prevention of stroke in symptomatic patients with ipsilateral carotid artery stenosis of <50%.

Among asymptomatic patients, ≥60% carotid artery stenosis is a proven indication for CEA, but only if the surgical risk is <3% and the patient's life expectancy is >5 years. If the surgical risk is 3%–5%, it is acceptable to recommend CEA in patients with *bilateral* carotid artery stenosis >75%. If the surgical risk is 5%–10%, asymptomatic patients should not be referred for CEA.

3. b Current indications for CEA are based largely on NASCET, ACAS, and other randomized trials that have compared CEA with medical therapy for the prevention of stroke. However, a large number of high-risk patients were excluded from these trials, and the results cannot, therefore, be extrapolated to a high-risk patient population. At this time, percutaneous carotid artery revascularization is only recommended in this high-risk cohort.

Medical and anatomic criteria that classify patients as "high surgical risk" include congestive heart failure (class III/IV) and/or known severe LV dysfunction (LVEF <30%), open heart surgery needed within 6 weeks, recent MI (>24 hours and <4 weeks), unstable angina (Canadian Cardiovascular Society class III/IV), severe pulmonary disease, contralateral carotid occlusion, contralateral laryngeal nerve palsy, radiation therapy to the neck, previous CEA with recurrent stenosis, cervical ICA lesions above C2 or common coronary artery lesions below the clavicle, severe tandem lesions, and age >80 years. Furthermore, it has been suggested that patients with pre-existing renal dysfunction may be at higher risk of perioperative stroke or hematoma after CEA [18]. While diabetes mellitus may predispose patients to CAD, which increases the risk of perioperative MI, diabetes mellitus has not been found to be an independent predictor of adverse perioperative outcomes after CEA [19–21].

Re-do CEA carries a 4.6%–10% major complication rate in large registries [22–24]. AbuRahma compared one surgeon's operative results on 124 patients undergoing re-do CEA for recurrent carotid stenosis with 265 primary CEAs performed on similar patients.

The 30-day rates of ipsilateral stroke and perioperative cranial nerve injuries were significantly higher among patients undergoing repeat surgery [23].

The risk of CEA increases with age: in a study of 2,089 Medicare patients undergoing CEA, operative mortality was 1.1% for patients <65 years, 2.8% for patients 70–74 years, 3.2% for patients 75–79 years, and 4.7% for patients >80 years [25]. Finally, patients with LV dysfunction are at increased risk for death or MI after CEA as compared with patients with normal ventricular function.

The SAPPHIRE trial enrolled high-risk patients ($N = 723$) with high-grade symptomatic ($\geq 50\%$) or asymptomatic ($\geq 80\%$) carotid stenosis and randomized 307 patients to carotid stenting ($n = 156$) or CEA ($n = 151$). Patients who were inoperable (as determined by a surgeon) due to co-morbid medical or anatomical considerations were entered into a stent registry ($n = 409$), and those who were at too high risk for intervention (as determined by an interventionist) were entered into a surgical registry ($n = 7$). In the randomized portion of the study, the 30-day composite endpoint of death, MI, or stroke was found to be significantly lower in the stenting arm as compared with the surgical arm (5.8% vs 12.6%, $P = 0.047$). The 30-day composite endpoint among patients in the stent registry was 7.8%, which was lower than the rate among patients in the surgical arm of the randomized cohort. The 1-year data for this trial are pending.

4. **d** The aortic arch can be classified into three types, based on the distance of the origin of the great vessels from the top of the arch (see **Figure 8**). The widest diameter of the left common carotid is used as a reference unit.

    - In a type I arch, all of the great vessels originate within one diameter length from the top of the arch.
    - In a type II arch, all of the great vessels originate within two diameter lengths from the top of the arch.
    - In a type III arch, the great vessels originate beyond two diameter lengths from the top of the arch.

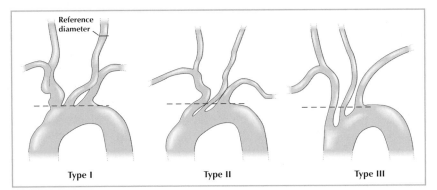

Figure 8. Classification of the aortic arch. Percutaneous interventions are technically more challenging to perform on type III arches than on type I arches.

Figure 9. Debris captured in an ANGIOGUARD™ XP (Cordis Corp.) distal emboli protection device after carotid stenting.

Type III arches are harder to access during percutaneous intervention than type I arches. The common carotid artery divides into the internal and external carotids at the C4–C5 level in 50% of patients. The bifurcation is higher in approximately 40% of patients, and lower in the remainder. The external carotid artery provides flow to the facial muscles and thyroid. It has a complex system of collaterals and symptoms from stenosis are therefore rare, although severe bilateral disease can cause jaw and tongue claudication.

Restenosis is uncommon after carotid artery stenting, and the incidence is lower than after coronary artery stenting. This is probably due to the larger final minimum luminal diameter after carotid artery stent deployment. Published rates of restenosis vary from 3.5% to 8% [26,27], and are lower than after carotid angioplasty alone [28–30].

Periprocedural stroke due to distal embolization is a serious complication of percutaneous carotid revascularization, with reported stroke rates ranging from 3% to 10% in series of patients who did not receive distal emboli protection [31–33]. With the use of transcranial Doppler, embolic signals have been detected in virtually all patients undergoing carotid artery revascularization, both percutaneous and surgical. Debris is routinely retrieved from emboli protection devices after almost all carotid interventions (see **Figure 9**). Histological analyses have found that such debris consists of fibrous and calcific plaque, lipid and cholesterol vacuoles, and fibrin and platelet clumps. While it is unlikely that a randomized trial will ever be performed to compare carotid stenting with and without distal emboli protection, the devastating nature of stroke makes it almost unethical to not use distal protection with all carotid interventions. In the SAPPHIRE trial, the rate of stroke was 3.8% among high-risk patients undergoing carotid stenting with distal protection. This is lower than most historical controls, and lower than the rate of stroke in the surgical arm of the study (3.8% vs 5.3%, not significant).

Figure 10. The ANGIOGUARD™ XP distal emboli protection device (Cordis Corp.).

There are three categories of distal emboli protection devices. The filter wires, of which the ANGIOGUARD™ XP (Cordis Corp., FL, USA) is the prototype (see **Figure 10**), consist of a porous umbrella-type filter at the end of a guidewire. They have the advantage of allowing distal blood flow during the procedure, but at the expense of some debris slipping through. Distal occlusion balloons, of which the PercuSurge GuardWire™ (Medtronic Inc., MN, USA) is the prototype, have the theoretical advantage of capturing more debris, but the disadvantage of occluding all distal flow. This is especially problematic in patients with contralateral occlusion. The third type of emboli protection device is the occlusion catheter, which is a guiding catheter with a distal balloon. When inflated in the common carotid artery, this creates a negative gradient distally and reverses flow. The lesion is then treated and debris extracted prior to deflating the balloon and re-establishing antegrade flow.

Transient hypotension and bradycardia due to pressure on the carotid sinus are common during carotid stenting, and can be readily treated with crystalloid infusion, atropine, and vasopressors as needed. Temporary pacemakers are rarely required.

There are few data on the use of GPIIb/IIIa inhibitors during carotid stenting. Initial observational data have suggested that their use may be safe [34]. However, with the potential risk of intracranial hemorrhage and without clear data in support of a benefit [35], GPIIb/IIIa inhibitors should not be routinely used until further studies better define their role during carotid stenting with emboli protection devices.

5. e The prevalence of renal artery stenosis (RAS) among patients undergoing cardiac catheterization has been estimated at 10%–25% [36–38]; in patients undergoing lower extremity angiography for claudication, the incidence of atherosclerotic RAS has been estimated at 20%–40% [39]. Among patients with refractory hypertension or episodes of congestive heart failure, the prevalence of ≥70% RAS at the time of cardiac catheterization has been reported to be 19% [40]. Atherosclerotic RAS is a component of renal dysfunction in approximately 15% of patients with end-stage renal disease, and approximately 30% of these patients have bilateral involvement. All patients with atherosclerotic RAS have a poor prognosis – mortality rates as high as 32% were

reported in a small series at 2 years' follow-up [41]. Progression of RAS is not infrequent, and is dependent on the baseline level of stenosis – the highest rates of progression are observed among patients with ≥60% stenosis at baseline [42,43]. Among all patients, the average rate of progression is approximately 7% per year [42].

Angiographic evaluation of the renal arteries during cardiac catheterization should be a planned procedure and the decision to proceed should be based upon clinical suspicion for the presence of RAS. While a fairly benign procedure in experienced hands, trauma to the ostium of the renal artery, contrast nephrotoxicity, and atheroembolism are all possible complications. Thus, routine renal angiography during all cardiac catheterizations is not recommended.

With strong clinical suspicion, however, renal angiography at the time of cardiac catheterization is a cost-effective diagnostic modality. The following are all reasons to suspect RAS, and therefore can justify renal angiography at the time of cardiac catheterization:

- onset of hypertension in a patient ≤30 years or ≥55 years
- refractory hypertension on three or more medications or with a diastolic blood pressure of ≥100 mm Hg
- rapidly accelerating hypertension in a patient with previously controlled hypertension
- malignant hypertension
- unexplained renal insufficiency (creatinine >1.5 mg/dL) or failure
- azotemia after initiation of an angiotensin-converting enzyme (ACE) inhibitor
- azotemia in a patient with evidence of atherosclerotic disease in other vascular beds and not due to any other known etiology
- presence of an abdominal or flank bruit
- asymmetrical kidney size, as assessed by noninvasive imaging

6. c **Figures 1a** and **1b** are angiograms of the left and right renal arteries, respectively. The left renal artery has a 90% atherosclerotic ostial lesion, while the right renal artery is angiographically normal.

Most RAS is caused by either atherosclerotic disease (90%) or fibromuscular dysplasia (FMD) (10%). FMD involves all three layers of the arterial wall and primarily affects young women between the ages of 15 and 50 years. It tends to affect the distal two thirds of the renal artery, and is readily identified on angiography by its beaded, aneurysmal appearance. Atherosclerotic RAS, on the other hand, tends to affect the ostium and proximal one third of the renal artery, and its incidence increases with age [44].

The percutaneous treatment of RAS is preferred to surgical revascularization. Perioperative mortality after surgical revascularization ranges between 2% and 13% in various reports [45–47]. In contrast, the mortality associated with percutaneous revascularization has been reported at <1% in several series [48–50].

It is generally accepted that angioplasty alone is a suitable therapy for FMD. RAS secondary to atherosclerotic disease, however, is probably better treated with stenting due to more consistent and durable short- and long-term results [50,51]. Stenting is definitely indicated for aorto-ostial lesions (due to a high incidence of elastic recoil), restenotic lesions, or when suboptimal results are achieved after angioplasty alone, including dissection, ≥30% residual stenosis, or a residual gradient of ≥15 mm Hg. Technical success rates for stenting have ranged from 97% to 100% [52–54], and rates of restenosis at 6 months to 1 year have been approximately 12%–25% [51,53]. As would be expected, the acute and long-term success rates of angioplasty alone are lower [51].

Indications for renal artery percutaneous revascularization are controversial as there are few data from randomized clinical trials, and trials comparing intervention with medical therapy have had relatively small sample sizes and inconsistent results [55–57]. Nevertheless, significant RAS (≥70% or ≥15 mm Hg pressure gradient) plus one of the following are generally accepted indications for renal artery intervention:

- recurrent episodes of congestive heart failure and acute pulmonary edema. Renal artery stenting has been found to decrease the frequency of congestive heart failure, flash pulmonary edema, and the need for hospitalization in patients with significant RAS [58]
- hypertension refractory to medical therapy with two or more antihypertensive agents
- hypertension refractory to medical therapy with a mean arterial blood pressure of ≥110 mm Hg
- chronic renal insufficiency, either non-oliguric or dialysis dependent, with no other etiology. It has been estimated that underlying atherosclerotic renovascular disease resulting in ischemic nephropathy is a major contributor to the etiology of renal failure in 20% of patients >50 years old on dialysis [59]
- acute renal failure, with normal-sized kidneys and a normal ultrasound appearance

The hypertension associated with FMD tends to be renovascular in etiology and is best medically treated with an ACE inhibitor, with percutaneous intervention reserved for refractory hypertension. After undergoing percutaneous revascularization, patients with FMD have an approximately 50%–80% chance of a cure [60,61]. The majority of hypertensive patients with atherosclerotic RAS, however, have essential hypertension.

Data on the effectiveness of renal artery revascularization in improving hypertension in patients with atherosclerotic disease have been conflicting, and randomized trials comparing percutaneous revascularization with optimal medical therapy for the treatment of hypertension have shown inconsistent results [55–57]. In one of the largest trials to date, van Jaarsveld et al. randomized 106 hypertensive patients to angioplasty or medical therapy and found comparable results with the two strategies [56]. This trial, however, has been criticized because patients with ≥50% RAS were included, and thus many of the patients may not have had renovascular hypertension. Furthermore, the crossover rate from medical therapy to angioplasty was >40% in an intention-to-treat analysis, probably resulting in a type II error.

It is generally agreed that percutaneous revascularization tends to reduce the severity of hypertension and the number of antihypertensive medications required in the majority of patients [62]. However, only a minority of patients (10%–20%) with atherosclerotic RAS are "cured" of their hypertension after percutaneous intervention. A mean pretreatment arterial blood pressure of ≥110 mm Hg and bilateral RAS are predictors of an improvement in blood pressure after renal artery revascularization [53].

In young patients, renal function should not be adversely affected unless there is bilateral RAS. However, in older patients with extensive systemic vascular disease, hypertension, or diabetes, unilateral stenosis may result in deterioration of renal function. Rocha-Singh et al. studied the long-term impact of renal artery stenting on renal function in patients with an abnormal baseline renal function (serum creatinine ≥1.5 mg/dL) and bilateral RAS of a solitary kidney. Revascularization therapy resulted in stabilization or improvement of renal function in 94% of patients, with only 7.3% eventually requiring hemodialysis [63]. Other studies have found similar results [51,52]. It is important to bear in mind, however, that these were all observational studies, and the data have not been confirmed by randomized trials. Predictors of adverse outcomes include baseline serum creatinine of ≥1.5 mg/dL, diabetes mellitus, and age ≥70 years [64].

7. **b** An accurate history and physical examination are very important for the diagnosis and treatment of lower extremity vascular disease. The two most important symptoms of obstructive iliofemoral disease are intermittent claudication and rest pain. Claudication and pseudoclaudication can be readily distinguished based upon the exertional nature of true claudication, which improves with rest. Pseudoclaudication tends to worsen with long periods of sitting, or with flexion at the hip joint while in an upright position. While the risk factors for lower extremity atherosclerotic disease are the same as those for CAD (diabetes, smoking, hypertension, hyperlipidemia, and family history of premature CAD), diabetics and smokers tend to have the fastest progression and the worst outcomes.

The most common cause of symptoms among patients <40 years old is aorto-iliac disease, while femoropopliteal disease is responsible for 65% of cases of claudication in patients >40 years old. The most common cause of death among patients with intermittent claudication is MI [65], and the risk of cardiac death among patients with PVD is about 6-fold that of the general population.

Among all patients with intermittent claudication, the rate of progression to severe ischemia or amputation is about 2% per year. Diabetics with intermittent claudication have a higher likelihood of progression to rest pain, and a higher risk of gangrene and limb loss [66,67]. Diabetics make up >50% of all patients who undergo lower extremity amputations due to vascular insufficiency. Similarly, smokers have faster rates of progression as compared with nonsmokers, and carry a 20% risk of limb loss if they continue to smoke after developing intermittent claudication. In one population-based study, approximately 20%–25% of patients who underwent a lower extremity revascularization procedure ultimately required ipsilateral amputation or another revascularization procedure [68].

The ABI is the ratio of the highest systolic blood pressure in the arm to the systolic blood pressure at the ankle. It is a sensitive test for the presence of lower extremity obstructive atherosclerotic disease. An ABI >1 is considered normal and an ABI <0.95 is considered abnormal. Patients with an ABI <0.8 usually have symptoms of intermittent claudication, and rest pain is usually present in patients with an ABI <0.4. A low ABI is a predictor of all-cause mortality [69,70], and the ABI should be determined in all patients with evidence of atherosclerosis in any vascular bed. Diabetics often have calcification of the media in the arterial walls of the lower limbs, and the resultant decrease in compliance can cause false elevations in ankle pressure measurements. However, the lower extremity digital arteries do not become calcified in diabetics, so toe systolic pressures can be used instead of ankle pressures.

Exercise ABIs are also valuable to assess the degree of lower extremity vascular compromise. In patients without lower extremity vascular disease, the ankle pressure can drop slightly after strenuous exercise, but returns to normal promptly upon cessation of exercise. In patients with PVD, the drop in pressure occurs much earlier and to a larger extent, and takes longer to return to normal. In general, the test should be considered abnormal if there is ≥20% drop in systolic pressure and it takes longer than 3 minutes to return to normal after exercise is stopped.

Pentoxifylline is a methylxanthine derivative that is thought to decrease platelet adhesiveness and blood viscosity. It is one of only two drugs currently approved by the Food and Drug Administration for the treatment of intermittent claudication. Despite initial optimism based on observational studies and randomized trials showing slight improvements in walking distance [71,72], pentoxifylline does not appear to be much superior to placebo. Cilostazol, a phosphodiesterase-III inhibitor, has shown significant improvements in pain-free walking distance as compared with placebo and pentoxifylline in randomized controlled trials [73–75]. However, it is contraindicated in patients with LV dysfunction, which has limited its use in a patient population with a high prevalence of CAD.

8. **b** The treatment goals for patients who develop intermittent claudication are improvement in pain-free walking distance, prevention of progression to resting ischemia, prevention of gangrene and limb loss, and prophylaxis against future cardiovascular events. A detailed risk factor evaluation is of utmost importance in achieving these goals. Special attention should be placed upon smoking cessation, control of diabetes mellitus, and weight loss. Regular, supervised exercise programs are the cornerstone of therapy for improving quality of life; several studies have shown exercise programs to be the most effective therapy for improving pain-free walking distance. While patients in some studies tripled their pain-free walking distance, most patients can expect a doubling of their symptom-free walking distance within the first few months of starting an exercise program [76].

Hypertension and hyperlipidemia, while not as strongly correlated with progression of symptoms as diabetes and smoking, should also be aggressively treated. All patients should receive statin therapy, especially given the high incidence of cardiovascular events among patients with PVD. Similarly, all patients should receive oral antiplatelet therapy consisting of either aspirin, clopidogrel, or both if warranted for cardiac reasons. Furthermore, some data suggest that aspirin may retard the angiographic progression of lower extremity disease, as well as progression to total occlusion. Warfarin should not be routinely used, but may have a role in the prevention of graft thrombosis following surgical bypass procedures.

With risk factor modification and optimal medical therapy, including an exercise program, most patients with intermittent claudication can expect their symptoms to stabilize or improve over time [77]. Diabetes mellitus and a deterioration in serial ABIs are independent predictors of progression of symptoms and angiographic disease.

According to American Heart Association (AHA) guidelines, endovascular interventions and surgical bypass procedures should be reserved for patients who have incapacitating intermittent claudication that is lifestyle-limiting or interferes with their work, and those who have progressed to rest pain, have nonhealing ulcers or gangrene, or are at risk for limb loss [78]. The definitions of "lifestyle-limiting" and "incapacitating" are, of course, subjective. One group has defined severe claudication as symptoms after <50 yards (55 m) of walking and >50 mm Hg postexercise drop in ankle systolic pressure [79]. In general, however, this varies from patient to patient, and individual circumstances must be taken into account when deciding a threshold for endovascular or surgical intervention for symptom relief. As the safety, efficacy, and long-term reliability of endovascular interventions improve, the threshold for such treatment will decrease. It must be kept in mind, however, that the majority of patients with angiographic evidence of obstructive lower extremity disease do not have symptoms of intermittent claudication, and the mere presence of disease does not justify referral for endovascular or surgical revascularization.

9. e The AHA guidelines for iliac artery and SFA intervention recommend percutaneous intervention for stenoses <5 cm in length. Stenoses of 5–10 cm can be treated with PTA, but surgery may have longer patency rates. Surgery is the recommended treatment for stenoses >10 cm long.

    Figure 3 shows a mid-SFA lesion that was found to be 8 mm in length by quantitative angiography and is suitable for percutaneous intervention. Several factors must be taken into account when assessing the suitability of a lesion for percutaneous intervention. Calcification, chronic total occlusions, distal disease, and poor distal run-off all adversely influence the acute success and long-term patency of lower extremity interventions. Current guidelines recommend balloon angioplasty, with provisional stenting only for suboptimal angiographic results (residual stenosis >30%), a gradient across the stenosis of ≥10 mm Hg, or for dissections. **Figure 2** shows a residual dissection after PTA in the distal SFA, and should be stented.

    Risk factors for restenosis after stenting include lesion length >30 mm, vessel diameter <7 mm, and multiple stents [80]. Lower extremity endovascular procedures can be divided into aorto-iliac interventions, femoropopliteal interventions, and infrapopliteal interventions. As a general rule, the more inferior the site of intervention, the worse the procedural success and long-term patency rate.

Figure 4 is an angiogram of an 80% lesion in the left external iliac artery. For iliac lesions amenable to percutaneous intervention, procedural success rates are >90% and long-term patency rates are similarly high at 80%–85%. Balloon angioplasty with provisional stenting results in similar long-term clinical and angiographic success rates as direct stenting [81]. With newer stent designs, drug-eluting stents, and improvements in adjunctive and chronic pharmacotherapy, stents may be directly deployed more often in the future.

Femoropopliteal interventions have slightly lower acute and long-term success rates than iliac interventions, with acute success rates of about 85% and 3-year patency of approximately 60%. In general, due to historically high restenosis rates, stenting should be used only provisionally when PTA results are suboptimal. The SIROCCO trial, which is not yet completed, randomized patients with SFA disease to bare metal stents ($n = 18$) or sirolimus-eluting stents ($n = 18$). A significantly different binary restenosis rate was found between the treatment and control arms (0% vs 23.5%, respectively) [82]. Duplication of these results in larger trials may increase enthusiasm for SFA stenting.

Other issues to be addressed include external forces to SFA stents, such as compression and torsion. While self-expanding stents are less vulnerable to external compression than balloon-expandable stents, torsion remains a problem. These obstacles will have to be successfully overcome to allow the widespread use of SFA stents.

Figure 5 is an infrapopliteal angiogram that shows severe disease in the anterior tibial artery, a totally occluded peroneal artery that reconstitutes by collaterals, and a completely occluded posterior tibial artery. AHA guidelines recommend infrapopliteal interventions only for critical limb ischemia, nonhealing ulcers, or limb salvage; in the absence of these issues, medical therapy and exercise programs are the initial treatments of choice. The success rates for percutaneous intervention, both acute and long term, are not as favorable as for iliac or SFA interventions. However, despite higher long-term restenosis rates, there is often a significant increase in distal blood flow immediately following PTA, and this is helpful for ulcer healing.

10. **e** Atherosclerotic disease of the upper extremities and the subclavian and innominate arteries is much less common than lower extremity disease. The risk factors include those for CAD, such as diabetes, smoking, and dyslipidemia. However, other etiologies must also be considered, including Takayasu's disease, giant cell arteritis, connective tissue disorders (eg, systemic lupus erythematosus and scleroderma), and radiation-induced vascular disease [83].

Symptoms include arm and hand claudication, subclavian steal syndrome, cold hands and secondary Raynaud's phenomenon, ischemic ulcerations and gangrene, angina pectoris in patients with prior LIMA bypass grafts (see **Figure 6**), dialysis arteriovenous fistula failure due to low flow, and thromboembolism. Data on upper extremity endovascular procedures are limited, and all of the above are accepted indications for percutaneous revascularization. However, a gradient across a stenosis without any symptoms or ischemia is not an acceptable indication for revascularization.

Stenting in the subclavian and innominate arteries is thought to be superior to angioplasty alone due to the elimination of elastic recoil, prevention of distal emboli, and a lower incidence of residual dissections [84]. Acute success rates and long-term patency rates following stenting appear to be excellent, with low periprocedural complication rates [85,86]. As such, in the absence of contraindications, angioplasty should generally be followed by stenting in the great vessels.

References

1. Hertzer NR, Beven EG, Young JR et al. Coronary artery disease in peripheral vascular patients. A classification of 1000 coronary angiograms and results of surgical management. *Ann Surg* 1984;199:223–33.
2. Marek J, Mills JL, Harvich J et al. Utility of routine carotid duplex screening in patients who have claudication. *J Vasc Surg* 1996;24:572–7; discussion 577–9.
3. Cheng SW, Wu LL, Ting AC et al. Screening for asymptomatic carotid stenosis in patients with peripheral vascular disease: a prospective study and risk factor analysis. *Cardiovasc Surg* 1999;7:303–9.
4. Executive Committee for the Asymptomatic Carotid Atherosclerosis Study. Endarterectomy for asymptomatic carotid artery stenosis. *JAMA* 1995;273:1421–8.
5. O'Holleran LW, Kennelly MM, McClurken M et al. Natural history of asymptomatic carotid plaque. Five year follow-up study. *Am J Surg* 1987;154:659–62.
6. Chambers BR, Norris JW. Outcome in patients with asymptomatic neck bruits. *N Engl J Med* 1986;315:860–5.
7. Naylor AR, Mehta Z, Rothwell PM et al. Carotid artery disease and stroke during coronary artery bypass: a critical review of the literature. *Eur J Vasc Endovasc Surg* 2002;23:283–94.
8. Wolf PA, Kannel WB, Sorlie P et al. Asymptomatic carotid bruit and risk of stroke. The Framingham study. *JAMA* 1981;245:1442–5.
9. Moore WS, Barnett HJ, Beebe HG et al. Guidelines for carotid endarterectomy. A multidisciplinary consensus statement from the Ad Hoc Committee, American Heart Association. *Circulation* 1995;91:566–79.
10. Herlitz J, Wognsen GB, Haglid M et al. Risk indicators for cerebrovascular complications after coronary artery bypass grafting. *Thorac Cardiovasc Surg* 1998;46:20–4.
11. Hamulu A, Yagdi T, Atay Y et al. Coronary artery bypass and carotid endarterectomy: combined approach. *Jpn Heart J* 2001;42:539–52.
12. Char D, Cuadra S, Ricotta J et al. Combined coronary artery bypass and carotid endarterectomy: long-term results. *Cardiovasc Surg* 2002;10:111–5.
13. Estes JM, Khabbaz KR, Barnatan M et al. Outcome after combined carotid endarterectomy and coronary artery bypass is related to patient selection. *J Vasc Surg* 2001;33:1179–84.
14. Borger MA, Fremes SE, Weisel RD et al. Coronary bypass and carotid endarterectomy: does a combined approach increase risk? A metaanalysis. *Ann Thorac Surg* 1999;68:14–20; discussion 21.
15. North American Symptomatic Carotid Endarterectomy Trial Collaborators. Beneficial effect of carotid endarterectomy in symptomatic patients with high-grade carotid stenosis. *N Engl J Med* 1991;325:445–53.
16. Barnett HJ, Taylor DW, Eliasziw M et al. Benefit of carotid endarterectomy in patients with symptomatic moderate or severe stenosis. North American Symptomatic Carotid Endarterectomy Trial Collaborators. *N Engl J Med* 1998;339:1415–25.
17. Hobson RW 2nd, Weiss DG, Fields WS et al. Efficacy of carotid endarterectomy for asymptomatic carotid stenosis. The Veterans Affairs Cooperative Study Group. *N Engl J Med* 1993;328:221–7.
18. Hamdan AD, Pomposelli FB Jr, Gibbons GW et al. Renal insufficiency and altered postoperative risk in carotid endarterectomy. *J Vasc Surg* 1999;29:1006–11.
19. Ahari A, Bergqvist D, Troeng T et al. Diabetes mellitus as a risk factor for early outcome after carotid endarterectomy – a population-based study. *Eur J Vasc Endovasc Surg* 1999;18:122–6.
20. Akbari CM, Pomposelli FB Jr, Gibbons GW et al. Diabetes mellitus: a risk factor for carotid endarterectomy? *J Vasc Surg* 1997;25:1070–5; discussion 1075–6.
21. Ballotta E, Da Giau G, Renon L. Is diabetes mellitus a risk factor for carotid endarterectomy? A prospective study. *Surgery* 2001;129:146–52.
22. AbuRahma AF, Snodgrass KR, Robinson PA et al. Safety and durability of redo carotid endarterectomy for recurrent carotid artery stenosis. *Am J Surg* 1994;168:175–8.

23. AbuRahma AF, Jennings TG, Wulu JT et al. Redo carotid endarterectomy versus primary carotid endarterectomy. *Stroke* 2001;32:2787–92.

24. Gagne PJ, Riles TS, Imparato AM et al. Redo endarterectomy for recurrent carotid artery stenosis. *Eur J Vasc Surg* 1991;5:135–40.

25. Fisher ES, Malenka DJ, Solomon NA et al. Risk of carotid endarterectomy in the elderly. *Am J Public Health* 1989;79:1617–20.

26. Willfort-Ehringer A, Ahmadi R, Gschwandtner ME et al. Single-center experience with carotid stent restenosis. *J Endovasc Ther* 2002;9:299–307.

27. Chakhtoura EY, Hobson RW 2nd, Goldstein J et al. In-stent restenosis after carotid angioplasty-stenting: incidence and management. *J Vasc Surg* 2001;33:220–5; discussion 225–6.

28. Munari LM, Belloni G, Perretti A et al. Carotid percutaneous angioplasty. *Neurol Res* 1992;14(2 Suppl.):156–8.

29. Gil-Peralta A, Mayol A, Marcos JR et al. Percutaneous transluminal angioplasty of the symptomatic atherosclerotic carotid arteries. Results, complications, and follow-up. *Stroke* 1996;27:2271–3.

30. Schoser BG, Becker VU, Eckert B et al. Clinical and ultrasonic long-term results of percutaneous transluminal carotid angioplasty. A prospective follow-up of 30 carotid angioplasties. *Cerebrovasc Dis* 1998;8:38–41.

31. Yadav JS, Roubin GS, Iyer S et al. Elective stenting of the extracranial carotid arteries. *Circulation* 1997;95:376–81.

32. Diethrich EB, Ndiaye M, Reid DB. Stenting in the carotid artery: initial experience in 110 patients. *J Endovasc Surg* 1996;3:42–62.

33. Theron JG, Payelle GG, Coskun O et al. Carotid artery stenosis: treatment with protected balloon angioplasty and stent placement. *Radiology* 1996;201:627–36.

34. Kapadia SR, Bajzer CT, Ziada KM et al. Initial experience of platelet glycoprotein IIb/IIIa inhibition with abciximab during carotid stenting: a safe and effective adjunctive therapy. *Stroke* 2001;32:2328–32.

35. Hofmann R, Kerschner K, Steinwender C et al. Abciximab bolus injection does not reduce cerebral ischemic complications of elective carotid artery stenting: a randomized study. *Stroke* 2002;33:725–7.

36. Ramirez G, Bugni W, Farber SM et al. Incidence of renal artery stenosis in a population having cardiac catheterization. *South Med J* 1987;80:734–7.

37. Harding MB, Smith LR, Himmelstein SI et al. Renal artery stenosis: prevalence and associated risk factors in patients undergoing routine cardiac catheterization. *J Am Soc Nephrol* 1992;2:1608–16.

38. Weber-Mzell D, Kotanko P, Schumacher M et al. Coronary anatomy predicts presence or absence of renal artery stenosis. A prospective study in patients undergoing cardiac catheterization for suspected coronary artery disease. *Eur Heart J* 2002;23:1684–91.

39. Olin JW, Melia M, Young JR et al. Prevalence of atherosclerotic renal artery stenosis in patients with atherosclerosis elsewhere. *Am J Med* 1990;88:46N–51N.

40. Khosla S, Kunjummen B, Manda R et al. Prevalence of renal artery stenosis requiring revascularization in patients initially referred for coronary angiography. *Catheter Cardiovasc Interv* 2003;58:400–3.

41. Pillay WR, Kan YM, Crinnion JN et al. Prospective multicentre study of the natural history of atherosclerotic renal artery stenosis in patients with peripheral vascular disease. *Br J Surg* 2002;89:737–40.

42. Zierler RE, Bergelin RO, Davidson RC et al. A prospective study of disease progression in patients with atherosclerotic renal artery stenosis. *Am J Hypertens* 1996;9:1055–61.

43. Caps MT, Perissinotto C, Zierler RE et al. Prospective study of atherosclerotic disease progression in the renal artery. *Circulation* 1998;98:2866–72.

44. Safian RD, Textor SC. Renal-artery stenosis. *N Engl J Med* 2001;344:431–42.

45. Cambria RP, Brewster DC, L'Italien GJ et al. The durability of different reconstructive techniques for atherosclerotic renal artery disease. *J Vasc Surg* 1994;20:76–85; discussion 86–7.

46. Cherr GS, Hansen KJ, Craven TE et al. Surgical management of atherosclerotic renovascular disease. *J Vasc Surg* 2002;35:236–45.

47. Novick AC, Ziegelbaum M, Vidt DG et al. Trends in surgical revascularization for renal artery disease. Ten years' experience. *JAMA* 1987;257:498–501.

48. Iannone LA, Underwood PL, Nath A et al. Effect of primary balloon expandable renal artery stents on long-term patency, renal function, and blood pressure in hypertensive and renal insufficient patients with renal artery stenosis. *Cathet Cardiovasc Diagn* 1996;37:243–50.

49. Dorros G, Jaff M, Jain A et al. Follow-up of primary Palmaz-Schatz stent placement for atherosclerotic renal artery stenosis. *Am J Cardiol* 1995;75:1051–5.

50. White CJ, Ramee SR, Collins TJ et al. Renal artery stent placement: utility in lesions difficult to treat with balloon angioplasty. *J Am Coll Cardiol* 1997;30:1445–50.

51. van de Ven PJ, Kaatee R, Beutler JJ et al. Arterial stenting and balloon angioplasty in ostial atherosclerotic renovascular disease: a randomised trial. *Lancet* 1999;353:282–6.

52. Rodriguez-Lopez JA, Werner A, Ray LI et al. Renal artery stenosis treated with stent deployment: indications, technique, and outcome for 108 patients. *J Vasc Surg* 1999;29:617–24.

53. Rocha-Singh KJ, Mishkel GJ, Katholi RE et al. Clinical predictors of improved long-term blood pressure control after successful stenting of hypertensive patients with obstructive renal artery atherosclerosis. *Catheter Cardiovasc Interv* 1999;47:167–72.

54. Burket MW, Cooper CJ, Kennedy DJ et al. Renal artery angioplasty and stent placement: predictors of a favorable outcome. *Am Heart J* 2000;139:64–71.

55. Plouin PF, Chatellier G, Darne B et al. Blood pressure outcome of angioplasty in atherosclerotic renal artery stenosis: a randomized trial. Essai Multicentrique Medicaments vs Angioplastie (EMMA) Study Group. *Hypertension* 1998;31:823–9.

56. van Jaarsveld BC, Krijnen P, Pieterman H et al. The effect of balloon angioplasty on hypertension in atherosclerotic renal-artery stenosis. Dutch Renal Artery Stenosis Intervention Cooperative Study Group. *N Engl J Med* 2000;342:1007–14.

57. Webster J, Marshall F, Abdalla M et al. Randomised comparison of percutaneous angioplasty vs continued medical therapy for hypertensive patients with atheromatous renal artery stenosis. Scottish and Newcastle Renal Artery Stenosis Collaborative Group. *J Hum Hypertens* 1998;12:329–35.

58. Gray BH, Olin JW, Childs MB et al. Clinical benefit of renal artery angioplasty with stenting for the control of recurrent and refractory congestive heart failure. *Vasc Med* 2002;7:275–9.

59. Appel RG, Bleyer AJ, Reavis S et al. Renovascular disease in older patients beginning renal replacement therapy. *Kidney Int* 1995;48:171–6.

60. Losinno F, Zuccala A, Busato F et al. Renal artery angioplasty for renovascular hypertension and preservation of renal function: long-term angiographic and clinical follow-up. *AJR Am J Roentgenol* 1994;162:853–7.

61. Miller GA, Ford KK, Braun SD et al. Percutaneous transluminal angioplasty vs. surgery for renovascular hypertension. *AJR Am J Roentgenol* 1985;144:447–50.

62. Nordmann AJ, Woo K, Parkes R et al. Balloon angioplasty or medical therapy for hypertensive patients with atherosclerotic renal artery stenosis? A meta-analysis of randomized controlled trials. *Am J Med* 2003;114:44–50.

63. Rocha-Singh KJ, Ahuja RK, Sung CH et al. Long-term renal function preservation after renal artery stenting in patients with progressive ischemic nephropathy. *Catheter Cardiovasc Interv* 2002;57:135–41.

64. Dorros G, Jaff M, Mathiak L et al. Four-year follow-up of Palmaz-Schatz stent revascularization as treatment for atherosclerotic renal artery stenosis. *Circulation* 1998;98:642–7.

65. Kannel WB, McGee DL. Update on some epidemiologic features of intermittent claudication: the Framingham Study. *J Am Geriatr Soc* 1985;33:13–8.

66. Jonason T, Ringqvist I. Factors of prognostic importance for subsequent rest pain in patients with intermittent claudication. *Acta Med Scand* 1985;218:27–33.

67. Jonason T, Ringqvist I. Diabetes mellitus and intermittent claudication. Relation between peripheral vascular complications and location of the occlusive atherosclerosis in the legs. *Acta Med Scand* 1985;218:217–21.

68. Farkouh ME, Rihal CS, Gersh BJ et al. Influence of coronary heart disease on morbidity and mortality after lower extremity revascularization surgery: a population-based study in Olmsted County, Minnesota (1970–1987). *J Am Coll Cardiol* 1994;24:1290–6.

69. Newman AB, Sutton-Tyrrell K, Vogt MT et al. Morbidity and mortality in hypertensive adults with a low ankle/arm blood pressure index. *JAMA* 1993;270:487–9.

70. Vogt MT, Cauley JA, Newman AB et al. Decreased ankle/arm blood pressure index and mortality in elderly women. *JAMA* 1993;270:465–9.

71. De Sanctis MT, Cesarone MR, Belcaro G et al. Treatment of intermittent claudication with pentoxifylline: a 12-month, randomized trial – walking distance and microcirculation. *Angiology* 2002;53(Suppl. 1):S7–12.

72. Belcaro G, Nicolaides AN, Griffin M et al. Intermittent claudication in diabetics: treatment with exercise and pentoxifylline – a 6-month, controlled, randomized trial. *Angiology* 2002;53(Suppl. 1):S39–43.

73. Strandness DE Jr, Dalman RL, Panian S et al. Effect of cilostazol in patients with intermittent claudication: a randomized, double-blind, placebo-controlled study. *Vasc Endovascular Surg* 2002;36:83–91.

74. Rendell M, Cariski AT, Hittel N et al. Cilostazol treatment of claudication in diabetic patients. *Curr Med Res Opin* 2002;18:479–87.

75. Dawson DL, Cutler BS, Hiatt WR et al. A comparison of cilostazol and pentoxifylline for treating intermittent claudication. *Am J Med* 2000;109:523–30.

76. Gardner AW, Poehlman ET. Exercise rehabilitation programs for the treatment of claudication pain. A meta-analysis. *JAMA* 1995;274:975–80.

77. McDaniel MD, Cronenwett JL. Basic data related to the natural history of intermittent claudication. *Ann Vasc Surg* 1989;3:273–7.

78. Weitz JI, Byrne J, Clagett GP et al. Diagnosis and treatment of chronic arterial insufficiency of the lower extremities: a critical review. *Circulation* 1996;94:3026–49.

79. Prepared by the Ad Hoc Committee on Reporting Standards, Society for Vascular Surgery/North American Chapter, International Society for Cardiovascular Surgery. Suggested standards for reports dealing with lower extremity ischemia. *J Vasc Surg* 1986;4:80–94.

80. Henry M, Amor M, Ethevenot G et al. Palmaz stent placement in iliac and femoropopliteal arteries: primary and secondary patency in 310 patients with 2–4-year follow-up. *Radiology* 1995;197:167–74.

81. Tetteroo E, van der Graaf Y, Bosch JL et al. Randomised comparison of primary stent placement versus primary angioplasty followed by selective stent placement in patients with iliac-artery occlusive disease. Dutch Iliac Stent Trial Study Group. *Lancet* 1998;351:1153–9.

82. Duda SH, Pusich B, Richter G et al. Sirolimus-eluting stents for the treatment of obstructive superficial femoral artery disease: six-month results. *Circulation* 2002;106:1505–9.

83. Greenfield LJ, Rajagopalan S, Olin JW. Upper extremity arterial disease. *Cardiol Clin* 2002;20:623–31.

84. Rodriguez-Lopez JA, Werner A, Martinez R et al. Stenting for atherosclerotic occlusive disease of the subclavian artery. *Ann Vasc Surg* 1999;13:254–60.

85. Al-Mubarak N, Liu MW, Dean LS et al. Immediate and late outcomes of subclavian artery stenting. *Catheter Cardiovasc Interv* 1999;46:169–72.

86. Sullivan TM, Gray BH, Bacharach JM et al. Angioplasty and primary stenting of the subclavian, innominate, and common carotid arteries in 83 patients. *J Vasc Surg* 1998;28:1059–65.

Chapter 11
Noncoronary cardiac intervention

Ivan P Casserly and E Murat Tuzcu

Questions

1. A 30-year-old asymptomatic female is referred for percutaneous atrial septal defect (ASD) closure. A surface echocardiogram demonstrates right atrial and right ventricular dilatation with normal systolic function. Transesophageal echo (TEE) shows a 5-mm atrial septal secundum defect. Prior to planned ASD closure in the catheterization laboratory, the following O_2 saturation measurements are obtained: left innominate 92%, right innominate 71%, superior vena cava (SVC) 91%, right atrium 85%, inferior vena cava (IVC) 77%, right ventricle 84%, main pulmonary artery 82%, left atrium 98%. What is the *most* appropriate next step in this patient's management?
   a. Proceed with percutaneous ASD closure using the CardioSEAL® (NMT Medical Inc., MA, USA) device
   b. Proceed with percutaneous ASD closure using the Amplatzer® (AGA Medical Corp., MN, USA) device
   c. Refer for surgical closure of ASD
   d. Obtain further cardiac imaging studies
   e. Advise medical therapy

2. An 84-year-old female presents with progressive shortness of breath on exertion (New York Health Association [NYHA] class III). Surface echo demonstrates moderate right atrial and right ventricular enlargement. A 2-cm ASD (secundum type) is observed on TEE. The patient is referred for percutaneous ASD closure. Right heart catheterization performed prior to closure provides the following information:

   • pulmonary artery pressure 74/29 mm Hg (mean 42 mm Hg)
   • mean left atrial pressure 22 mm Hg
   • aortic pressure 190/91 mm Hg (mean 128 mm Hg)
   • O_2 saturations: SVC 64%, IVC 68%, pulmonary artery 86%, left atrium 96%, left femoral artery 95%
   • O_2 consumption 110 mL/m²
   • body surface area (BSA) 2.0 m²
   • hemoglobin (Hb) 13.0 g/dL

   Based on these data, which of the following statements is *true*?
   a. This degree of pulmonary hypertension precludes ASD closure
   b. Pulmonary vascular resistance is 6 Wood units
   c. The O_2 saturation measurements indicate right-to-left shunting and are therefore a contraindication for ASD closure
   d. The pulmonary blood flow (Qp) to systemic blood flow (Qs) ratio is 1.8
   e. It is reasonable to proceed with ASD closure

3. A 62-year-old female is referred for percutaneous patent foramen ovale (PFO) closure. She has a history of multiple strokes without any residual neurologic deficit. A TEE was performed during her most recent stroke, 3 months ago. This demonstrated a 1-cm x 1.5-cm mobile mass in the right atrium, which had the appearance of a thrombus, and a PFO with evidence of right-to-left flow during a bubble study. There was no other evidence of structural heart disease. The patient commenced coumadin at that time and now presents to your clinic. A TEE was performed 1 week prior to the clinic visit. Figure 1 is a representative picture of the right atrium from that study. What is the *most* appropriate management of this patient?

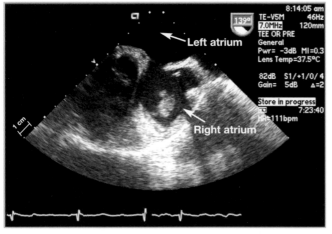

Figure 1. Transesophageal echocardiogram.

   a. Continue coumadin therapy for a further month and repeat TEE to confirm resolution of the thrombus before proceeding with PFO closure

   b. Admit the patient for heparin therapy for 24 hours prior to PFO closure, and continue coumadin therapy for 6 months following the procedure

   c. Proceed with PFO closure and coumadin therapy for 6 months following the procedure

   d. Proceed with PFO closure and administer aspirin and clopidogrel for 6 months following the procedure

   e. Refer the patient for cardiothoracic surgery

4. Regarding the design of the Amplatzer PFO occluder and Amplatzer septal occluder devices, which of the following statements is *false*?

   a. The waist diameter of the PFO occluder is fixed

   b. In the PFO occluder, the left atrial disc is larger than the right atrial disc

   c. The size of the PFO occluder is based on disc size, while that of the ASD occluder is based on waist size

   d. In the ASD occluder, the left atrial disc is larger than the right atrial disc

   e. The maximum size of the ASD occluder is 38 mm

5. A 45-year-old male develops acute onset of left arm and left leg hemiparesis, which resolves over the following 48 hours. Subsequent magnetic resonance imaging/magnetic resonance angiography of the carotid and intracerebral vessels are normal. TEE shows a PFO and an interatrial septal aneurysm. Hematologic evaluation is normal. Which of the following statements is *true* regarding the optimal therapy for this patient?
    a. Percutaneous closure of the PFO is more effective than medical therapy with either aspirin or coumadin
    b. Percutaneous closure of the PFO is equally effective as medical therapy with coumadin
    c. Aspirin is superior to coumadin in preventing future transient ischemic attacks or stroke
    d. In most series, surgical closure is associated with recurrence rates of 1%–2% for transient ischemic attack and stroke at 2 years follow-up
    e. None of the above

6. A 48-year-old male is referred for alcohol septal ablation. The patient has a 2-year history of chest pressure and shortness of breath on exertion. His current medical therapy includes atenolol 100 mg bid and verapamil slow-release 360 mg qd. A DDD pacemaker was placed 1 year previously, with minimal improvement in symptoms. Surface echo shows a resting gradient of 95 mm Hg, systolic anterior motion of the mitral leaflet, anterior jet of 3+ mitral regurgitation, and asymmetric septal hypertrophy with a septal thickness of 3.8 cm. A posteroanterior (PA) cranial view of the left anterior descending artery (LAD) is shown in Figure 2. Which of the following therapies would you recommend?

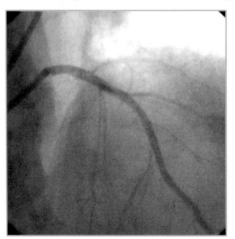

Figure 2. Posteroanterior cranial view of the left coronary artery.

    a. Surgical myectomy
    b. Alcohol septal ablation
    c. More aggressive medical therapy
    d. Adjustment of DDD pacemaker settings
    e. None of the above

7. You have decided to proceed with alcohol septal ablation on a 45-year-old female patient whose angiogram of the left coronary artery (PA cranial view) is shown in Figure 3. The baseline electrocardiogram is entirely normal and without conduction abnormalities. Which of the following statements is *true* with regards to the appropriate choice of equipment/therapy for this procedure?

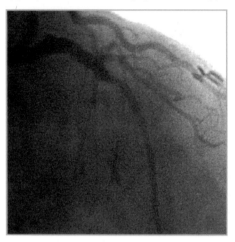

Figure 3. Posteroanterior cranial view of the left coronary artery.

a. Alcohol ablation of the first septal perforator should be performed
b. You would plan to inject 5–6 cc of absolute alcohol into the septal perforator
c. You would use a 2.5-mm diameter × 20-mm length over-the-wire balloon
d. You would use a 2.0-mm diameter × 9-mm length Monorail balloon
e. A temporary pacemaker is not required as the baseline electrocardiogram is normal
f. None of the above

8. A 69-year-old white female presents with chest discomfort and shortness of breath. Surface echo is significant for the presence of a resting dynamic outflow tract gradient of 71 mm Hg and thickening of the upper septum (measuring 3.1 cm). Comorbid medical conditions preclude the patient as a surgical candidate. She is referred for consideration of alcohol septal ablation. Figure 4 demonstrates the PA cranial view angiogram obtained on injection of the left main coronary artery. The decision is made to stent the mid-LAD in conjunction with an alcohol septal ablation of the first septal perforator branch. What is the *most* appropriate interventional approach?

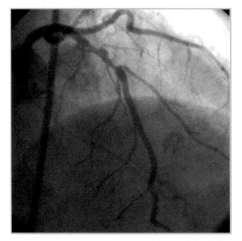

Figure 4. Posteroanterior cranial view of the left main coronary artery.

a. Wire the first septal perforator branch, perform alcohol septal ablation using a 1.5-mm x 15-mm Monorail balloon, then stent the LAD
b. Wire the LAD and first septal perforator branch, perform alcohol ablation using a 1.5-mm x 9-mm Monorail balloon, then stent the LAD
c. Wire the LAD and first septal perforator branch, perform alcohol ablation using a 1.5-mm x 9-mm over-the-wire balloon, then stent the LAD
d. Wire the LAD and first septal perforator branch, perform alcohol ablation using a 2.5-mm x 15-mm over-the-wire balloon, then stent the LAD
e. Stent the LAD and then perform alcohol ablation of the first septal perforator branch using a 1.5-mm x 15-mm over-the-wire balloon

9. Regarding clinical and echo outcomes following alcohol septal ablation, which of the following statements is *false*?
   a. Alcohol septal ablation is associated with a significant reduction in septal thickness at 1–2 years follow-up
   b. Nonseptal left ventricular hypertrophy is unaffected by the procedure
   c. Alcohol septal ablation is associated with a significant improvement in exercise tolerance at 1-year follow-up
   d. If subsequent myectomy is required, the risk of complete heart block is increased
   e. Approximately 10%–15% of patients do not achieve an adequate clinical response to alcohol septal ablation

10. During alcohol septal ablation, the hemodynamic tracing shown in Figure 5 was recorded. Which of the following maneuvers/interventions is a *possible* explanation for the observed effect?

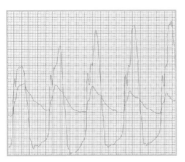

Figure 5. A hemodynamic tracing obtained during alcohol septal ablation.

   a. Forced exhalation against a closed glottis
   b. Isometric handgrip
   c. Elevation of both lower extremities
   d. Intravenous (IV) phenylephrine
   e. IV methoxamine

11. An 85-year-old female remains ventilator-dependent secondary to critical aortic stenosis (aortic valve area 0.5 cm²) after a carotid endarterectomy performed 2 weeks previously. Her history is significant for prior bypass surgery, moderate left ventricular systolic dysfunction (ejection fraction 30%), and dual chamber pacemaker implantation for complete heart block. She is referred for aortic valvuloplasty. Figure 6 shows the pressure waveforms obtained from the left ventricle and aorta at baseline and following aortic valvuloplasty (using a 20-mm diameter x 55-mm length Owen balloon).

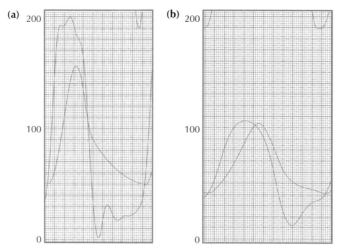

Figure 6. Pressure waveforms obtained from the left ventricle and aorta at **(a)** baseline and **(b)** postaortic valvuloplasty.

i. These tracings demonstrate *which* of the following?
   a. Optimal aortic valvuloplasty result with elimination of aortic valve gradient
   b. Residual aortic valve gradient with requirement for further balloon valvuloplasty using a larger balloon
   c. Severe aortic regurgitation

ii. What therapy would cause the change in the hemodynamic tracings in Figure 7 (overleaf)?
   a. Decrease in pacemaker rate to 50 bpm
   b. Decrease in pacemaker rate to 50 bpm and IV nitroprusside
   c. Increase in pacemaker rate to 100 bpm and IV neosynephrine
   d. Increase in pacemaker rate to 100 bpm and IV nitroprusside
   e. Insertion of an intra-aortic balloon pump (IABP)

12. What is the *most common* major procedural complication associated with aortic valvuloplasty?
   a. Death
   b. Stroke
   c. Complete heart block
   d. Surgical femoral arterial access complications
   e. Severe aortic regurgitation

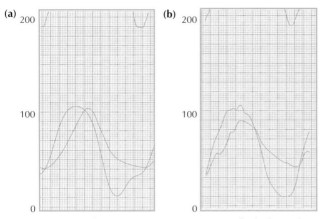

Figure 7. Hemodynamic tracings **(a)** postaortic valvuloplasty; **(b)** posttherapy.

13. A 69-year-old male presents with recurrent episodes of exertional syncope. Evaluation reveals severe calcific aortic stenosis with peak and mean aortic valve gradients of 55 mm Hg and 39 mm Hg, respectively, an aortic valve area of 0.6 cm², and an ejection fraction of 35%. He has had prior bypass surgery. Cardiac catheterization demonstrates patent grafts, but severe diffuse native vessel disease. He has been referred for percutaneous aortic valvuloplasty. Hemodynamic tracings at baseline and following inflation with a 23-mm diameter x 55-mm length Owens balloon are shown in Figure 8. Which of the following statements is *true* regarding this patient?

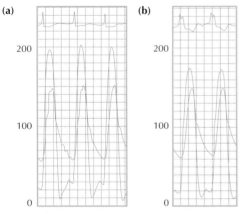

Figure 8. Hemodynamic tracings at **(a)** baseline and **(b)** postinflation with a 23-mm x 55-mm Owens balloon.

a. A further inflation with a 25-mm diameter Owens balloon is indicated

b. Compared with medical therapy, balloon valvuloplasty has a proven beneficial effect on survival in patients with these indications

c. The mechanism of action of valvuloplasty in calcific aortic stenosis is similar to that in rheumatic mitral stenosis

d. Restenosis of the aortic valve is inevitable, regardless of the acute procedural outcome

e. The aortic valve area achieved acutely with optimal valvuloplasty is marginally worse than that achieved with a valvular prosthesis

14. A 53-year-old female is referred for percutaneous mitral valvuloplasty. She has a 6-month history of fatigue and exertional dyspnea. Echo demonstrates features consistent with rheumatic mitral stenosis, +1 mitral regurgitation, and an echo score (Wilkin's) of 7/16. A transseptal puncture is performed in the catheterization laboratory and simultaneous left atrial and left ventricular pressures are recorded (see Figure 9). Right heart catheterization reveals a cardiac output of 4 L/min and a pulmonary artery pressure of 70/30 mm Hg (mean 44 mm Hg). The mean mitral valve gradient was 14 mm Hg.

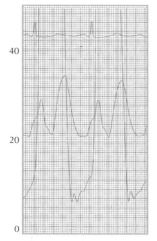

Figure 9. Simultaneous left atrial and left ventricular pressure recordings. Paper speed 50 mm/s.

i. What is the mitral valve area (MVA)?
a. There are insufficient data to allow calculation of the MVA
b. 2.0 cm²
c. 1.5 cm²
d. 1.0 cm²
e. None of the above

ii. Which of the following is the *most* reasonable management strategy for this patient?
a. Consider alternative causes for the patient's clinical symptoms
b. Medical therapy
c. Percutaneous mitral valvuloplasty
d. Mitral valve replacement with a mechanical valve
e. Mitral valve replacement with a bioprosthetic valve
f. Surgical commissurotomy

15. Which of the following is *not* an accepted criterion to help decide the endpoint of a percutaneous mitral commissurotomy procedure?
a. Valve area >1.0 cm²/m² of BSA
b. Complete opening of at least one commissure
c. Appearance or increment of regurgitation to grade 2 in Seller's 0–4 classification
d. Decrease in mean pulmonary and left atrial pressures by 50%
e. None of the above

16. A 65-year-old female presents with slowly progressive shortness of breath and decreased energy over the past 2–3 years. She has had a cardiac murmur since childhood. Her medical history is significant for chronic atrial fibrillation and hypertension. Surface echo and TEE demonstrate a peak and mean gradient of 60 mm Hg and 30 mm Hg, respectively, across the pulmonic valve, and a small ASD of 5-mm diameter. A biplane pulmonary angiogram is performed (see Figure 10). Which of the following statements is *true* regarding this patient?

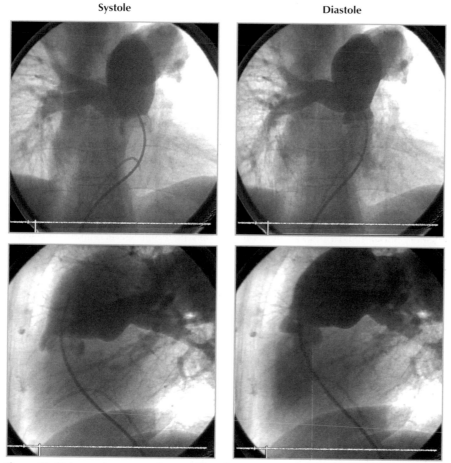

Figure 10. Biplane pulmonary angiogram. PA: pulmonary artery.

a. Balloon valvuloplasty is contraindicated
b. Left-to-right shunting across the ASD would be worsened by relief of the pulmonary gradient
c. The pulmonary gradient is probably subvalvular in location
d. There is no associated pulmonary regurgitation
e. This gradient is unlikely to account for the patient's symptoms

Answers

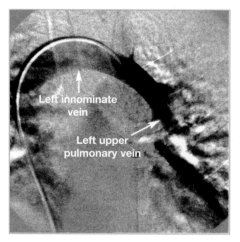

Figure 11. A partial anomalous pulmonary venous connection.

1. **d** It is mandatory to exclude the presence of partial anomalous pulmonary venous
connection (PAPVC) in adult patients prior to ASD closure. In this particular case, right
atrial and right ventricular dilatation in the presence of an ASD with a small diameter
(5 mm) warrants the search for additional factors that could contribute to right-sided
chamber dilatation. Forms of PAPVC include drainage of the right upper pulmonary
artery into the SVC or right atrium (most common); a right lower pulmonary vein into
the IVC (rare); and a left upper pulmonary vein into the left innominate vein (rare)
(see **Figure 11**).

In the catheterization laboratory, PAPVC can be detected by measuring O_2 saturations
from the right-sided venous structures and cardiac chambers. In this patient, high O_2
saturation measurements in the left innominate vein and SVC, with normal O_2 saturation
in the right innominate vein, suggest drainage of a left pulmonary vein into the left
innominate vein. A step-up between the upper and lower SVC would suggest an
anomalous connection between a right upper pulmonary vein and the SVC. Additionally,
PAPVC may be indicated on a pulmonary angiogram by right atrial filling either before,
or simultaneous with, left atrial filling. An imaging study (TEE, magnetic resonance
imaging, computed tomography) that demonstrates the precise anatomic connections
of the anomalously draining veins will definitively diagnose PAPVC.

2. **e** $Qp = O_2$ consumption (BSA m^2 x O_2 consumption mL/m^2) / (Hb [g/dL] x 1.36 [mL O_2/g
   of hemoglobin] x 10 x [PV O_2 – PA O_2]) (substitute LA O_2 sat for PV O_2 sat)
   = 2 x 110 / (13 x 10 x 1.36 x [0.96 – 0.86])
   = 12.4 L/min

   PVR = mean pulmonary artery pressure – mean left atrial pressure / Qp
   = 42 – 22 / 12.4
   = 1.6 Wood units

   Systemic venous O_2 saturation = ([SVC x 3] + [IVC x 1] / 4)

   Qp/Qs ratio = systemic arterial O_2 – systemic venous O_2 / pulmonary venous O_2 –
   pulmonary arterial O_2 (substitute LA for PV O_2 saturation)
   = 0.95 – 0.65 / 0.96 – 0.86
   = 3.0

   Closing an ASD in a patient with severe pulmonary hypertension presents some concern. This is based on the fear of right ventricular failure, particularly if there is evidence of right-to-left shunting across the defect. Closure of the ASD in this circumstance may result in further elevation of pulmonary pressures and worsening of the patient's clinical condition. In this patient, the O_2 saturation of 96% in the left atrium and the arterial systemic O_2 concentration of 95% argue against the presence of a large right-to-left shunt at the level of the atria. While the patient does indeed have severe pulmonary hypertension, this is not a contraindication to ASD closure *per se*. A careful assessment of pulmonary vascular resistance reveals a near normal value, further arguing in favor of closure. The high Qp/Qs ratio suggests a highly significant left-to-right shunt, which, together with the evidence of right-sided volume overload, argues in favor of ASD closure.

3. **e** This case illustrates the importance of an open-minded approach when presented with patients referred for percutaneous closure of a PFO. This patient was diagnosed with a right atrial thrombus and a PFO at the time of her recent stroke, and paradoxical embolism from the thrombus was presumed to be the likely mechanism. However, one should be aware that isolated right atrial thrombosis in the absence of structural heart disease is a very rare event. In addition, the right atrial mass has not diminished in size over a 3-month period of coumadin therapy. This should raise suspicion that the right atrial mass is a tumor and not a thrombus.

   The mechanism of recurrent stroke in this patient is most likely the formation of microthrombi on the surface of the right atrial tumor with paradoxical embolism to the cerebral circulation. The most appropriate management of this patient is referral to cardiothoracic surgery for removal of the right atrial mass. Surgical closure of the PFO during the procedure is technically straightforward, using either a suture or, rarely, a pericardial patch to close the PFO. Pathological examination of the right atrial mass in this patient revealed histology consistent with the diagnosis of an atrial myxoma.

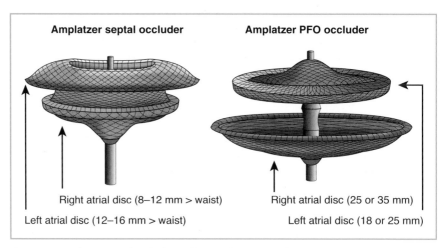

Amplatzer septal occluder	Amplatzer PFO occluder
Right atrial disc (8–12 mm > waist)	Right atrial disc (25 or 35 mm)
Left atrial disc (12–16 mm > waist)	Left atrial disc (18 or 25 mm)

Figure 12. Amplatzer septal occluder and patent foramen ovale (PFO) occluder devices.

4. **b** It is important to be aware of the specific design features of the available devices used to percutaneously close PFO and ASD. The Amplatzer PFO and septal occluder devices are currently the most popular devices. They serve to highlight some of the fundamental design differences between PFO and ASD closure devices (see **Figure 12**).

   The PFO occluder has a narrow waist that sits in the communication between the right and left atria. Currently, the PFO occluder comes in two sizes, based on the size of the right and left atrial discs, with a fixed waist size. Importantly, the right atrial disc is larger than the left atrial disc, which facilitates closure of the PFO.

   The Amplatzer septal occluder device comes in various sizes, based on the waist of the device (4–38 mm). The right disc is 8–12 mm (depending on device size) larger than the waist, and the left disc is 12–16 mm larger than the waist. The largest defect that can be closed using the septal occluder device is 38 mm.

5. **e** This patient is presenting with a cryptogenic stroke – ie, a stroke in the absence of any definable cause. This is the most common reason for referral for percutaneous closure of a PFO. In order to adequately advise patients regarding this procedure, it is important to be familiar with the data regarding the efficacy of the various therapeutic options currently available. There have been no randomized trials comparing medical versus surgical or percutaneous closure of PFO in the treatment of patients with cryptogenic stroke and a PFO.

   Randomized data comparing aspirin with coumadin are also lacking, but a *post hoc* analysis of the WARSS (Warfarin or Aspirin for Recurrent Stroke Study) trial suggested a trend toward reduced incidence of stroke or death at 2 years in patients treated with coumadin (9.5% vs. 17.9%, $P = 0.28$) [1]. A meta-analysis of several observational studies reported a reduced incidence of cerebrovascular ischemic events in patients treated with coumadin versus antiplatelet monotherapy over a mean follow-up period of 42 months (odds ratio 0.37, confidence interval 0.23–0.60) [2]. The risk of recurrent cerebrovascular ischemic events in surgical series varied between 0% and 18% at 1–2 years follow-up [3,4]. This may be explained by the unique clinical characteristics of this group, and emphasizes the point that recurrent cerebrovascular ischemic events may occur despite complete closure of the PFO in a cryptogenic stroke population.

6. a   This question highlights some of the important considerations in the preprocedural evaluation of patients for alcohol septal ablation. The patient is a middle-aged man who has symptomatic hypertrophic cardiomyopathy that has failed to improve with DDD pacing and maximal medical therapy. The therapeutic options therefore include alcohol septal ablation and surgical myectomy.

In the largest reports of outcomes following alcohol septal ablation, the mean septal thickness is usually in the range of 2.0–2.5 cm [5]. The use of alcohol septal ablation for septal thickness >3 cm frequently results in incomplete resolution of the left ventricular outflow tract gradient. Moreover, in experienced centers, surgical myectomy results are excellent in this subset of patients.

The severity and direction of the mitral regurgitation jet raises the strong suspicion of structural mitral valve disease. Systolic anterior motion of the mitral valve should produce a posterior, rather than an anterior, jet of mitral regurgitation. The ability to treat structural mitral valve disease at the same time as surgical myectomy favors the surgical option.

The angiogram demonstrates a normal LAD with a prominent bifurcating septal perforating branch. In the absence of the above considerations, this anatomy would be suitable for alcohol septal ablation.

7. f   The angiogram shows the LAD with a diminutive first septal perforator branch and a medium-sized second septal perforator branch (see **Figure 13**). This second perforator is the probable target for alcohol ablation. The area of the septal myocardium perfused by the second septal perforator can be assessed by injecting echo contrast. Ideally, the proximal septum is targeted, including the endocardial surface against which the anterior mitral leaflet makes contact. This highlights the importance of echo contrast injection in making the final decision as to the choice of septal perforator injected.

Over-the-wire balloons are required for this procedure, since alcohol must be injected through the balloon lumen. The balloon diameter should be sized to the diameter of the septal perforator branch, which is generally in the range of 1.5–2.0 mm. It is also important to use short, 9-mm balloons to maximize the area of the septum that is infarcted. In this case, using a long balloon would make it difficult to infarct the septum supplied by both divisions of the septal perforator branch. We have found that it is important to use a flexible balloon (eg, Maverick) to aid delivery of the balloon past the typically acute angulation at the junction of the septal perforator and LAD.

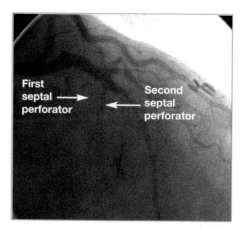

Figure 13. Posteroanterior cranial view of the left coronary artery (as in **Figure 5**).

The volume and speed of alcohol injection appear to predict the risk of complete heart block. More recently, the trend has been to slowly inject (approximately 1 mL/min) smaller volumes (2–3 mL) of alcohol in an effort to minimize this risk. It is important to be aware that alcohol septal ablation is associated with a frequency of complete heart block of 30%–60% in published series. Most cases are transient and occur within 48 hours of the procedure, although rare cases of unheralded complete heart block have occurred 3–9 days postprocedure. Therefore, it is standard practice to insert a temporary pacemaker into the right ventricular apex in all patients at the start of the procedure, except those with a permanent pacemaker or implantable cardioverter defibrillator with pacing capabilities, and to keep the pacemaker in place for a minimum of 48–72 hours. The rate of permanent pacemaker implantation following alcohol septal ablation is between 10%–30% in various reports [6,7], although the introduction of the modified alcohol injection technique and the use of adjunctive contrast echo suggest that this figure should now be closer to 5%.

8. c This case presents the challenge of managing a significant mid-LAD lesion in addition to performing alcohol ablation of the first septal perforating branch. The safest approach is to initially wire both the LAD and the first septal perforating branch (see **Figure 14**). Stenting the LAD first will probably result in significant plaque shift into the first septal perforating branch compromising wire access, and hence should be performed following alcohol ablation. It is wise to maintain wire access to the LAD during the alcohol ablation procedure in case of unforeseen complications. As explained in Question 7, the use of appropriately sized (to the dimensions of the septal perforator), short (generally <9 mm) over-the-wire balloons is mandatory during alcohol septal ablation procedures. In this case, a 2.5-mm balloon would clearly be oversized for the septal perforator branch and would increase the risk of vessel perforation.

9. b Outcomes following alcohol septal ablation have been reported in several case series with variable lengths of follow-up. These have consistently demonstrated the efficacy of alcohol septal ablation in significantly reducing septal thickness, left ventricular

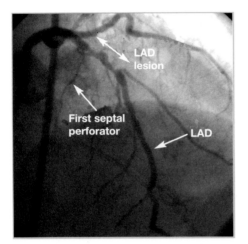

Figure 14. Posteroanterior cranial view of the left coronary artery. LAD: left anterior descending.

outflow tract gradient, and exercise tolerance. Regression of nonseptal left ventricular hypertrophy has also been shown at 1–2 years follow-up after alcohol ablation.

The clinical response to alcohol ablation is inadequate in 10%–15% of patients, resulting in a repeat alcohol ablation procedure or referral for myectomy. In those patients referred for myectomy following alcohol ablation, the risk of complete heart block is greatly increased. This is because right bundle branch block is present in ~70% of patients following alcohol ablation, while myectomy is associated with left bundle branch block in a majority of cases. Except for this consideration, subsequent myectomy is equally successful as *de novo* myectomy.

10. a The tracing shows a progressive increase in outflow tract gradient (left ventricular cavity to aortic) associated with a decrease in systolic pressure and pulse pressure. In hypertrophic cardiomyopathy, the dynamic outflow tract gradient will be increased by maneuvers/interventions that: (a) decrease preload (ie, amyl nitrate, forced expiration against closed glottis [Valsalva maneuver]); (b) increase myocardial contractility (ie, dobutamine); or (c) decrease afterload. Elevation of the lower extremities will increase preload; isometric handgrip, IV phenylephrine, and methoxamine will increase afterload and will therefore decrease the dynamic outflow tract gradient.

11i. c

11ii.d This case illustrates one of the potential complications of aortic valvuloplasty (the reported frequency of which is 0%–1.6%). Following aortic valvuloplasty, this patient developed severe aortic regurgitation, which was hemodynamically manifested by equalization of the left ventricular end diastolic and aortic diastolic pressures. At this point, therapy is aimed at minimizing the diastolic filling period by increasing the pacemaker rate and by reducing afterload using IV nitroprusside. Therapies that increase afterload (eg, neosynephrine), lengthen diastolic filling period (eg, decreased pacemaker rate), or increase diastolic pressures (eg, IABP) will exacerbate aortic regurgitation.

12. d The reported procedural complications and outcomes following aortic valvuloplasty vary according to the patient population studied and the experience of the operators. Generally, the populations studied are old (>80 years) and have significant comorbidities that preclude them from aortic valve replacement. The most common major morbidity is related to vascular access. Our practice is to use a 12-Fr Cook sheath in the femoral artery to allow delivery of an Owen dilatation balloon. Other operators use a 14-Fr sheath and either a double- or single-balloon technique for valve dilatation.

The large sheath size, together with the high incidence of peripheral vascular disease, is responsible for major surgical femoral complications (~5%). In these cases, the use of 8-Fr or 10-Fr Perclose devices to preclose the arteriotomy prior to insertion of the larger sheaths has helped to reduce the bleeding risk associated with the large arteriotomy. In 148 consecutive patients aged >80 years undergoing aortic valvuloplasty, the incidence of death, stroke, and persistent complete heart block was 2.7%, 2.0%, and 1.4%, respectively [8]. Particular attention should be paid to the presence of pre-existing right bundle branch block, which increases the risk of complete heart block due to trauma to the left bundle during balloon inflation.

13. d This patient was not deemed a candidate for operative aortic valve replacement due to the presence of severe coronary artery disease that was not amenable to revascularization. Therefore, aortic valvuloplasty was offered for palliation. The effect on survival of aortic valvuloplasty compared with medical therapy is controversial. Available data are based entirely on retrospective analyses that are subject to considerable selection bias. Any early benefit is assumed to be lost due to poor durability of the procedure secondary to inevitable restenosis. This has resulted in the doctrine that aortic valvuloplasty should only be used as a bridge to aortic valve replacement.

In this case, there was a decrease in the mean aortic valve gradient from 50 mm Hg to 23 mm Hg. A halving of the gradient is an acceptable result. A doubling of the valve area from approximately 0.5 cm^2 to 1 cm^2 is also considered an acceptable valvuloplasty result. This is still significantly inferior to the valve area achieved with a valvular prosthesis, which is usually >1.5 cm^2.

Valvuloplasty in calcific aortic stenosis is thought to act by fracturing calcium deposits in the leaflets, with a smaller contribution from separation of fused commissures and stretching of the valve structures. This is distinct from the mechanism of action in patients with rheumatic mitral stenosis, where separation of fused commissures is the predominant mechanism.

14i. d The calculation of the MVA is shown in **Figure 15**. It is important to note the variables that are required for calculation of the MVA (heart rate, cardiac output [in mL], diastolic ejection period, and the computer-generated mitral valve gradient). These are all potential sources of error that may enter into the calculation of MVA based on hemodynamic assessment.

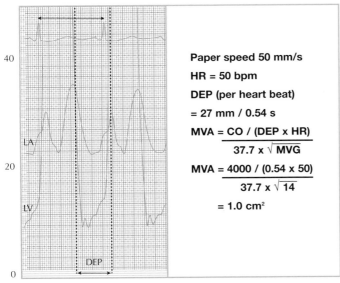

Paper speed 50 mm/s

HR = 50 bpm

DEP (per heart beat)

= 27 mm / 0.54 s

$$MVA = \frac{CO / (DEP \times HR)}{37.7 \times \sqrt{MVG}}$$

$$MVA = \frac{4000 / (0.54 \times 50)}{37.7 \times \sqrt{14}}$$

= 1.0 cm²

Figure 15. Calculation of the mitral valve area (MVA). CO: cardiac output; DEP: diastolic ejection period; HR: heart rate; LA: left atrium; LV: left ventricle; MVG: mitral valve gradient.

14ii.c This patient is a good candidate for percutaneous mitral valvuloplasty. She has clearly defined symptoms attributable to mitral stenosis. Traditionally, it has been felt that elevated left atrial pressure contributes to exertional symptoms, while low cardiac output explains the symptoms of fatigue. There is trivial mitral regurgitation by echo, and the echo score based on leaflet mobility, leaflet thickening, subvalvular disease, and calcification is within the acceptable range (<10) for proceeding with percutaneous valvuloplasty.

15. d Determining the endpoint of mitral valvuloplasty is one of the most challenging aspects of the procedure. Different operators use different criteria based on their own experience. In general, however, the synthesis of hemodynamic and echo data offers the best chance of achieving an optimal result with the minimum risk of complications. The multiple elements required to calculate the MVA by hemodynamic assessment can introduce significant error (see Question 14), which can be amplified by hemodynamic alterations during the procedure. In the acute setting, echo assessment of MVA is best performed by manual planimetry of the valve using either surface echo (parasternal short axis view) or TEE (transgastric short axis view). A decrease in pulmonary and left atrial pressures helps to guide the response to valvuloplasty, but should not be used as an absolute endpoint.

The most critical element is to determine splitting of the mitral commissures. This is best achieved using either surface echo (parasternal short axis view) or TEE (transgastric short axis view). Splitting of the commissures will be accompanied by commissural regurgitation. Ideally, both commissures should be split, but adequate results can be achieved provided at least one of the commissures is split. Worsening mitral regurgitation is suggested by an increasing "V" wave on the left atrial tracing, but, again, echo assessment is probably superior for this determination. No further valvuloplasty should

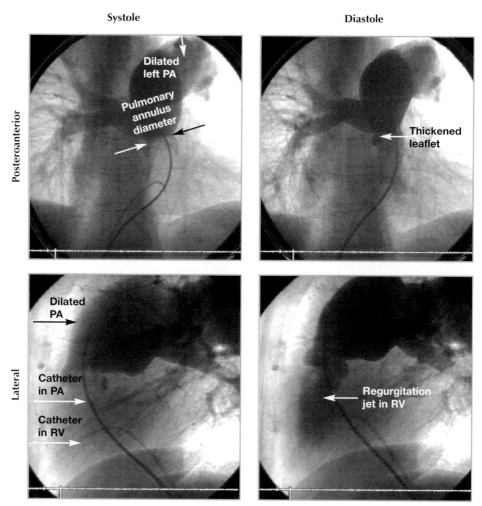

Figure 16. Biplane pulmonary angiogram. PA: pulmonary artery; RV: right ventricle.

be performed if the degree of mitral regurgitation reaches +2 at any point during the procedure, as the risk of severe mitral regurgitation becomes significant with further valvuloplasty.

16. b This patient has hemodynamically significant pulmonary stenosis (ie, >50 mm Hg peak gradient), which is probably the source of her symptoms. The pulmonary angiogram in **Figure 16** demonstrates some of the classic findings in pulmonary valvular stenosis: the leaflets are clearly thickened, and there is marked poststenotic dilatation of the pulmonary artery (more prominent in the left pulmonary artery). The diastolic frames also demonstrate some regurgitation across the pulmonic valve. Hemodynamic assessment (gradual withdrawal of Goodale–Lubin catheter across the pulmonary valve with simultaneous recording from the right ventricular cavity) found that the gradient was largely at the level of the pulmonary valve.

Pulmonary valvuloplasty has been successfully completed in adults (>21 years), with excellent acute hemodynamic outcomes. This is also true for older patients (such as in this case) and even in those with calcification of the pulmonary valve leaflets. Most centers use a single-balloon technique, although some use a double-balloon technique if the pulmonary annulus is >20 mm in diameter. Long-term outcome of pulmonary valvuloplasty in adults is not as well described as that in children. In the largest series of adolescent and adult patients treated with pulmonary valvuloplasty (*N* = 53), the mean follow-up was 6.9 years, and the procedure had a durable benefit [9].

References

1.   Homma S, Sacco RL, Di Tullio MR et al. Effect of medical treatment in stroke patients with patent foramen ovale: patent foramen ovale in Cryptogenic Stroke Study. *Circulation* 2002;105:2625–31.

2.   Orgera MA, O'Malley PG, Taylor AJ. Secondary prevention of cerebral ischemia in patent foramen ovale: systematic review and meta-analysis. *South Med J* 2001;94:699–703.

3.   Homma S, Di Tullio MR, Sacco RL et al. Surgical closure of patent foramen ovale in cryptogenic stroke patients. *Stroke* 1997;28:2376–81.

4.   Dearani JA, Ugurlu BS, Danielson GK et al. Surgical patent foramen ovale closure for prevention of paradoxical embolism-related cerebrovascular ischemic events. *Circulation* 1999;100(19 Suppl.):II171–5.

5.   Lakkis NM, Nagueh SF, Dunn JK et al. Nonsurgical septal reduction therapy for hypertrophic obstructive cardiomyopathy: one-year follow-up. *J Am Coll Cardiol* 2000;36:852–5.

6.   Chang SM, Nagueh SF, Spencer WH 3rd et al. Complete heart block: determinants and clinical impact in patients with hypertrophic obstructive cardiomyopathy undergoing nonsurgical septal reduction therapy. *J Am Coll Cardiol* 2003;42:296–300.

7.   Kern MJ, Holmes DG, Simpson C et al. Delayed occurrence of complete heart block without warning after alcohol septal ablation for hypertrophic obstructive cardiomyopathy. *Catheter Cardiovasc Interv* 2002;56:503–7.

8.   Eltchaninoff H, Cribier A, Tron C et al. Balloon aortic valvuloplasty in elderly patients at high risk for surgery, or inoperable. Immediate and mid-term results. *Eur Heart J* 1995;16:1079–84.

9.   Chen CR, Cheng TO, Huang T et al. Percutaneous balloon valvuloplasty for pulmonic stenosis in adolescents and adults. *N Engl J Med* 1996;335:21–5.

Abbreviations

Abbreviations

ABI	ankle brachial index
ACC	American College of Cardiology
ACE	angiotensin-converting enzyme
ACEI	angiotensin-converting enzyme inhibitor
ACS	acute coronary syndromes
ACT	activated clotting time
AHA	American Heart Association
AL	Amplatz left
AP	anteroposterior
AR	Amplatz right
ASD	atrial septal defect
ATP	adenosine triphosphate
bid	twice a day (Latin: *bis in die*)
bpm	beats per minute
BSA	body surface area
CABG	coronary artery bypass graft
CAD	coronary artery disease
CCR	chemokine receptor
CCS	Canadian Cardiovascular Society
CEA	carotid endarterectomy
CI	confidence interval
CKMB	creatine kinase MB
COPD	chronic obstructive pulmonary disease
CPB	cardiopulmonary bypass
CRP	C-reactive protein
CT	computed tomography
CTO	coronary total occlusion
CVR	coronary flow velocity reserve
DCA	directional coronary atherectomy
EBU	extra-backup
ECG	electrocardiogram
ECM	extracellular matrix
EF	ejection fraction
ELG	El Gamal
FEV	forced expiratory volume
FFR	fractional flow reserve
FMD	fibromuscular dysplasia
FVC	forced vital capacity
GISE	Italian Society for Invasive Cardiology
GL	geometric left
GP	glycoprotein
Hb	hemoglobin
HR	hazard ratio
IABP	intra-aortic balloon pump
ICA	internal carotid artery
IL	interleukin
IMA	internal mammary artery
IV	intravenous
IVC	inferior vena cava
IVUS	intravascular ultrasound

JL	Judkins left
LAD	left anterior descending
LAO	left anterior oblique
LBBB	left bundle branch block
LCX	left circumflex
LDL	low-density lipoprotein
LIMA	left internal mammary artery
LMWH	low molecular weight heparin
LV	left ventricular
LVEDP	left ventricular end diastolic pressure
LVEF	left ventricular ejection fraction
MACE	major adverse coronary event
mAo	mean pressure of the aorta
MBG	myocardial blush grade
MCP	monocyte chemoattractant protein
M-CSF	macrophage colony-stimulating factor
MI	myocardial infarction
MO$_2$	myocardial oxygen
mP2	mean pressure of the distal vessel
MRI	magnetic resonance imaging
MVA	mitral valve area
MVG	mitral valve gradient
NIH	National Institutes of Health
NYHA	New York Health Association
OM	obtuse marginal
OPCAB	off-pump coronary artery bypass surgery
OR	odds ratio
oxLDL	oxidized low-density lipoprotein
PA	posteroanterior
PAPVC	partial anomalous pulmonary venous connection
PAWP	pulmonary arterial wedge pressure
PCI	percutaneous coronary intervention
PDA	posterior descending artery
PDGF	platelet-derived growth factor
PFO	patent foramen ovale
PO	by mouth (Latin: *per os*)
POBA	plain old balloon angioplasty
PTA	percutaneous transluminal angioplasty
PTCA	percutaneous transluminal coronary angioplasty
PTFE	polyterafluoroethylene
PVD	peripheral vascular disease
QCA	quantitative coronary angiography
qd	once a day (Latin: *quaque die*)
Qp	pulmonary blood flow
Qs	systemic blood flow
RAO	right anterior oblique
RAS	renal artery stenosis
RCA	right coronary artery
RCB	right coronary bypass
rCVR	relative coronary flow velocity reserve

Abbreviations

RHC	right heart catheterization
RIMA	right internal mammary artery
RR	relative risk
RV	right ventricular
SC	subcutaneous
SCR	shepherd's crook right
SFA	superficial femoral artery
STEMI	ST-segment elevation myocardial infarction
SVC	superior vena cava
TEE	transesophageal echocardiogram
TIA	transient ischemic attack
tid	three times a day (Latin: *ter in die*)
TIMI	Thrombolysis in Myocardial Infarction
TLR	target-lesion revascularization
TVR	target-vessel revascularization
UFH	unfractionated heparin
VL	Voda left
XB	extra-backup

Trial acronyms

ACAS	Asymptomatic Carotid Atherosclerosis Study
ACIP	Asymptomatic Cardiac Ischemia Pilot
ACME	Angioplasty Compared to Medicine
ADMIRAL	Abciximab before Direct Angioplasty and Stenting in Myocardial Infarction regarding Acute and Long-term Follow-up
ADVANCE-MI	Addressing the Value of Facilitated Angioplasty after Combination Therapy or Eptifibatide Monotherapy in Acute Myocardial Infarction
ARTS	Arterial Revascularization Therapies Study
ASSENT	Assessment of the Safety and Efficacy of a New Thrombolytic
AVERT	Atorvastatin versus Revascularization Treatment
AWESOME	Angina with Extremely Serious Operative Mortality Evaluation
BARI	Bypass Angioplasty Revascularization Investigation
BARI 2D	BARI 2 Trial in Diabetics
BENESTENT	Belgian Netherlands Stent Study
BOAT	Balloon versus Optimal Atherectomy Trial
BRAVO	Blockade of the GPIIb/IIIa Receptor to Avoid Vascular Occlusion
CADILLAC	Controlled Abciximab and Device Investigation to Lower Late Angioplasty Complications
CAPTURE	c7E3 Fab Antiplatelet Therapy in Unstable Refractory Angina
CAVEAT	Coronary Angioplasty versus Excisional Atherectomy Trial
CREDO	Clopidogrel for the Reduction of Events During Observation
CRUISE	Coronary Revascularization Utilizing Integrilin and Single-bolus Enoxaparin
CURE	Clopidogrel in Unstable Angina to Prevent Recurrent Ischemic Events
DEBATE	Doppler Endpoints Balloon Angioplasty Trial Europe
DESTINI	Doppler Endpoint Stenting International Investigation
ECST	European Carotid Stenosis Trial
EPIC	Evaluation of c7E3 for Prevention of Ischemic Complications
EPILOG	Evaluation in PTCA to Improve Long-term Outcome with Abciximab GPIIb/IIIa Blockade
EPISTENT	Evaluation of Platelet IIb/IIIa Inhibitor for Stenting Trial
ERACI	Argentine Randomized Trial of Percutaneous Transluminal Coronary Angioplasty versus Coronary Artery Bypass Surgery in Multivessel Disease
ESSENCE	Efficacy and Safety of Subcutaneous Enoxaparin in Non-Q-wave Coronary Events
EXCITE	Evaluation of Oral Xemilofiban in Controlling Thrombotic Events
FINESSE	Facilitated Intervention with Enhanced Speed to Stop Events
FIRE	FilterWire During Transluminal Intervention of Saphenous Vein Grafts
FRISC	Fast Revascularization during Instability in Coronary Artery Disease
GISSOC	Gruppo Italiano di Studio sullo Stent nelle Occlusioni Coronariche
GUSTO	Global use of Strategies to Open Coronary Arteries
MASS	Medicine, Angioplasty, or Surgery Study
NASCET	North American Symptomatic Carotid Endarterectomy Trial
NICE	National Investigators Collaborating on Enoxaparin
NRMI	National Registry of Myocardial Infarction
OPUS-TIMI	Orbofiban in Patients with Unstable Coronary Syndromes–Thrombolysis in Myocardial Infarction
PACT	Plasminogen-activator Angioplasty Compatibility Trial
PAMI	Primary Angioplasty in Myocardial Infarction
PARAGON	Platelet IIb/IIIa Antagonist for the Reduction of Acute Coronary Syndrome Events in a Global Organization Network

PEPCI	Pharmacokinetic Study of Enoxaparin in Patients Undergoing Percutaneous Coronary Intervention
PRISM	Platelet Receptor Inhibition for Ischemic Syndrome Management
PRISM-PLUS	PRISM in Patients Limited by Unstable Signs and Symptoms
PURSUIT	Platelet Glycoprotein IIb/IIIa in Unstable Angina: Receptor Suppression using Integrilin Therapy
RAVEL	Randomized Comparison of a Sirolimus-eluting Stent with a Standard Stent for Coronary Revascularization
REPLACE	Randomized Evaluation in PCI Linking Angiomax to Reduced Clinical Events
RITA	Randomized Intervention Treatment of Angina
SAFER	Saphenous Vein Graft Angioplasty Free of Emboli Randomized
SAPPHIRE	Stenting and Angioplasty with Protection in Patients at High-risk for Endarterectomy
SAVED	Saphenous Vein *De Novo*
SHOCK	Should we Emergently Revascularize Occluded Coronaries in Cardiogenic Shock
SICCO	Stenting in Chronic Coronary Occlusion
SIRIUS	Sirolimus-eluting Stent in *De Novo* Native Coronary Lesions
SoS	Stent or Surgery
SPEED	Strategies for Patency Enhancement in the Emergency Department
STRESS	Stent Restenosis Study
SYMPHONY	Sibrafiban versus Aspirin to Yield Maximum Protection from Ischemic Heart Events Post-acute Coronary Syndromes
TACTICS	Treat Angina with Aggrastat and Determine Cost of Therapy with an Invasive or Conservative Strategy
TIME	Trial of Invasive versus Medical Therapy in Elderly Patients with Chronic CAD
TIMI	Thrombolysis in Myocardial Infarction
TOAST	Total Occlusion Angioplasty Study
VEGAS	Vein Graft AngioJet Studies
WARSS	Warfarin or Aspirin for Recurrent Stroke Study